Jokob Mulner, member of the Nuremberg Twelve Brothers House, making clay models for moulds from which brass-founders cast single and three-branched candlesticks, 1471. Nürnberg Stadtarchiv 75-Amb-2-317-90-r. © Nuremberg City Archive

Brass from the past

Brass made, used and traded from prehistoric times to 1800

Vanda Morton

Archaeopress Publishing Ltd
Summertown Pavilion
18-24 Middle Way
Summertown
Oxford OX2 7LG

www.archaeopress.com

ISBN 978-1-78969-156-6
ISBN 978-1-78969-157-3 (e-Pdf)

© V Morton and Archaeopress 2019

All rights reserved. No part of this book may be reproduced, or transmitted, in any form or by any means, electronic, mechanical, photocopying or otherwise, without the prior written permission of the copyright owners.

This book is available direct from Archaeopress or from our website www.archaeopress.com

Contents

Chapter 1 Experiment and emergence 1

Chapter 2 Medieval Europe and far beyond 29

Chapter 3 The Sacred and the Salesmen 69

Chapter 4 Age of Discovery 101

Chapter 5 Merchants and migrants 139

Chapter 6 Continuity and conflict in Europe 179

Chapter 7 Trade and technology 221

Chapter 8 The turning tide 265

Glossary 307

Bibliography 311

Appendix Metallurgical tables https://bit.ly/2HgrCwi 339

Index 341

List of Figures

Colour Illustrations

Jokob Mulner, member of the Nuremberg Twelve Brothers House, making clay models for moulds from which brass-founders cast single and three-branched candlesticks, 1471. Nürnberg Stadtarchiv 75-Amb-2-317-90-r. © Nuremberg City Archive

Franco-roman glass horn with interlaced brass strips, AD 4-5. From Samson excavations, south of Namur, photo © author, courtesy of Musée Archéologique de Namur, Collection Fondation SAN

Brass currency bars from Hedeby, northernmost port of the Carolingian empire, freshly cast from metal ores mined in the Balkans. 16-25cm long. *Top*, photo © Wikinger Museum, Haithabu, Schleswig; *Bottom*, photo © author, by courtesy of the Viking Museum, Haithabu, Schleswig

Ninth-century brass astrolabe cast in (today's) Syria by Khalīf, apprentice to 'Alī ibn 'Īsā. Its plates cover latitudes 33° to 41°. Added inscriptions show that it was used in Armenia. Astrolabe Inv. 47632, © History of Science Museum, University of Oxford

Brass candlestick inlaid with gold, silver and a black substance, 1340s. The side shown is thought to depict Tashi Khatun, Mongolian regent and mother of Sheik Abu Ishak of Shiraz. She is being offered fruit and a book. Inv. 47632. © Museum of Islamic Art, Doha, Qatar

Detail of brass candlestick inlaid with gold, silver and a black substance, 1340s. The side shown is thought to depict Tashi Khatun, Mongolian regent and mother of Sheik Abu Ishak of Shiraz. She is being offered fruit and a book. Inv. 47632. © Museum of Islamic Art, Doha, Qatar

Pestle and mortar cast in Bristol, 18th century. Inv. 3243, photo © author, courtesy of Blaise Castle Museum, Bristol

Tobacco box, 1762, made in Iserlohn, a source of zinc carbonate ore. The lid shows the conquest of Martinique in 1762, the base bears the initials of King George III of England. © Jan R. Schäfer, courtesy of Iserlohn Stadtmuseum

Cast thimble and bell, excavated from a mid-eighteenth-century domestic site in the colonial settlement at Williamsburg, Virginia. photo © author, courtesy of Collections of Colonial Williamsburg Foundation's Department of Archaeology. Inv. thimble: ER1339.19B,00030-19BB, bell: ER1346.19B,00024-19BB

Byzantine or Coptic low-zinc, leaded brass bowl, probably made in Egypt in the 5th-6th century, one of several high-status burial goods in a solo, pagan Anglo-Saxon burial dating to the 6th-7th century. This elite burial was recently found by chance near Snettisham, eastern England. photo © Ian Cartwright, School of Archaeology, University of Oxford

Twisted-stem brass candlestick cast at Skultuna brass works c.1700, photo © author, courtesy of Skultuna Bruks Museum, Sweden

Illustrations in the text

Figure 1. Diagram of temperatures during cementation 7
Figure 2. Miniature brass helmet from a royal grave at Ur 8
Figure 3. Dagger from Umm-an-Nar 10
Figure 4. The Middle East to India, map showing some sites mentioned in the text 11
Figure 5. *above*, horse figurine, Kachbulag; *below*, brass arrowhead, Sary Tepe 13
Figure 6. Brass bowl from Taxila 15
Figure 7. Transcaucasian belt-clasp 20
Figure 8. Map, early medieval European sites mentioned in the text 23
Figure 9. Roman face-mask vizor sports helmet, Ribchester Roman fort, British Museum 1814,0705.1 24
Figure 10. Roman brooches, *left* Aucissa type fibula, c.10 BC-AD 50; *right* Hod Hill type, AD 44-80, Alchester Roman camp 25
Figure 11. 8-10-century ewer, Khurasan, 44cm high 30
Figure 12. Celestial attendant, Kashmir, tenth century, height 6.03 cm 31
Figure 13. Padmapani, god of compassion, in sorrowful thought, 16cm high 33
Figure 14. Talismanic plaques, Tibet: horses, AD 701-900; peacocks, AD 801-1000, 7.8cm high .. 34
Figure 15. Middle East and Asia, map showing some sites mentioned in the text 39
Figure 16. Fatimid domestic vessels, early 11 century, Eretz Israel Museum, Tel Aviv 43
Figure 17. Europe, map showing some sites mentioned in the text 45
Figure 18. Beaded-rim bowl from Norway, 5 to 7 century AD, 140 cm wide, Bergen University Museum B2892 47
Figure 19. Deer hunt engraved on the lock-plate of a late Roman casket, Bonn, Römisch-Germanisches Museum, Cologne, after photo by Dr Michael Schmauder, with his kind permission 48
Figure 20. Rhine/Meuse area, map showing some sites mentioned in the text 51
Figure 21. Russian rivers, map showing some sites mentioned in the text 59
Figure 22. Comparative lengths of brass bars, *left to right*, Ma'den Iafen, Myrvälde, Kamanget, Hedeby, des Jarres 62
Figure 23. Northern Africa, map showing some sites mentioned in the text 64
Figure 24. Koi Gourrey figurines: *above*, hornbill; *centre*, female crocodile; *below*, male crocodile, after illustration by Dr Laurence Garenne-Marot, with her generous permission 66
Figure 25. Ewer, Western Iran, c.1220-40; *right*, detail showing a musician 71
Figure 26. South-East Asia and India, map showing some sites mentioned in the text 73
Figure 27. Diagram showing *left*, Chinese method, *right* Indian method distillation retorts 75
Figure 28. Golden Hall temple, Wudang Mountain, China, 1416 77
Figure 29. Brass market stall, Champeaux market, Paris, 1403-4, after Bibliotheque Nationale de France 12559 f.167r 79
Figure 30. Wool cards in use 80

Figure 31. A batteur at work, after a woodcut by Jost Amman, 1568 84
Figure 32. Europe, map showing some sites mentioned in the text 87
Figure 33. Peter Vischer the elder with his hammer, St Sebald tomb, Nuremberg 91
Figure 34. Brass figure of Theodoric by Peter Vischer the younger, c.1519 92
Figure 35. Veneto-Saracenic style ewer, c.1500, Poldi Pezzoli Museum, Milan No.1656 96
Figure 36. Asia, map showing some sites mentioned in the text 103
Figure 37. Europe and northern Africa, map showing some sites mentioned in the text 106
Figure 38. Seventeenth-century brass pan with lizard decoration, British Museum 1955 AF 225 108
Figure 39. Austria, map showing some sites mentioned in the text 111
Figure 40. Astrolabe made and engraved in Nuremberg by Georg Hartmann, c.1540s 114
Figure 41. Everyday products of the Nuremberg Twelve Brothers: *hanging up, left:* oil lamps, *left to right.* 1525, 1518, *hanging up, right:* key rings *left to right:* 1586 and 1528; foreground, *left to right,* 2 candlesticks 1526; ewer 1544; tankard 1518; two arquebus barrels 1528; tap 1586; candlesticks *left to right:* 1525 and 1518; a brass-worker rasping a cannon barrel, 1518 115
Figure 42. Map of England and Wales, showing some sites mentioned in the text 123
Figure 43. Isleworth mills and Brode's (Monsieur le Broade's) house, 1635 127
Figure 44. Ratzeburg and its lake in 1586, looking north (Bäk is at the top left corner), after *Civitatis Orbis Terrarum* V, photo Ozgar Tafekai 129
Figure 45. Stockholm Castle interior, 1616, with brass and copper floor tiles. The king is receiving the Dutch ambassador 135
Figure 46. Asia, map showing some sites mentioned in the text 140
Figure 47. Chandelier by Winant Nacken, 1633, in Skultuna church 145
Figure 48. Sweden, map showing some sites mentioned in the text 148
Figure 49. Map of Ratzeburg lake area, showing some sites mentioned in the text 151
Figure 50. Figure of a Portuguese soldier, 15 to 16 century 154
Figure 51. Examples of kuduos, *top right:* kuduo, royal grave, Kumasi, Ghana; *lower left and right:* Ashanti kuduos from the Gold Coast 156
Figure 52. Taynton field names, including Brass Mill Field 159
Figure 53. Diagram of a reverberatory furnace 160
Figure 54. Map of the Bohemia area showing some sites mentioned in the text 168
Figure 55. Lienz brass battery works in ruins after the fire of 8 April 1609 174
Figure 56. Map of Asia, showing some sites mentioned in the text 180
Figure 57. Chinese imperial palace, equatorial sphere: diameter *c.*1m. 1669-1688 180
Figure 58. Bidri pandan inlaid with brass leaves and silver flowers, Bidar, 17 century 182
Figure 59. Islamic brass; *background,* eighteenth-century brass alams held in the royal Shi'a house of mourning, Hyderabad; *foreground,* cast-brass calligraphic dragon's head finial, 1650-1750 (10.7cm wide) 183
Figure 60. Map of England and Wales, showing some sites mentioned in the text 187

Figure 61. Manillas from the three corners of the slave trade. *left*: manilla excavated at King Street, Bristol; *centre*: manilla excavated beside the former slave market site, Nevis, Caribbean; *right*: manillas from the kingdom of Benin 190

Figure 62. Drawer handles, *top left* tear-drop handle with key-plate c.1690; *top right* bat-wing plate c.1720; *lower left* bat-wing style, later mid-1700s; *bottom right* loop handle with disc plates 193

Figure 63. Ember brass mill interior, 1689-91, after a drawing by Eric Odelstierna, National Archive, Stockholm, Bergskollegium Huvularkivet E 111 5 (relating to '4', folio 611) .194

Figure 64. Musical instruments; author's drawings, *top*, bass trombone, 1612, Isaac Ehe, Nuremberg; *below, left* French horn in F, 1700-25, Christian Bennet, London; *upper right* basset horn, late 18 century, Johan Heinrich Grendel, Dresden; *lower right* natural trumpet, 1667, Simon Beale, London, Germanisches Nationalmuseum, Nuremberg M1168; Bate Collection, University of Oxford, 603, 487, x78 197

Figure 65. Part of Holstein, map showing some sites mentioned in the text 205

Figure 66. Bettenhausen brass works, west façade, 1679, 50 metres wide, after Alois Holmeyer, 1923, plate 360 212

Figure 67. Horse and rider, cast brass, Achenrain brass works c.1650-60, 43.5 cm. high, Tiroler Landesmuseum B43 217

Figure 68. Oba Ewuakpe with attendants, Benin...224

Figure 69. West Africa, map showing some sites mentioned in the text 226

Figure 70. Kitchen wares: *top left* three-footed English cauldron, *top right* skimmer; *centre* domestic pan; *lower left* 'frying pan' or skillet; *lower right* pestle and mortar, Blaise Castle Museum, Bristol, TA 6900, T 5018, T260; Vienna Technical Museum 38612.......228

Figure 71. Plan of Baptist Mills, based on a plan by de Wilstar, 1750..238

Figure 72. Brass candlesticks, *left to right*, Netherlands type with tulip-shaped cup, 1680; Spanish type, c.1650; English type 1700-1720 240

Figure 73. Cherub spandrels, on an eighteenth-century long-case clock, Richard Rooker 241

Figure 74. Brass flat-iron, cast at Skultuna, with decorative dolphin features, 1726................244

Figure 75. Plan of Hegermühle brass works, Finow-Eberswalde, Brandenburg 1784 245

Figure 76. Surviving Hegermühle buildings: *above*, remains of the 1739 furnace-house walls; *below*, 1724 workers' housing. 246

Figure 77. Europe, map showing some sites mentioned in the text 249

Figure 78. Battery hammers, late 18 century: *back*, forging strips for wire-drawing, *centre*, hollowing out cauldrons, *near*, flattening plate brass for shaping into drinking vessels. Krunitz Encyclopedia 89, page 400, inset figure 5206.2..251

Figure 79. Reichraming brass mill, 1763: 1 furnace house; 2 calamine mill; 3 wire-drawing; 4 brass battery; 5 smelting; 6 charcoal; 7 stables; 8 manager's mansion; 9 workers' dormitories; 10 channel supplying water to mill wheels, Seitenstetten Abbey, by kind permission 255

Figure 80. Sweden, map showing some sites mentioned in the text 258

Figure 81. British naval brass equipment: *left,* late eighteenth-century sextant; *centre,* barrel-spigot (tap) with bucket-hook, 1780-1815; *right,* careening block, c.1809, with central reinforcing sheave (plate) and brass coak lining the rope-groove, National Maritime Museum, Greenwich, NAV 1140; ZBA 0439; AEL 0453 265

Figure 82. Ancestral masks from Temne, Sierra Leone, with applied brass strips 268

Figure 83. Decorated bowl or container from Borneo, thought to be Dayak, Museum Nasional Jakarta 25269..270

Figure 84. Eighteenth-century brass harness bosses from Colonial Williamsburg, Virginia, CWF 2.4; 2.16; 0185 ANH; 0786 ANH 270

Figure 85. The restored Geddy Foundry, Colonial Williamsburg, Virginia 272

Figure 86. England and Wales, map showing some sites mentioned in the text 275

Figure 87. Candlesticks: *right,* late eighteenth century square-based candlestick: *left,* Georgian petal-based type candlesticks, *c.* 1780 and *c.* 1760 276

Figure 88. Greenfield Mills, Holywell, Flintshire, north-east Wales, after a painting *Gweithfeydd Copr yn Nyffryn Maesglas*, 1792 286

Figure 89. Turner's Brass House, Birmingham, 1753, after S. and N. Buck engraving, *East Prospect of Birmingham* 288

Figure 90. Plan of Norrköping brass works, based on a plan by Pontelius, 1751 295

Figure 91. Brass keyhole plate cast at Norrköping c.1780, based on an insert between pages 128-9, Helmfrid 1954, *Holmenöden under fyra sekler*, Östergotlands Museum, Sweden 296

Brass currency bars from Hedeby, northernmost port of the Carolingian empire, freshly cast from metal ores mined in the Balkans. 16-25cm long. *Top*, photo © Wikinger Museum, Haithabu, Schleswig; *Bottom*, photo © author, by courtesy of the Viking Museum, Haithabu, Schleswig

Chapter 1

Experiment and emergence

From prehistory to the Roman Empire – *c*.2500 BC-AD 500

Brass, a gold-coloured, metal alloy, shows an engaging aptitude to follow the same course of history as its human makers. The glint of flames on brass ornaments on the shelves of elderly relatives may evoke memories, but brass, in the past, was for more than just trinkets. It is an exciting glistening alloy with a turbulent history, ranging from the exotic to the sordid, and it was often at the centre of political, social or economic interactions. Brass was long seen as resembling gold, which is a metal that can occur in nature, whereas brass, like bronze, must be alloyed. Brass-making, however, was more difficult than normal metal-alloying, caused pollution, and used up valuable fuel resources – yet new brass-alloying processes and decorative techniques were steadily developed over time, to meet increasing demand, so what did people make from it, how, and why? Since it was a somewhat difficult alloy to make, why take the trouble to make it at all? Why would people want brass, and who were they?

Brass, being an alloy, is a mixture of metals that have been selected and modified, and it is alloyed from copper and zinc whereas bronze is alloyed from copper and tin, so why might brass sometimes be chosen in preference to bronze? They had some differences, brass (8% zinc) has a desirable rich yellow-golden colour much like that of gold and it was tougher, harder and stronger than bronze. Brass is also more malleable (easier to hammer) than bronze and more ductile (it could be drawn out into fine wire). This was because the crystal structure of brass consists of positively-charged atoms[1] of copper and zinc that are virtually equal in size, allowing them to form a tightly-packed solid cubic crystal lattice of metal atoms with a cloud of negatively-charged electrons moving freely throughout the lattice. This arrangement allows the smooth surfaces of layers within this lattice to slide against each other when hammered, so the alloy is hard and holds together well under stress. Another useful property of brass is its low friction, making it useful for moving parts like locks and hinges. This contrasts with bronze, in which the atoms of copper and tin forming its crystal lattice are very different in size, so the surfaces of the layers are more corrugated, meaning that hammering or stretching forces its atoms out of position more readily. These dislocations distort the bronze crystal lattice sooner, producing a more brittle metal that is prone to cracking.

Another great advantage of brass over bronze is that not only are zinc sources far more widespread than the tin sources needed for bronze-making, but zinc ores often

---
[1] strictly-speaking, ions

occur in the same region as copper ores. Early on in the prehistoric period, pure or 'native' copper ore often lay near the surface, and native copper, a malleable, easily hammered material, was among Neolithic exchange goods handled before 8000 BC at Hallan Çenü and at Çayönü settlements in eastern Anatolia,[2] where small artefacts made from locally-sourced native copper date from 7400 BC onwards. In Anatolia, at Catal Höyük (double mound), native copper beads accompanied female burials dated to *c.*5900 BC.[3] In China, copper artefacts have also been found, dating back to at least the 4th millennium BC[4]. The extensive Kargaly copper mines in today's northern Kazakh steppes, date from the late 4th to the early 3rd millennium BC.[5] Copper is also thought to have been worked from the 4th millennium BC in Nubia, south of Egypt, and around 1200 BC at three sites with furnaces at Agadès, south-east of Azelick, Niger.[6] An important late Egyptian and Roman copper source was in Cyprus (source of the Latin term *cuprum* – copper), where copper was locally exploited from 3500 BC, intensifying from 2500 BC upon cultural contact with Anatolia.[7]

In Europe, the copper-exploiting, or Chalcolithic, Age started at different times depending on the area. Nearer to the Carpathian Mountains, the Vinča culture (*c.*5700-4500 BC), embracing most of the Balkan states, produced a copper axe dating to 5500 BC. Later Vinča-culture burial mounds, or 'cities of the dead', near Pločnik in today's Bulgaria, have revealed copper tools, weapons and personal adornment such as armbands[8] and, by about 4800 BC, nearby sites produced needles, awls, beads and earrings.[9] The Bulgarian Varna culture (*c.*4400-4000 BC) used copper axe-heads, spearheads and chisels. The Copper Age started about 4200 BC in Austria and Bavaria, and about 3300 BC (the date of the copper axe of Ötzi, the ice-man) in the rest of southern Europe.[10]

Smelting, using heat to release metals from their ores, made it possible to alloy one metal with another. Lead is simple to smelt from ore, so its onset is hard to trace, but evidence exists from the 6th millennium BC.[11] Copper was probably smelted in the late 6th to early 5th millennium BC, in the Balkans and perhaps Iran,[12] by which time metal workers could experiment with alloying copper with other, different, metals, wondering what each new alloy had to offer. Knowledge of the history of these ancient alloys is growing all the time thanks to the expertise and technology of

---

[2] Özdoğan 1999: 54; Sharp-Jonkowsky 2002: 88
[3] Mellaart 1967: 22, 52, Radivojević et al 2017: 108-9
[4] Reade 1991:34
[5] O'Brien 2015: 187-8
[6] Herbert 1984: 7, 16-17
[7] Steel 2004: 3, 83, 121-1266, 114, 143-144
[8] Jovandovič 1990: 55-56
[9] Gale *et al* 2003: 124-125
[10] Heyd 2008: 24
[11] Craddock 1995, 125, 205
[12] Radivojevic et al 2017, 121

archaeo-metallurgists, those scientists who analyse and study archaeological metal objects. The deliberate alloying of copper and tin to make bronze signalled the start of what we call the Bronze Age. The origin of most early copper alloys is not known, but in north Georgia in the Caucasus region (between the Black Sea and the Caspian Sea), copper was first alloyed with locally available arsenic in the 5th millennium BC,[13] and arsenic, antimony, nickel and lead ores were subsequently all alloyed with copper there[14]. A few bronze examples date to about 4300 BC from Aphrodisias, western Anatolia[15] and a bronze spiral ring containing 11% tin, from Talin, Armenia, dates to around 3500-3300 BC.[16]

The origins of brass making are hard to clarify, but small-scale developments evidently occurred earlier than previously thought. For example, in the eastern Black Sea region of the south Caucasus (now western Georgia), small outcrops of polymetallic ores containing copper, zinc and iron[17] are very common.[18] People of the agricultural Kura-Araxes culture (about 3500 to 2500 BC) alloyed and worked metals in Armenia, the south Caucasus, north-west Iran and eastern Anatolia. From the early 3rd millennium BC, rare small copper-alloy artefacts, particularly beads, occur in this area, containing up to five to six per cent of zinc, as well as arsenic and tin.[19] The contemporary northern-Caucasus Maikop culture used these alloys in high-status burial goods, one of which was made of zinc metal.[20] In Armenia, a few early brass objects were made from the late 3rd millennium BC,[21] and were widespread in South Georgia and East Anatolia during the 2nd millennium BC.[22]

China, too, had zinc-rich copper ores from which early brass was perhaps sometimes accidentally produced,[23] under the right conditions, enough to make a few objects. Its zinc ores stretched from the T'ai-Hang Mountains of Shaanxi province in the north,[24] through the central provinces,[25] and southwards to Guangdong and to Pa Niu (now just inside Thailand).[26] Rare, scattered reports of fortuitous Chinese brass include a hairpin from Shaanxi province claimed to date to about 3000 BC, and an awl from

---

[13] Meliksetian *et al* 2010: 201
[14] Ghambashidze 1919: 42, 47
[15] Sharp-Jonkovsky 2002:143
[16] Sagona 2018: 268-9
[17] Zinc-rich chalchopyrite
[18] Ghambashidze 1919: 42, 47
[19] Abesadze: 1969: 207, 253, 282-283
[20] Chernykh, 1992: 66-67, 74
[21] Meliksetian et al 2011: 205, 209
[22] Iskili and Altunaynak: 2014, 78
[23] Fan Xiaopan, personal communication
[24] Golas 1999: 138, note 375; Song Yingxing 1966, 471
[25] Chen Jianli *et al* 2005: 49; Smith 1918: 68
[26] Needham 1974: volume 5, part 2, 212;

a site in Shandong province claimed to date to the 3 millennium.[27] However, dating and stratification may be a problem for these isolated claims, especially given that the widespread adoption of brass in China did not come until the 15 century AD. Local craftsmen in either the Caucasus or China may have produced low-zinc objects accidentally, rather than intentionally, because in practice their yield would mostly have been copper, with up to about 1% of zinc, most of the zinc having vaporised, and most of the iron having ended up in slag (the waste product of smelting).[28] The main problem was that, despite availability of materials, brass was not straightforward to alloy in the prehistoric era, and the difficulties may have taken a long time to solve.

To alloy copper with zinc, metal workers had to overcome two daunting technological problems. The simplest way to make an alloy is to mix the two metals together, metal with metal, but zinc is more difficult to handle in this way, and this will be discussed below. The other reason is that zinc metal does not occur on its own in nature – it combines too readily with other elements – but it is common as a composite ore, occuring as zinc carbonate or zinc sulphide.[29] Even copper is rarely found as a native metal on its own, so making making brass deliberately by combining copper ore directly, even with zinc carbonate, was not likely in early prehistory.

The next question is: who first made brass intentionally, and where? An alert prehistoric copper craftsman might occasionally have noticed unusual, slightly golden, effects when smelting ores containing copper and zinc, and have repeated the experiment. The earliest evidence for deliberate brass-making seems to emerge in a very few places in the Caucasus, the Middle East and neighbouring areas in the 3 millennium BC. Later on, it is rare for a copper-zinc brass to contain less than about 8% zinc. Therefore, for this account, brass will be assumed to contain at least 8% zinc, by weight, and the objects described in this chapter as brass have all been analysed and found to be copper containing at least 8% zinc, together with only small amounts of other ingredients.

During the prehistoric period, the story relies on scarce archaeological finds of individual brass fragments, but a picture is nevertheless emerging. As farming led towards a settled lifestyle, communities gradually became organised into simple villages. Farming – herding and grain-cultivation – brought the chance to produce more food than was needed for subsistence alone. Surplus grain or meat raised the prosperity and status of the communities, allowing agricultural and herding leaders to emerge. Surpluses allowed these privileged members of family society to profit from the work of those producing the excess food, and perhaps to pay or maintain certain craft specialists like the makers of the special new metal, brass.

---

[27] Zhou Wenli 2016: 9
[28] Gilmour et al forthcoming and B. Gilmour personal communication
[29] $ZnCO_3$, smithsonite or ZnS, sphalerite

The practice of crafts like making brass would have been determined by the overall customs of the culture, but the golden appearance of the alloy would arrest the attention of ambitious leaders. Over time, stronger leaders and more organised societies emerged, developing from nomadic pastoralism to villages, city-states and kingdoms and finally to empires with fluctuating, fluid boundaries. Shifting boundaries gave excellent opportunities for contact between peoples, and this kind of interaction allowed the diffusion of knowledge needed to identify and find copper or zinc carbonate ores and to learn brass-making technology from practising craftsmen. Archaeological evidence shows that leaders gained prestige and power by circulating rare gifts that carried a magic aura. Early brass came into this category, and those who knew the mysteries of making it would have been worth enticing or capturing. The rare surviving prehistoric brass items suggest that they were intended for handling by those higher up the social ladder. Skilled craftsmen who made brass are likely to have depended on those who could repay them and perhaps lift them from subsistence living and keep them while they worked.

The skill of these very early, possibly itinerant, brass craftsmen is astonishing, because of the complexity of making brass as opposed to bronze (for which copper and tin can simply be melted together). There are two major problems. One, as already mentioned, is that zinc does not occur in nature as a pure metal. It is too volatile and reactive – combines too readily with other elements – and only occurs in compound ores such as zinc sulphide (ZnS) or zinc carbonate ($ZnCO_3$). The other problem is that zinc is so volatile that it vaporises and escapes into the atmosphere when heated above 907°C, a temperature below the melting-point of copper (1084°C). Therefore, since zinc and copper cannot both be in a molten state at the same time, it was no use trying to melt them together. The answer was to produce zinc oxide, as a first step, because zinc oxide is more stable (less volatile) than zinc on its own. To produce it, zinc carbonate ore ($ZnCO_3$) was crushed, layered with powdered charcoal (which is carbon, a reducing agent) and left to roast at about 700°C in a partially enclosed space. Carbon (C) combines with oxygen (O_2) and escapes as carbon dioxide gas (CO_2), leaving behind zinc oxide (ZnO).[30]

The most practical zinc ore for the cementation brass-making process was zinc carbonate ore, commonly called calamine. However, in many mountainous areas, zinc sulphide ores predominated, so before reaching the zinc oxide stage, unwanted sulphur had first to be partly purged from the ore by slow roasting with charcoal until sulphurous gases escaped. Upland metal-workers in the Middle East discovered that they could produce zinc oxide by heating powdered zinc sulphide ore (ZnS) over a fire until it vaporised, combining with oxygen in the air to form white clouds of zinc-oxide vapour. This method eliminated most of the unwanted sulphur, and white powdery zinc-oxide encrustations could be collected on cold metal or clay bars above the open

---

[30] $ZnCO_3 \rightarrow ZnO + CO_2$

fire, as the vapour cooled.[31] Flakes of the zinc oxide could then be scraped off and collected either for making brass or for medicinal eye-ointments. An Indian medical treatise dating to about 500 BC mentions the product of this process, and its use for treating sore eyes and open wounds.[32] The zinc oxide flakes could be used for brass-making by the cementation process to be described below, during which, as the brass began to form, the temperature could be lowered and a bit more zinc oxide powder quickly sprinkled onto the surface.[33]

In order to actually make brass, prehistoric metal-workers would have needed (besides copper and zinc oxide) wood for making charcoal – which provided both heat and a reducing atmosphere – the heat of a fire, and some small containers to act as crucibles to hold the molten metal. Pottery was already culturally important, and fire- and heat-proof ceramic vessels were a natural choice as containers for molten or liquid material because they were already used at early copper-working sites for heating ore to extract copper metal.[34]

The cementation method mainly used by prehistoric brass-makers involved placing powdered zinc oxide and crushed charcoal (carbon) together inside a vessel (crucible), together with finely divided copper fragments. Zinc oxide reacted with carbon inside the crucible to produce zinc vapour plus, mainly, carbon dioxide.[35] A tightly-fitting crucible lid could prevent zinc vapour from escaping into the atmosphere. Once the zinc vapour started diffusing in through the surfaces of the warm but still-solid copper fragments, brass began to form. The temperature always had to be kept just below the melting-point of copper (1084°C) to make sure the small, hot copper pieces remained solid, because if copper melts it will sink into a pool at the bottom of the crucible and present too little surface area to absorb enough zinc vapour to form brass. Experiments suggest that 950-1000°C is the temperature for maximum uptake of zinc vapour by copper,[36] but the optimum temperature varies as the process proceeds.

During the brass-making process, as copper absorbs zinc, the temperature will drop as the solid alloy becomes fully liquid (the *liquidus* phase). Once 20% zinc has been absorbed, for example, this liquid phase will have lowered from 1084°C, to 1000°C. Once nearly 30% zinc is absorbed, the alloy will become liquid at only 919°C. Since this is barely above the temperature at which zinc vaporises enough to diffuse into the copper, the reaction soon halts. This explains why early cementation brass did not normally contain more than just under about 30% zinc[37]. Ideally, therefore, the

---

[31] Allan 1979: 38-9; Reade 1991: 34
[32] Craddock, Freestone *et al* 2000: 29
[33] Craddock and Cowell 2000: 123-124
[34] Jovandovic 1990: 55-56
[35] $2\,ZnO + C = 2\,Zn + CO_2$
[36] Bourgarit and Bauchau 2010: 51; Newbury *et al* 2005: 75
[37] Rehren and Martinón-Torres 2008: 168

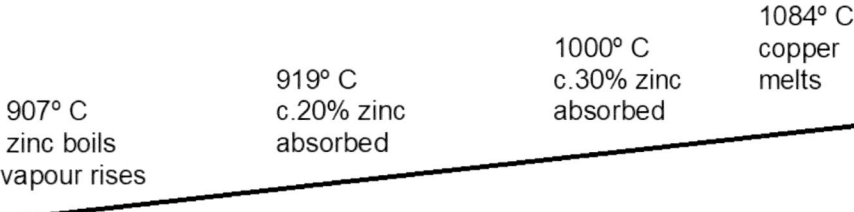

Figure 1. Diagram of temperatures during cementation

crucible temperature should be controlled to start at around 1000-1100°C and be gradually lowered to nearer 907°C (the boiling point of zinc),[38] not easy to regulate without thermometers, and perhaps seldom achieved. Once alloyed, the brass could be heated slightly further, to make sure it was evenly mixed, after which it could be cast – poured out into a mould.

Brass can contain varying amounts of zinc. To find out whether an early copper alloy metal object is made of brass, it is necessary to know the ratio, by weight, of copper to zinc and to any other elements present. As already mentioned in this account, only copper alloys containing at least eight per cent zinc (by weight) will here be counted as brass. For the alloy to be a true brass, other elements such as iron, lead or tin should only be present in much lower percentages than zinc – generally up to two or three per cent, but this was almost certainly uncontrollable in prehistoric times. Lead, tin or iron lower the melting point, reducing the amount of zinc that copper can absorb, so that when the molten brass temperature drops to 907°C it will contain less zinc than with a purer mix. Too high a percentage of the elements arsenic and antimony makes brass brittle to work.

Although there is very little evidence for brass-making in the early Bronze Age – the early to mid 3rd millennium BC – what does survive gives an idea of which levels of society saw the fascination of brass and could afford it. At Thermi on the island of Lesbos, just across the water from ancient Troy, the few excavated brass objects include a knife, pin, ornament and pierced disc,[39] dating probably to the mid to late 3rd millennium BC. These were luxury items in use during the first and second phases of Troy, where a powerful leader, with his family, retinue and servants, ruled from a citadel, surrounded by farming families, perhaps dependent on him for protection. Thermi, on Lesbos, which had an organised layout of long rectangular buildings, had

---

[38] Rehren 1999: 1083-1085
[39] Begemann *et al* 1992: 224; Thornton and Ehlers 2003: 4

the same stratified society as Troy itself, so it is not hard to guess to which social level the users of the four golden-looking brass objects belonged.

More evidence for Middle Eastern brass in the mid to late 3rd millennium BC comes from a high-status grave at the cemetery at Ur, which contained part of a miniature 9 cm-high brass helmet,[40] too small to have been worn into battle but a clear statement of power and influence. Ur was the world's largest city of its day, and capital of the great and progressive Sumerian Empire (southern Iraq). It imported luxury goods from Afghanistan, India and Asia Minor, and was the leading port on the Persian Gulf, which then extended further north. Brass grave-goods at Ur, including the helmet, a dagger, spear-tip (10.7 cm long)[41] and bowls, imply the presence there of a wealthy elite, and were perhaps originally used as revered objects for gift exchange, or produced solely to accompany important burials. Another Sumerian city, Kish (Kiš) on the Euphrates, covering several square kilometres of land about 80 km south of today's Baghdad, had forty tells (habitation mounds) and a simple palace – a sure sign of a stratified society. A brass bowl and a toilet knife found at Kish seem to have been intended for elite users.[42] Also dating to the early Bronze Age (the mid 3rd millennium BC) they too suggest which levels of society desired brass – and could afford it.

Figure 2. Miniature brass helmet from a royal grave at Ur

In the late 3rd millennium BC, the evidence suggests that brass was exploited by powerful members of society to achieve their own ends. On the steppes of Kalmykia (now western Kazakhstan), the elaborate catacomb graves of the settled farming community at Ergeni (Ergeninski) north of the Caspian Sea contained a brass knife and hook.[43] Other excavated material from this site shows that the pastoral horse-riding nomads of Kalmykia dressed in multi-coloured clothing adorned with long strings of bronze and paste-glass beads. In many prehistoric contexts, exotic materials possessed by powerful leaders reinforced their status, and unusual objects gained in

---

[40] Hauptmann and Pernicka 2004: 82, and figure 127
[41] Hauptmann and Pernicka 2004: figure 103
[42] Hauptmann and Pernicka 2004: 27 and figure 35
[43] Thornton 2007: 126

prestige as they passed around the exchange networks, imparting their value to the giver and receiver. A leader who participated in the exchange system stood far higher on the social ranking ladder than the craftsman who made and worked the brass.[44] At Ergeni in Kalmykia, and contemporary sites, objects made of unusual metals were used in complex burial ceremonies and as symbolic talismans for reinforcing territorial control. Around 2500-2200 BC, shortly before the Kalmykia steppes dried up for a millennium,[45] brass and locally-mined copper were cast in moulds. Casting brass in a mould was less straightforward than casting bronze but solutions were available and experiments would have shown craftsmen how to cast brass successfully. The rare brass items excavated from such sites must represent a small minority of what originally existed, many more having been lost or looted, so never available for analysis. At two sites in today's Georgia, besides third-millennium copper-working evidence, copper tools and weapons, two objects, a dagger blade from Nuli in today's northern Georgia and a disc-shaped pin from Telebi (Telavi), east Georgia, each contained five to six per cent of zinc, which might (for this early date) be called a low-zinc brass.[46]

Societies were starting to urbanise, but the few surviving brass objects from millennia ago may represent the way in which many more were used and valued. From the late 3rd millennium BC, excavated material from two Middle Eastern urban settlements includes three different kinds of brass artefact. Two of them, a needle and seal of pure brass were excavated at Namazga (in today's southern Turkmenistan), a 60-hectare nascent urban community on a strip of land where rivers descend to the plain from the dramatic eastern Iranian slopes.[47] The third artefact, a dagger, also of pure brass[48] was excavated at Umm-an-Nar,[49] a different contemporary urban settlement on an island next to Abu Dhabi on the Persian Gulf (now United Arab Emirates).[50] It has been calculated that circulated prestige goods, like the brass objects described, must have facilitated the creation of political and economic hierarchies by imparting authority to them – an effect that radiated from the impressive rarity of the brass objects, their symbolism, possible invested spirituality and the elevated strata of society within which they circulated.[51] With urbanisation, however, brass making would gradually begin to lose its aura of mystery and become more utilitarian.

The stone buildings of the Umm-an-Nar culture, which lasted from 2600-2000 BC, included one extra-large house with many rooms for the dominant leader, and beehive-shaped burial cairns constructed of carefully fitted stone. Copper-working

---

[44] Frachetti, 2006: 27
[45] Anthony 2009: 48
[46] Kavtaradze 1999: 73, 91, 95
[47] 37°37 N 59°55 E
[48] 90% copper, 10% zinc
[49] artefact 1011, A1, Grave II
[50] Frifelt 1991: 98, 100-101
[51] Weeks 2003: 189

Figure 3. Dagger from Umm-an-Nar

took place at this site,[52] whose inhabitants drew their prosperity from fishing and maritime contacts, and from trading their local copper to Mesopotamia (Iraq). It is possible that the brass dagger found there was made in Mesopotamia as an object for exchange connected with the copper trade. It was found in a stone-built grave, twelve metres in diameter, furnished with picture-stones, one bearing the image of a camel in low relief. At Nuzi, in Akkadia, east of the River Tigris in northern Mesopotamia (north-eastern Iraq), two excavated brass rings date from the late 3rd millennium BC.[53] Nuzi was a provincial town full of Hurrian people who originated from Anatolia where zinc ores occur,[54] so they may have arrived already knowing how to alloy brass.

This was becoming more than just a period of emerging city-states; it merged into a time of ruthless rulers pressing into new territories, forming empires whose boundaries surged back and forth with the winners and losers. Contact across shifting boundaries diffused technical knowledge. As one powerful state or empire shrank towards oblivion, crafts such as brass-making would remain or travel with expert craftsmen, to be taken up and exploited by a newly emerging power. The development of better agriculture and herding meant that cattle- or grain-owning communities were prospering, to the envy of their neighbours. Dating from shortly after 2000 BC, at Dalversin, an influential state in the fertile, prosperous Amu Darya (Oxus) valley (in today's southern Uzbekistan) a brass ring and pin were excavated,[55] suggesting interaction with neighbouring states.

Within the boundaries of the new emerging states, more complex hierarchical societies were developing. In the early 2nd millennium BC, at Tepe Yahya, in the Soghun valley, in Kerman province (in today's southern Iran, 1,200 metres above sea level and several days' walk east of the Persian Gulf) three fragments of brass, and part of a bracelet dating to the early 2nd millennium BC were excavated.[56] A '*tepe*' was a group of *tells* or great mounds created by generations of inhabitants building their homes upon those that had been there before. This early town had defensive walls and buttresses,[57] and the rooms containing the bracelet and other finds show that its houses and other buildings belonged to a ranked society, and the bracelet found there was probably worn by a well-to-do woman.

---

[52] Frifelt 1991: 188. Oriental Department, Moesgard Museum, Jutland
[53] Website: news.harvard.edu/gazette/1998/05.14/Fragments.of.a.Forgotten.Past
[54] Website 2003: bu.edu/phpbin/researchbriefs/display.php?id=121
[55] Thornton 2007: 126
[56] Thornton 2007: 126; Thornton and Ehlers 2003: 4-5
[57] Lamberg-Karlovsky and Magee 2001: 128, 180, 302-322

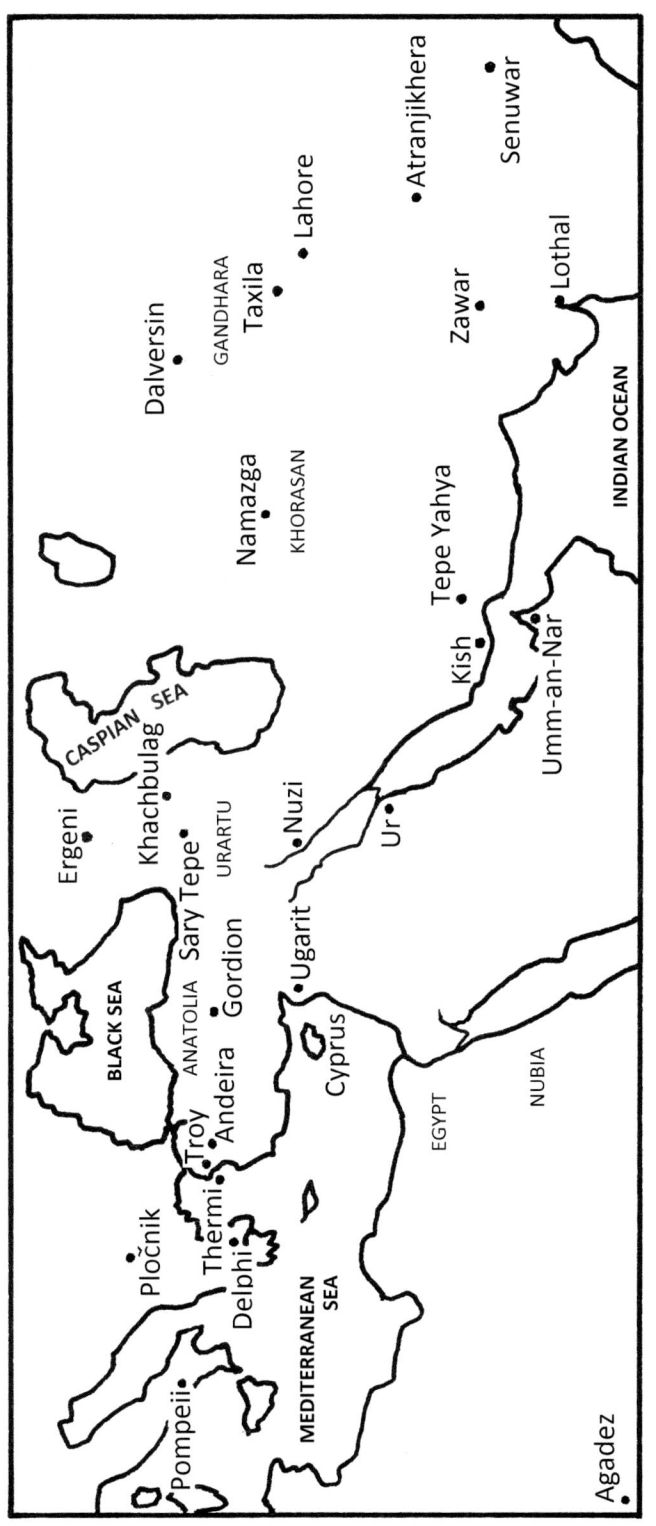

Figure 4. The Middle East to India, map showing some sites mentioned in the text

During this city-state period, when trade contact was widening, the metal contents, and therefore qualities, of brass varied. Metal finds from middle or late Bronze Age sites in the Caucasus, for example, tended to be less pure than in the 3 millennium BC, being mostly of mixed copper alloy with varying amounts of zinc, tin and lead and sometimes arsenic or antimony. Ugarit, a Mediterranean headland port (now in Syria) stood on the south-north route from Egypt to Anatolia – a rich source of zinc carbonate ore.[58] Pottery and metal finds show that Ugarit had strong links with contemporary Egypt and with Cyprus, one source of the copper needed for making brass. A brass statuette and brass ring from Ugarit, dating to around 1400-1300 BC had a lower zinc content but higher lead content than earlier objects, which suggests a change in the way the metal was being used.

By the later 2nd millennium BC, Middle-Eastern brass contained greater amounts of metals like lead and tin, perhaps due, in part, to casting rather than hammering the alloy. Casting brass in a mould would prove less straightforward than casting bronze, but a deliberately-added extra two per cent of lead lowers the melting point of the copper and so makes a more fluid alloy – allowing the molten metal to be poured easily and flow into all the corners of a casting-mould. On the other hand, the increased lead may have resulted from the use of locally available lead-rich copper ores, whilst increased tin could have resulted from recycling old bronze and brass objects together to make new ones.

Rising prosperity in the emerging empires and city-states is revealed by archaeological evidence for brass items possessed mainly by the elite of the Elamite, Sumerian, Babylonian, Assyrian and neighbouring territories, wealthy enough to pay for their production. Such leaders would have valued brass objects, possibly already charged with superstition and mystical religious belief, and used them as talismans to protect their households or to present as charmed gifts to leaders of other city-states. Not only could brass look rather like gold, but it may have had a cachet of its own due to the little-understood processes required to make it. The Azerbaijan coastline, along the east side of the Caspian Sea, brought the region into frequent contact with other cultures and made it a popular target for expansionist empire-builders.[59]

During the earlier 1 millennium BC the Assyrians were being forced out of the region between the Black Sea and Caspian Sea by a rival dynasty, the Urartians, whose domains covered today's eastern Turkey, western Azerbaijan and Iran. In 764-735 BC, Sarduri II of Urartu (a state at the time called Biainili) built Sary Tepe, an elite settlement. Strung along a rocky ridge, the site (in today's Armenia, north-east of today's Dilijan), it still shows clear traces of buildings, from which a brass bracelet and an arrowhead of almost pure brass were excavated. Brass would have made an effective sharp tip to

---

[58] Website: metmuseum.org/toah/ugar/hd_ugar.htm
[59] Babaev *et al* 2007: 31-32

an arrowhead, but no other brass arrowheads have been analysed from Sary Tepe (just one bronze one), so it may have been created as a rare prestige item rather than for normal use. Also from a Urartian settlement is a figurine of a small horse, with a loop perhaps intended for hanging it from a necklet as a talisman – an early example of 'gunmetal', a term meaning copper containing similar amounts of zinc and tin.[60] Another luxury item of dress-adornment was an 11th- to 12th-century BC brass bracelet fragment excavated from Khachbulag[61], lying between the south-east corner of the Black Sea and the Caspian Sea in western Azerbaijan (south-west of today's Mingecevir).

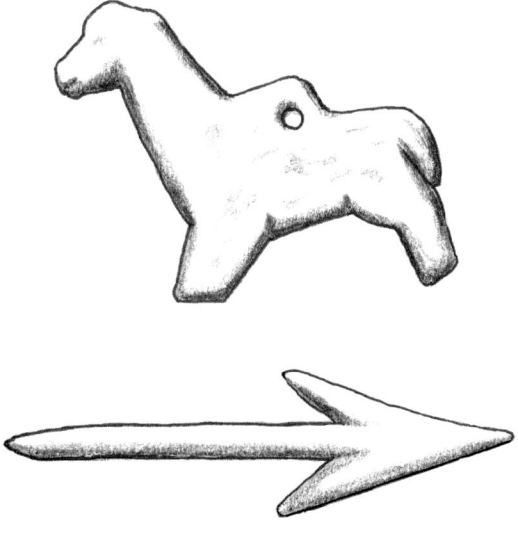

Figure 5. *above*, horse figurine, Kachbulag; *below*, brass arrowhead, Sary Tepe

Although recorded sporadically up to this point, brass became more widespread from the 8 century BC, when brass objects are known from Urartu, Phrygia and Assyria.[62] The metal craftsmen of Urartu used hammering, casting, embossing and engraving techniques. Mineral-rich Anatolia (in today's north-east Turkey, just south of the eastern Black Sea) steps into the picture in the 8 century BC as a source of calamine and a centre of brass-making. A brass bracelet dating to that period was excavated from Cavustepe, south-east of Lake Van in north-east Anatolia.[63] In central Anatolia, by the 8 century BC, Gordion, part of Phrygia, the state ruled by the powerful Midas, had fortification walls and formally planned buildings along a broad main street.[64] Farm produce was abundant around Gordion, and its people were in close contact with eastern Mediterranean trade. Besides metal-working, its farmers and craftsmen practised textile-weaving and pottery.[65] A brass bowl was found there, as well as four early brooches made of bronze, but each containing a little zinc – the copper alloy now termed gunmetal. Brass craftsmen are likely to have worked among agricultural communities where iron and bronze tools were made. Some may have moved to gain the support and protection of a wealthy patron, perhaps transferring from one to another, allowing technical knowledge of brass-making to diffuse over time.

---

[60] Kashkai and Selimkanov 1973: 222, plate 20, figures 46 and 45
[61] Kashkai and Selimkanov 1973: 221, plate 199, figure 37
[62] Kroll *et al* 2010: 36
[63] Kroll *et al* 2010: 17
[64] Thornton, 2007: 127; Young 1981: 287
[65] Kealhofer 2005: 1 and 31

The excavated and analysed brass samples so far described as brass in this chapter contain an average of 14.4% zinc, and most of them come from sites representing social hierarchies where a more powerful leader could provide a livelihood for skilled craftsmen such as brass-workers but brass was not yet an everyday utilitarian metal. Many excavated copper-alloy items contain too little zinc or too much lead or tin to count as the copper/zinc alloy defined here as brass, so early brass-making was far from a standardised process. Metal ores in some areas contained widely varying amounts of different metals, for instance zinc or copper ores from the Indian sub-continent commonly included iron and more than three per cent of lead. The pure metal known as 'native copper' (Cu), still occurs in the Deccan region of India – the triangular raised plateau spanning central and southern India.[66] At the far north-western corner of the Deccan, an early second-millennium BC site at Prakash produced a brass object found to contain over 25% zinc,[67] a fairly high percentage for the time.

About six leaded copper-alloy fragments containing up to 6% zinc, dating to around 1500 BC, were excavated from Lothal beside the Indian Ocean (in today's Gujarat).[68] This organised port had a well-built dock, ruler's residence, merchant houses and customs house. Lothal had been settled by Harappan people from the developed Indus Valley civilisation further north, near sources of zinc-bearing copper ore.[69] The civilisation stretched from the Himalayas down to Gujarat, and its merchants traded to Mesopotamia and up and down the east African coast.[70]

Copper alloys of the Indus Valley, though they may have contained zinc, also held a considerable percentage of lead and tin – so they were quaternary alloys – combinations of four metals – with a variable balance of copper, zinc, lead and tin. They were therefore not brass by the definition used here, but may represent an increasing use of leaded gunmetal.[71] The Indus Valley civilisation ended by about 1500 BC. By the 6th century BC, the Ganga (Ganges) plain was becoming urbanised. Increasingly high percentages (by weight) of zinc began to be included in early Indian leaded copper alloys. At Atranjikhera, Uttar Pradesh, a settlement on the Kali Nadi river, a tributary of the River Ganges (Ganga), two pin fragments dating to around 600-500 BC were found, made from leaded gunmetal.[72]

The clouds of change hung over the Indus Valley region. The powerful Persian leader Darius (521-486 BC) advanced, conquered and added the north-western Indian states to

---

[66] Chakrabati and Nayanjot 1996: 18
[67] Kharakwal 2012, http//infinityfoundation.com/mandala/t_pr_khara_zinc.htm, 4
[68] Lothal analysis, number 4169
[69] Kharakwal and Gurjar 2006: www.ancient-asia-journal.com/rt/printerFriendly/aa.06112/23, volume 1, 5
[70] Taxila, 1999: http;//archeology.about.com/od/thers/g/taxila.htm
[71] Kharakwal 2012: http//infinityfoundation.com/mandala/t_pr_khara_zinc.htm, 4
[72] Gaur 1983: 231, 233 and figure 69/4; 444 and figure 130/5; Thornton 2007, 127

his vast empire, which already included Anatolia and Persia (Iran). His conquests brought these diverse areas into trading contact, and Darius himself was later reported to own an Indian cup which looked like gold but gave off a slightly offensive smell – characteristic of brass.[73] Another powerful and warlike ruler, Alexander the Great of Macedonia (356-323 BC), may have helped to diffuse brass technology,[74] by sweeping in to take much of Egypt, Mesopotamia, the entire Near East and Asia Minor – which fell to him after nearly 200 years of Persian rule. From 332 BC, Alexander the Great also conquered lands bordered by the River Indus and the Arabian Sea, moving north to Taxila beneath the Himalayas. Leaded-brass coins became expressions of power in India, and many were produced there to honour successive rulers, a few in the 4 and 3 centuries BC, but most from the 2 and 1 centuries BC.[75]

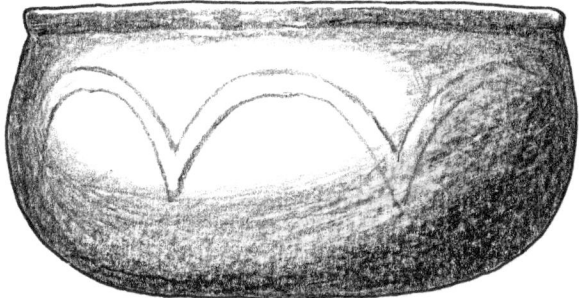

Figure 6. Brass bowl from Taxila

Taxila was already a famed Buddhist centre of monasteries and education, at the north end of the Indus Valley (about 32 km north of today's Rawalpindi, Pakistan). One of the most exciting third- to second-century BC finds from Taxila is a high-zinc brass bowl or vase (34.34% zinc), excavated from Bhir mound, a large settlement with many winding streets and closely packed dwellings.[76] Brass containing such a high percentage of zinc implies that it was made by a process different from the cementation process described above. Another high-zinc brass artefact (35.52% zinc) dated to between 700 and 200 BC,[77] was excavated at Senuwar, much further east near the borders of Bihar and Bangladesh.[78]

The existence of this high-zinc Indian brass can be explained. The great Mauryan Buddhist dynasty (321-185 BC) had been taking in territories throughout northern India. Chandragupta, the first Mauryan emperor, had a strong and ruthless advisor named Kautilya, a scholar educated at Taxila. Between 321 and 296 BC, Kautilya wrote of brass-making as part of an organised Indian metal-working industry. He specified that not only should the superintendent of mines understand 'the science of dealing with copper and other minerals', but should also have experience in the 'art of

---

[73] Hoover and Hoover 1950: 403, translation of Pseudo-Aristotle, *On Marvellous Things Heard* (4-3 century BC)
[74] Craddock and Eckstein 2003: 217
[75] Kharakwal and Gurjar 2006: http//infinityfoundation.com/mandala/t_pr_khara_zinc.htm, 7-8
[76] Marshall 1951, vol. 2: 568, table IV, plate 175, No 284; Marshall 1951, vol. 3: plates 2-7
[77] Kharakhwal and Gurjar, 2006: http//infinityfoundation.com/mandala/t_pr_khara_zinc.htm, 7
[78] *c.*110 km east-south-east of Varanasi, *c.*140 km south-west of Patna

distillation and condensation of mercury' and be responsible for 'the manufacture of brass (*arakuta*)' and for producing commodities from it and from other metals.[79] The art of distillation mentioned by Kautilya brought a key advance in brass making, since distilled zinc metal allowed more precise control of the proportion of zinc to copper and made it possible to include a higher percentage of zinc than was possible with the cementation process.

Clearly, the Senuwar brass artefact and the earlier Taxila bowl both have especially high zinc ratios, and were therefore unlikely to have been produced by the cementation process already discussed. Cementation, which relied on diffusing zinc vapour into copper, and produced brass containing a maximum of *c*.30-31% zinc, was unlikely to have produced objects containing as much as 34.34% and 35.52% zinc. The answer lay with zinc distillation. The Aravalli Hills of Rajasthan, north-west India, which partly consist of good copper ore known as dolomite, are also rich in zinc sulphide, mined there from the late 1st millennium BC onwards.[80] Mining of zinc sulphide ores around Zawar (now Jalore) in the Tiri River valley in the Aravalli Hills was already active during the 4th to 1st centuries BC,[81] and more recent tree-ring dating of timber suggests that it started there by the 9th century BC.[82] Zinc sulphide ores, besides containing sulphur, often held too much iron, which, during the cementation process, would also be reduced to the metallic state and would dissolve into the copper. The alloy produced would thus contain more iron and therefore less zinc, which meant that a different approach had to be taken. Ores were first prepared just as for the cementation process, by crushing and calcining (slow roasting) to produce zinc oxide. Then an extra process was added.

Powdered zinc oxide (ZnO) and charcoal (C) were distilled – heated together in a retort – to remove oxygen, iron and all other impurities, leaving only zinc metal vapour (Zn). As it heated, the vapour rose up the retort and trickled down a cooler tube as molten metallic zinc, which then solidified. The resulting zinc metal could be added directly to molten copper to make brass by mixing them together. The volatile zinc metal was thrown into the molten copper and quickly shoved beneath the surface before it could vaporise – very hot work. Direct mixing of distilled zinc metal now made it possible to produce high-zinc brasses containing more than 32% zinc. The main advantage of mixing distilled zinc directly with copper was that the amount of added metallic zinc could be regulated, so the craftsman could plan the type of brass to be produced. Small amounts of zinc metal could also be added to molten cementation brass, to raise its zinc content and make it purer. Distilled zinc metal must have been involved in producing the two high-zinc brass objects from Taxila and Senuwar.

---

[79] Ray 1956: 51, translating Kautilya
[80] Craddock and Eckstein 2003: 211
[81] Craddock et al 1990: 38
[82] Kharakhwal and Gurjar, 2006: http//infinityfoundation.com/mandala/t_pr_khara_zinc.htm, 7

Later, the archaeological evidence for brass making was backed up by the written word, bringing it into the historic period. Writing in about 44-33 BC, the Greek writer Strabo quoted a fourth-century historian, Theopompus, who described brass-making, referring back to the mid-7th century BC when Greeks were in the Byzantium (Constantinople) area. He mentioned an ore found in the neighbourhood of Andeira in Anatolia (today's Turkey), which was first 'burned' and then 'when heated in a furnace with a certain earth, distils mock silver; and this, with the addition of copper, makes the so-called 'mixture', which by some is called mountain copper' – *oreichalkon*.[83] Strabo, it has been suggested, might have been referring to Andreida near Balya Maden silver mines, located near Balikesir, south of the Sea of Marmara. In Asia Minor, the Hittites too had a word meaning 'mountain copper',[84] believed to have meant brass. The distilled 'mock silver' may have been droplets of pure zinc, condensed and deposited together with powdered zinc oxide. A seventh-century BC anonymous Greek poem[85] mentions lower-leg armour of shining *oreichalkos* (brass)[86]. The Romans would later adopt, or rather corrupt and adapt, *oreichalkos,* the Greek word for brass, by spelling it *aurichalcum* (golden copper) to mean brass.

Brass was therefore known to the Greeks, even though they preferred gold to brass for prestigious objects, so brass artefacts occurred mainly in Anatolia, their Black Sea settlement.[87] Anatolia may have been the source of 47 brass ingots recovered in 1988 by divers, from a ship wrecked in the first half of the 6th century BC near the prosperous Greek city of Gela, southern Sicily. The narrow ingots, two of which were about 40 cm long, one pointed at each end, formed part of a mixed cargo on a Greek vessel trading to Greek Mediterranean ports.[88] A roll of almost pure (therefore distilled) sheet zinc metal was excavated from a second-century-BC level of the Athenian market place (*agora*), so Greek merchants also handled distilled zinc metal, perhaps from Zawar in Rajasthan, northern India.

The Phoenicians were probably key distributors of zinc metal. In the 9 century BC, when the near-eastern Phoenician city of Tyre held power, Phoenician trading networks already included Anatolia, and they traded copper from Cyprus to the Levant and Egypt and probably helped to spread the knowledge of brass-making around the Mediterranean area, so their alliances are relevant. The Phoenicians had set up city-states around the Mediterranean shores for trading items like glass, purple dye and hunting-dogs in exchange for goods like copper, silver and tin. Sicilian historian, Diodorus Siculus, wrote, around 60-30 BC, that Phoenicians had

---

[83] Jones 1929: 115, translation of Strabo *Geography,* book 13.1, paragraph 56
[84] Craddock and Eckstein 2003: 211
[85] Shield of Herakles: lines 121-2
[86] Craddock 1995: 293-294,
[87] Caley 1963: 68
[88] Sebastiano Tusa, Soprintendenza per i beni culturali e ambientali del mare, Palermo. Inventory Nos. 4239, 4240

traded since time immemorial to Iberia (Spain), where they exploited rich, silver-bearing copper mines from the 8th to the 6th centuries BC.[89] In the 7 century BC, the Assyrians threatened Phoenician territory by raiding nearby northern Mesopotamia. Under threat from Assyrian aggression, the Phoenicians allied themselves with the newly united (North and South) Egypt, helping it to invade Judah and Israel and sack Jerusalem. The Assyrians and Babylonians meanwhile harried the Phoenicians' eastern Mediterranean home ports. When the Babylonians won, the Phoenicians changed allegiance and traded enthusiastically with Babylonian regions. However, when Babylon was captured by the Persians in 539 BC, the Phoenicians were quick to place their fleets at the Persians' disposal, and helped them to invade Greece (485-465 BC).[90] At this period, therefore, Phoenician traders were in a prime position to spread brass, its ingredients and technology around the Mediterranean.

The Phoenicians also traded with the Etruscans, who lived around the copper-rich Tuscan *Colline Metalliferae* in today's northern Italy and were prolific, lively and imaginative creators of tiny high-tin bronze figures.[91] From the 5th century BC, the Etruscans understood how to work calamine and other zinc and copper ores,[92] but the few analysed brass objects thought to date from the 4th to 2nd centuries BC, mainly figurines, appear more Roman in style than Etruscan.[93] During the Greek period, zinc oxide had an alternative ophthalmic use, and zinc oxide pills, thought to be intended for eye treatment, have been found in a tin pill-box from the wrecked second-century BC ship '*Relitto del Pozzino*'. Judging by the cargo, the vessel was travelling from the eastern Mediterranean, and sank near the busy Etruscan east-west trading port of Populonia (today's Piombino), Tuscany.[94] This is probably the earliest physical evidence for zinc oxide used for anything other than making brass. The Romans defeated the Etruscans in the 2nd century BC.

From the 5th century BC, mine workers at Castello di Parre in the Italian alpine foothills east of Bergamo, produced copper alloys with varying percentages of copper, tin, lead and zinc, Their ancestors are understood to have come west from the Balkans and also to have had links with the Etruscans. Nearby Gorno (a bulgar place-name) in the Val del Riso had exceptionally rich and extensive zinc carbonate (calamine) deposits and the region gradually became criss-crossed with mule-tracks to bring down ore from the high valleys to streams where it could be washed.[95] The Romans first gained control of the alpine foothills area and their zinc-mining activities around 222 BC, only to lose it two years later to local tribes who supported Hannibal, but the Romans recovered

---

[89] Diodorus Siculus 60-30 BC: paragraphs 20 and 35
[90] Markoe 2000: 37-50
[91] Riederer 2002: 132-152
[92] Stos-Gale 1993: 101
[93] Riederer 2000: 147; Craddock, 1986: 237-238
[94] Giachi *et al*, 2013: 1193
[95] Furia 2012: 23-25, 30-31

this mining region between 198 and 191 BC, as Cisalpine Gaul, incorporating it in 42 BC into their Italian empire and subjecting it to Roman law in 16-15 BC. A mule track (the Vià di Góren) linked the mineral-rich valleys and others led down to the north Italian plains, but so far the earliest evidence for brass workshops comes from a settlement just outside the Roman city walls of Milan (*Mediolanum*), dating from the 1st century BC to early 2nd century AD.[96] A first-century AD condenser with deeply impregnated evidence for the preparation of zinc oxide for brass cementation was recently found there, by the great canal that carried water into Milan from the River Ticino. The zinc ore is very likely to have come from Gorno.[97]

Brass objects began to be traded further west along the Mediterranean coasts. A fourth- to third-century BC monumental Phoenician tomb excavated near Cadiz in southern Spain, for example, yielded a four-centimetre brass pin with a round head. A votive hoard of the same period, found in a pit at El Amajaro (near Albacete, southeast Spain), included a fragment of brass sheet.[98]

Earlier on, brass had been made in Asia Minor rather than in the Mediterranean area, but after Alexander the Great of Macedonia took his army into Anatolia in 334 BC his troops may have discovered centres of brass-making and spread the knowledge to the eastern Mediterranean, where evidence for the use of brass started to increase. The Greek, Theophrastus of Lesbos (372-287 BC), wrote of brass-making, reporting that one particular ore, mixed with copper, could turn it yellow,[99] and a later Greek physician, Dioscorides (*c*.40-*c*.90 AD), described how powdery white zinc oxide (*pomphylox*) and zinc metal were collected from silver-smelting furnaces. The metal specks (zinc metal), he added, were lighter and brighter than silver, and used by brass-makers to add to their crucibles.[100]

Brass is also mentioned by Plutarch, born in Pontus and later a priest at Delphi. In about 44-33 BC, he wrote, in his life of the fifth-century BC Greek hero Pericles, that a brass wolf stood in the Temple of Apollo at Delphi at the time when Pericles fought to win back possession of the Oracle (from the Laodaemonians). Pericles had his victory inscribed on the brass wolf before returning Delphi to its former owners (the Phocians), thereby gaining himself and his fellow-Athenians priority access to the predictions of the Delphi oracle.[101] Although his account of Pericles' victory in the 5th century BC is a legend, Plutarch himself was evidently familiar with brass.

---

[96] Grassi 2015: 156-157
[97] Tizzi 1996: 115-119
[98] Montero-Ruis and Pereira 2007: 136-137
[99] Furia 2012: 30, 33
[100] Dioscorides Book V, section 84, in Gunther 1933, 624
[101] Plutarch 75 AD: Dryden online translation: http//classics.mit.edu/plutarch

Figure 7. Transcaucasian belt-clasp

During the 7th century BC, the early Greek period, the Iron Age Hallstatt 'Celtic' period had started in Europe. The term 'Celtic' describes a stylistic and linguistic culture enjoyed by peoples living north and south of the Alps, active as far afield as tin-producing Britain and the Mediterranean world.[102] Bronze craftsmanship would reach a peak with the succeeding La Tène culture, with the production of decorative brooches and the widespread use of lost-wax casting (described in the next chapter). In Britain, a second-century BC sword has been found in the River Thames at Isleworth, west London, decorated with La Tène-style maker's brass stamps, hammered to a thin foil. The pure brass stamps (80% copper to 20% zinc) appeared very golden when retrieved, and have a quite similar metal content to contemporary Middle-Eastern brass coins.[103] The absence of lead is significant – lead is not a useful addition in hammered (forged) brass, because the small lead droplets present throughout the solid brass make the alloy more likely to crack under the hammer.[104] Foil is so delicate and fine that no lead should be present in the brass beaten out to produce it. Though the decoration of the foils on the sword was La Tène in style, it is thought that the brass originated from the mines of Andreida, or Andeira, in Anatolia, where relatively high-zinc copper-alloys were produced at that time[105]. A very decorative belt clasp from the Caucasus (Georgia), also of true brass, dates from the 1 or 2 century AD and features curiously-shaped, curvaceous animals.[106]

Brass coins were minted for Mithradates VI (126-63 BC) of Pontus, northern Anatolia, during the early 1st century BC, and 'Celtic' warriors went to Anatolia in the 3rd and 2nd centuries BC.[107] Tribespeople moving westwards through Galicia from Anatolia are thought to have traded brass items and perhaps introduced brass technology

[102] Cunliffe 1997: 1-3
[103] Craddock *et al* 2004: 340
[104] Dungworth 1997: 902
[105] Craddock and Cowell 2000: 124
[106] Curtis 2002: 50-51, 117, plate 9c, 96, table 2, Alistair Pike analysis 137 of British Museum ANE 1921-6-28,1
[107] Pollard and Heron 2008: 198; Zn 13-26%

from Anatolia to Gaul and to other culturally 'Celtic' areas of Europe.[108] Surprisingly enough, most late Iron Age and early Roman brass artefacts have average zinc contents considerably higher than most of those produced in the later Roman period. Skilfully made late Iron Age decorative brass brooches and horse harness are found at various sites in northern England,[109] the metal alloy perhaps imported through Roman-occupied Gaul.[110] Zinc occurs in the coinage of native British tribes, for example in a coin of the Dobunni tribe and two coins of the Trinovantes.[111] In fact, indigenous brass-makers lived in specific areas of Europe both before and during the early Roman occupation.

Dating to 100-50 BC, three excavated fragments of brass brooches or horse-harness have been found at the Titelberg Iron Age hill-top settlement in Gaul (now in Luxembourg). The archaeologists suggest that they or their technology were introduced from Anatolia by the 2nd century BC,[112] especially since the resident Treveri were a tribe of La Tène Celts whose wider territories extended west to the River Danube trade route to the Black Sea. Brass coins bearing 'Celtic-style' designs dating from the mid-2nd century BC onwards were excavated around huts that served as coin-mints. Among 121 other metal objects excavated from the Titelberg hill fort, dating from about 50 BC to AD 300, 45 were made of brass, including several pins for fastening hair or clothing, a buckle, a tool and 29 brooches (fibulae). The highest zinc contents occur from the years AD 1-70.[113]

In the mid-first century BC, local craftsmen in alpine north-eastern Italy designed and made brass coins and brooches. Their region, lying on trade routes from Italy to the Danube basin and the Balkans, was by this time under Roman control, so most brass fibulae found there followed a northern Italian design intended for Roman soldiers. Hinged brass brooches to fasten Roman military cloaks occur from about 60 BC.[114] Alesia-type brooches dating to around 53-51 BC were found at a hill settlement at Alise in Gaul (50 km north-east of Dijon), thought to have been captured by Julius Caesar in 52 BC, but a site in the Jura has also been suggested. 14 out of 17 analysed Alesia-type brooches were of brass.[115] A sword scabbard with impressive brass fittings has been retrieved from the River Lubljanica in Slovenia,[116] and three fragments of brass brooches, dating to around 50 BC, from the Roman-controlled Zerovnišček Iron-Age hill-fort (Slovenia) located alongside part of the Baltic amber trade route.[117]

---

[108] Craddock *et al* 2004: 340, 343
[109] averaging around 20% zinc to 76% copper
[110] Dungworth 1996: 414-421
[111] Northover 1992: 264, 292-293, 295
[112] Hamilton 1996: 59
[113] Hamilton 1996: 43
[114] Istenic 2015: 41
[115] Istenič and Šmit 2007: 140-142, 145
[116] Istenič 2015: 40
[117] Laharnar 2009: 103

Roman brass items containing more than twenty per cent zinc began to circulate only during the rise of Rome towards the end of the 1st century BC, when the Romans began to dominate in Europe and beyond. They include a first-century-BC-style statuette of Hermes, from Egypt.[118] Late first-century-BC brass ingots, recovered from a Roman shipwreck off Corsica, were cast by pouring molten brass into an oval hollow in sand. A similar first-century-BC Roman brass ingot was found in the Upper Thames Valley in Britain.[119]

The Romans made brass by the cementation process, using calamine (zinc carbonate ore, $ZnCO_3$ or smithsonite), so they needed calamine sources. Pliny the Elder, in about 77-9 BC, mentioned *auricalcum* (brass) and wrote that calamine (*cadmeum*) was abundant in Asia Minor, had been found in Campania and was now being recovered near Bergamo (Italian Alps). Pliny, writing a century after these alpine valleys had been settled, describes their zinc ore sources as well-known abroad (*celebri trans marea*), so the mining had evidently started there earlier.[120] The source near Bergamo is confirmed by Roman coins dated AD 68-9 found at Gorno, the richest nearby zinc carbonate source then accessible near the surface.

Pliny had also heard reports of calamine being recently found in the province of Germania, presumably Roman *Germania Inferiore* (now Belgium and northern France).[121] In this region the Romans made brass between the Rivers Rhine and Meuse, near the great Vieille Montagne (or Altenberg) calamine source at Moresnet, south-west of Aachen (Aix-la-Chapelle). There is evidence for calamine mining near Stolberg, seven kilometres east of Aachen, even before the Roman occupation.[122] Roman brass-makers formed industrial settlements around Stolberg Roman villa for making brass by the cementation method. Nineteenth-century archaeological finds from the Stolberg area reportedly included figurines and horse harness, though little has survived.[123] Most brass-working sites of this period have been built over, but at Roman Anthée, near Dinant, crucibles, furnaces, and many brooches and other items of jewellery were found before the site disappeared.[124]

Brass was more common and higher in zinc content in the late British Iron Age and early Roman centuries than in subsequent Roman reigns.[125] In the period AD 50-100, forty per cent of 111 analysed enamelled brooches from Roman Germania Inferiore

---

[118] Craddock 1977: 107-8, discussing British Museum 836
[119] Weisgerber 2007: 148-150, 154
[120] Furia, 2012: 29-30
[121] Pliny, 77-9 BC, 1961 translation: Book 34, Chapter 1, section 2, *Vena que dictum est modo foditur ignique perficitur. fit et e lapide aeroso, quem vocant cadmeum, celebri trans maria et quondam in Campania, nunc et in Bergomatum agro extrema parte Italiae, ferunt nuper etiam in Germania provincia repertum*, 126-127
[122] Gechter 1993: 165
[123] Roderburg 1927: 9
[124] Peltzer 1909: 15-16
[125] Bayley 1984: 42

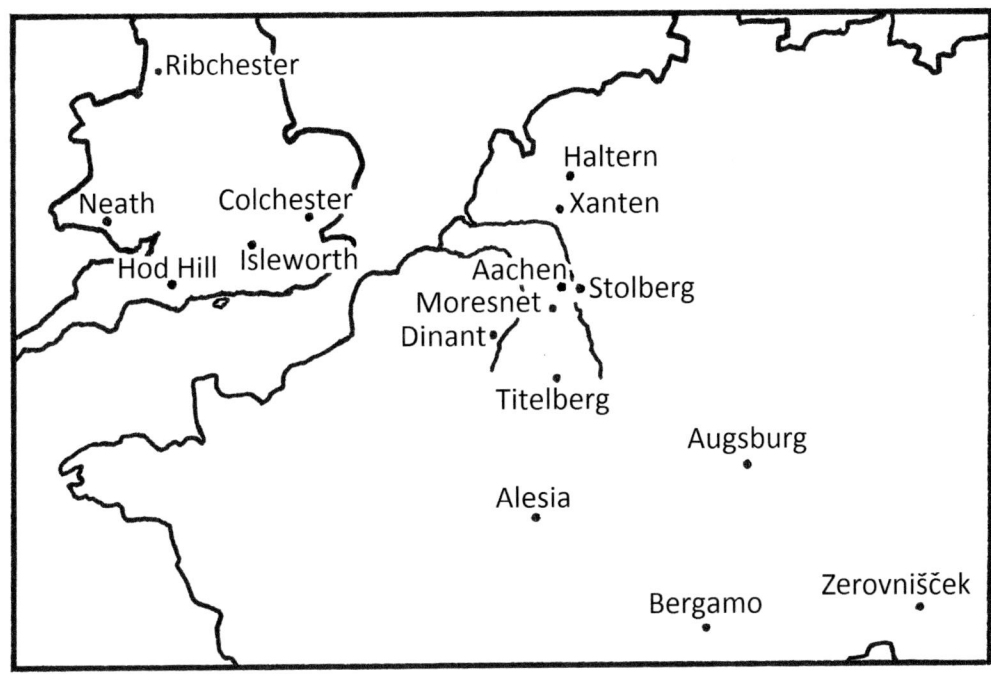

Figure 8. Map, early medieval European sites mentioned in the text

and Gallia Belgica were made of brass, but over the next century it declined to ten per cent. The use of mixed alloys of copper zinc, tin and lead increased proportionally, perhaps due to re-casting or to cost-cutting under financial duress.[126]

Before the Romans invaded, Britain's brass is thought to have been imported from Gaul, including three-quarters of brooches dating to before AD 70 from Hod Hill Iron Age fort (Dorset).[127] In the 1st century AD, thirty-seven per cent of all Romano-British copper-alloy artefacts were brass, compared to only four per cent by the 4 century AD. During recycling by melting down and re-casting, brass loses zinc through evaporation and may be diluted by the addition of bronze objects. In more remote first-century northern Britain, 'Celtic'-style, brass items (mainly horse trappings) contained slightly less zinc than Roman first-century brass made in Gaul, from which it was perhaps re-cycled. However, the brass used for Romano-British rural items normally resembles late Iron Age alloys more than Roman military alloys. For the 1 century AD, more brass brooches are excavated from remote Romano-British farming settlements than from towns.[128]

---

[126] Callewaert, Maxime *et al*, 2013. Elementary analysis of Roman enamelled brooches in Gallia Belgica and Germania, unpublished preliminary dissertation results. Historical Metallurgy Society conference, London, 2013
[127] Bayley and Butcher 1980: 31
[128] Dungworth 1996: 410, and 1997: 907-908

Earlier on in Roman Britain a much greater proportion of objects were of brass (as opposed to bronze) than was the case later. In the early 1st century AD, one third of the Roman brooches found in Britain were made of brass,[129] but from then onwards the average amount of zinc in their brass objects declined. Roman military horse trappings were usually bronze, but a late first- to second-century AD hoard found near Ribchester Roman fort (Ribble Valley, Lancashire) included a fine brass vizor sports-helmet, eye-guards, medals, horse-harness fragments, brooches and brass wire loops linking scale armour.[130] Roman artisans refined their copper by oxidising it, which allowed them to cast such complex objects.[131] At the first-century British Roman forts of both Colchester and Canterbury, brass was made in lidded brass-making crucibles (with zinc residues remaining on their inner surfaces). At Colchester, the inner surfaces of 24 crucible fragments had high levels of zinc (plus some copper, and occasionally lead), suggesting long high-pressure exposure to zinc vapour during cementation. A first-century brass sheet was found at Colchester, and several Roman military sites in Britain have yielded droplets of molten brass, traces of brass on crucibles and part-finished brass artefacts.[132] Nine late first-century brass brooches, found together with crucibles at the pre-Roman and Roman settlement at Baldock, average over twenty per cent zinc.[133]

Figure 9. Roman face-mask vizor sports helmet, Ribchester Roman fort

Many small, capped, brass-alloying crucibles, dating to the start of the 1 century AD, have also been excavated from Xanten Roman military camp in the Rhine-Meuse zone,[134] and much larger ones at Lyon.[135] Military equipment and coins were the most

---

[129] Bayley and Butcher 2004: 15
[130] Jackson and Craddock 1995: 78-79, 92, 97-98
[131] Peter Northover, personal communication
[132] Bayley 1984: 42; Bayley 1990, 7
[133] Stead 1986: 110, figure 4; 111, figure 42; 114, figure 43
[134] copper age 1999 (a): 1083-1084
[135] Rehren and Martinon-Torres 2008: 170

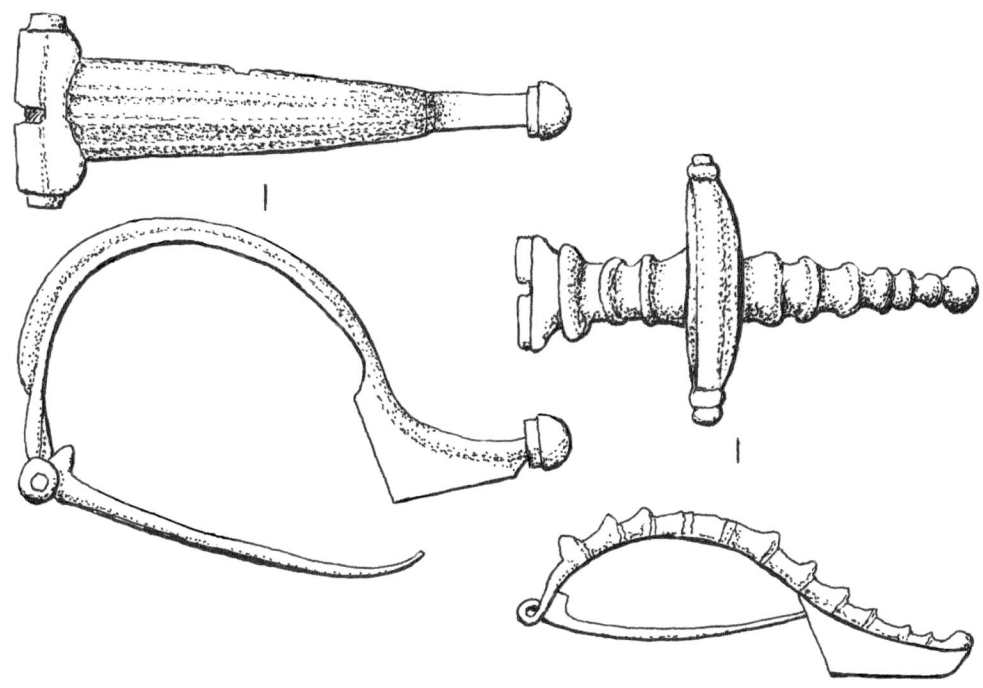

Figure 10. Roman brooches, *left* Aucissa type fibula, c.10 BC-AD 50; *right* Hod Hill type, AD 44-80, Alchester Roman camp

consistently produced early Roman brass items, and Roman coins show the overall pattern of changes in zinc content throughout the period of the Western Roman Empire. Starting from cementation brass using calamine and copper, the content moved to more adulterated alloys. After the relatively high-zinc early Roman coins of Julius Caesar, issued in Macedonia and northern Italy in 44-45 BC, brass coins gradually declined in zinc content.[136] Under the Emperors Augustus, Tiberius and Caligula (27 BC-AD 41), brass coins contained around twenty per cent zinc and a little tin and lead; but under Claudius (AD 41-54) average zinc contents dropped, while lead increased.[137]

Did the composition of Roman brass coins change because lead was added to improve viscosity for fine casting? As already discussed, two per cent of added lead lowers the melting point of copper, rendering the molten brass fluid enough to reach into the extremities of complicated moulds to produce cast metal without the tiny hydrogen-gas bubble-holes released from water-vapour in molten copper.[138] Although extra lead produced a brass that was more fluid for casting and was less liable to shrinkage as it cooled, leaded brass cracked more readily when hammered. Roman Aucissa and Hod Hill type brooches were found at the briefly-occupied Alchester Roman fort (7

---

[136] Istenič and Smit 2007: 140
[137] Riederer 2001: 220-225
[138] Bayley 2004: 16

km north of Oxford), whose gate-tower corner-posts date to AD 43-44. From after AD 70-80, some Roman brooches cast at sites like Colchester were produced from British-made high-lead bronze (copper, lead and tin), rather than brass.[139]

Metal casting grew more common in Roman Britain from the 1 century onwards, and each time brass was re-melted for casting, more zinc escaped as vapour. Up to about 10% zinc may be lost during re-casting, which helps archaeo-metallurgists to notice where this has happened. Objects recycled from Roman brass artefacts include a Romano-British cast-brass torc (necklet) with 'Celtic' decoration, dating to AD 50-150.[140] A good part of the mid-first-century Welsh hoard found north of Neath, South Wales, consisted of brass military horse trappings carrying traditional Welsh Iron-Age geometric designs, together with the handles of tankards and some foundry evidence for casting. Most of the horse pendants, terrets (rein rings) and strap-ends were of brass, perhaps recycled.[141] Impressive first-century Roman cast objects include helmets with applied brass features, brass fibulae, hinges and horse trappings, many excavated along the Rhine valley where the Roman army was long deployed, for example at Augsburg and Haltern, and by the River Oder (today's German-Polish border).[142] Roman brooches (*Augenfibeln*) found at Augsburg mostly contained 13% to 24% zinc.[143]

After Claudius, the amount of silver in Roman *denarii* was reduced, so the content of brass coins may have been adjusted to correlate their value with the debased silver coins.[144] A *dupondius* brass coin weighed and measured the same as a silver *ass* coin, but was worth twice as much.[145] Under Nero (AD 54-68) little brass was produced, though a sample of nine analysed brass coins and a *dupondius* coin contained almost no impurities, so they were almost pure brass. In Roman provincial Asia Minor, however, brass coins were still made well after Nero's death, their zinc content varying more by where they were made rather than when, suggesting that they were produced in individual local centres independent of Rome. They consistently contained around twenty per cent zinc,[146] some perhaps made from zinc oxide produced by the Middle Eastern mountain tradition (described above) of heating zinc ore to produce thick clouds of zinc vapour in order to collect the resultant zinc oxide flakes on bars over the fires.

---

[139] Bayley and Butcher 1980: 31
[140] British Museum: P&EE 4-7 1
[141] Davis and Gwilt 2008: 151-153
[142] Riederer 2001: 237-238
[143] Riederer 2002: 118-120
[144] Bayley and Butcher 2004: 49
[145] Furia 2012: 29
[146] Riederer 2001: 211-224

Zinc content in brass is significantly varied by factors like temperature control, the time the process lasts, the mass of the ingredients and the initial ratio of copper to zinc.[147] These factors could not be fully standardised in the Roman Empire, nor could the metal impurities be fully eliminated. Roman horse trappings were nearly all made of brass, but everyday Roman brass objects are rarely found. In fact common utilitarian items from volcano-struck Pompeii (AD 63) included no brass at all.[148] Most Roman statuettes were bronze, but occasional brass examples depict classical subjects like centaurs, gods and goddesses.[149] By this time, villa life had developed in rural areas, and smaller villa workshops may have cast low-zinc or bronze objects rather than attempting to forge higher-zinc sheet brass.

Overall, looking back to the Middle East it can be seen that from the mid-3rd millennium BC through to the later 1st millennium BC, brass-making was a sparsely-scattered, localised activity, with a few craftsmen probably carrying out all the processes. Works in brass may have been very highly valued and therefore mainly exchanged as diplomatic gifts between leaders – prompted by their golden appearance as well as their rarity. Expertise gradually diffused through long-distance cultural contact and later by trade, until Roman brass was used not only for luxury adornments but for military objects ranging from helmets and medals to horse-trappings. Over time, more brass was made, and the way it was viewed and used slowly but constantly changed, and would soon change further, with brass-making becoming more organised and arguably more industrial in approach.

---

[147] Bourgarit *et al* 2010: 51-52
[148] Riederer 2001: 177-197
[149] Riederer 2001: 160, 165, 173

Franco-roman glass horn with interlaced brass strips, AD 4-5. From Samson excavations, south of Namur, photo © author, courtesy of Musée Archéologique de Namur, Collection Fondation SAN.

Chapter 2

Medieval Europe and far beyond
c.500-1250

The expression 'Dark Ages' does suggest that we might not expect brass to be made immediately following the end of the classical Western Roman period. However, there is evidence that brass production continued to be a significant activity in various areas at the time, with contemporary religions inspiring its creative use. Exuberant, beautifully crafted brass items were made especially for sacred places and holy ritual. Brass objects were common in religious contexts and were rarely in everyday use outside wealthy households. During the immediate post-Roman centuries, however, the art of making powerfully evocative religious images flowered not in the West, but in Central Asia and on the Indian sub-continent, where the tradition set by the earlier Buddhist brass-makers of Central Asia lived on.

At this period, where Buddhism flourished, brass-making frequently flourished too. Prosperous Buddhist rulers whose monks and merchants travelled far afield could have spread awareness of the process. Brass coins were used in India in the 1 century AD, and a second-century brass casket has been excavated from Mānikyāla, a Buddhist site near today's Rawalpindi. In the later 2 century, brass was made in the Indian monastery of Kosala, near today's Nagpur, central India. The philosopher and alchemist, Nāgārjuna (c.AD 150-250), disclosed that, when the local Sātavāhana[1] ruler ran short of royal gold, he himself had turned stones (ore) into 'gold' using a divine recipe. He added that four statues of the Buddha had been created from his decoction. Nāgārjuna made it widely known that he produced his yellow-gold metal by heating calamine three times with copper to turn the copper into gold (brass).[2] A fifth-century brass Buddha image has survived from Gandhara[3], as well as a fifth-century cup-shaped octagonal bowl dredged from the Krishna delta, Andhra Pradesh, which contains zinc together with other elements.[4] This bowl, with eight small oil pans around its rim, has a lotus petal foot above a base of elephants and cobras placed on the back of a mythological tortoise.[5] Several sixth- to seventh-century brass statues survive in India, including Buddhist figures from Akota and Mahudi in Gujarat. A Chinese Buddhist monk, Xuanzang (602-664), who travelled throughout India in AD 629-645, saw brass Buddha statues in many provinces, some cast in several pieces for

---

[1] A pre-Mauryan dynasty ruling the Deccan
[2] Biswas 1996: 131-132, quoting the Rasa Ratnākara of alchemist Nāgārjuna,
[3] Biswas 1993: 311
[4] Srinavasan 2008: 383-389
[5] Victoria and Albert Museum, London: IM-9-1924

assembly later. He also found brass statues of Śiva in Varanasi.[6]

In the Indian to Afghan region, the third Kushan Emperor, Kanishka I (late 1st to early 2nd century AD), a great patron of learning and the arts, encouraged both religious tolerance and Buddhist cast brass imagery. His mausoleum contained an inscribed cylindrical brass casket.[7] The Kushan dynasty, thought to descend from the ancient nomadic Yuezhi tribe from north-western China, held sway from approximately AD 150-300. From his capital near the Khyber Pass, Kanishka ruled over most of today's Afghanistan, Pakistan and northern India. Afghanistan, which has evidence for early mining and metal-working, was rich in copper, zinc carbonate ore and tin, so Afghan craftsmen could source metals for their alloys locally.[8] Kushan survivors include a small Afghan brass seated male figure, dated to the 1st to 3rd century,[9] and a fourth- to fifth-century 41 cm-high image of the Buddha Śākyamuni,[10] the embodiment of enlightenment and the end of suffering. Slightly later images from the Swat Valley (along the eastern Afghan border) include a fifth- to sixth-century figure of the Buddha Śākyamuni,[11] and another of Avalokiteśvara.[12] During Kanishka's reign, Buddhism diffused westwards and eastwards along the Silk Road, whose ancient track-ways carried a lively trade with the Roman Empire and with two other great contemporary empires – Parthia (today's north-eastern Iran) and China.

Figure 11. 8-10-century ewer, Khurasan, 44cm high

From the 3rd century AD, mineral-rich Himalayan Central Asia was a leading Buddhist brass-producing area which included historic Khorasan – land of the rising sun. Khorasan flourished as part of the Sāsānian Empire, and included parts of today's eastern Iran, most of Afghanistan and slices of several other modern Central Asian states. In 651-2, the region was conquered by Arabs but the name lived on as Khurasan,

---

[6] Biswas 1993: 311, 318
[7] Biswas 1993: 315
[8] Reedy 1997: 105
[9] Reedy 1997: 133, figure A1(private collection)
[10] British Museum: 1958,0714.1
[11] Reedy 1997: figure A7, 136
[12] Reedy 1997: figure A9, 137

Figure 12. Celestial attendant, Kashmir, tenth century, 6.03 cm high

from which elegant eighth-century brass ewers survive.[13] Afghanistan with its own zinc source,[14] together with Kashmir, produced the highest-zinc brasses during the early Buddhist period,[15] Afghan brasses being richer in zinc than those from Pakistan or Kashmir. Two of the brass figures surviving from Kashmir are a ninth-century image of Buddha Śākyamuni and a tenth-century 'celestial attendant'.[16]

The Buddhist cast-brass images now held in museums convey something of the devout nature of these mountain people as well as their technology. The brass figures mainly represent the Buddha or a Bodhisattva (a figure embodying a particular positive quality of the Buddha, to free followers from suffering and inspire them to reach enlightenment). Collected museum objects, however, rarely derive from the secure archaeological contexts that could tell us so much more about the people who made or used them. Museum artefacts, although we are lucky still to have them, have often been the victims of past looting and obsessive collecting, rather than finds discovered through targeted research.

The 'Silk Road', the long-distance system of Central Asian routes connecting Afghanistan, Kashmir, Tibet and China, was used by merchants, itinerant craftsmen, Buddhist pilgrims and other travellers. Chinese travellers who visited the Swat valley (northern Pakistan), then a flourishing cultural and religious centre, reported no less than five hundred Buddhist monasteries, which would have contained countless brass images, amongst which some sixth- to eighth-century Buddhist figures from Swat survive.[17] Nearby Nepal, which exported copper, also produced religious statuettes – a surviving late fifth-century brass Buddha is thought to be Nepalese.[18]

Gandhara, a centre of religious and intellectual life at the northern tip of today's Pakistan, and including neighbouring parts of Afghanistan, was pivotal to the spread

---

[13] Victoria and Albert Museum, London: 434-1906
[14] Reedy 2003: 133
[15] Twilley 2003: 144
[16] Reedy 1997: 158, figure K 52; 16enlightenment3, figure K 66, Los Angeles County Museum M,81.25
[17] Reedy 2003: 133, 136, fig, 11;
[18] Sharma 2000: fig. 31

of Buddhism into Central and Eastern Asia. At Gandhara the technology of leaded brass made of the three metals, copper, zinc and lead, was emerging – a fifth-century brass Buddha from Gandhara contained over twenty per cent zinc.[19] Copper in this area naturally contained a certain amount of lead, which, in small amounts, does not present problems when casting objects, since two per cent of lead helps the brass to flow into the corners of the mould and reduces surface pitting. After the Buddhist monasteries of the Gandhara area were destroyed in the mid-5th century, the closely connected Kashmir region continued the brass-casting tradition, and gradually took over as the northern Indian hub of Buddhism.

A quarter of a group of analysed Himalayan statues prove to be made of brass, another quarter are copper, rather less than a quarter are of leaded brass, while the remainder are made of bronze or mixed (quaternary) alloys. In Kashmiri statues, brass and leaded brass also predominate over bronze, and well over half of surviving Central Tibetan statues are of brass. Statues from the Himalayan area appear to have higher zinc content than contemporary figures from the rest of the Indian sub-continent. Buddhist brass imagery still flourished in Kashmir in the 7th and especially the 8th centuries, under the Karakota dynasty (625-1003). A monastery in the Ladakh mountain range, Kashmir, housed a fifth-century brass figure of the Buddha Shakyamuni.[20] Analysed Kashmiri brass and leaded-brass religious images dating from the late 8th to the 11th century contained an average of 23.5% zinc and very variable amounts of lead.[21] Kashmiri brass figures, however, had an uncertain life-span, subject to the whims of the current ruler. A twelfth-century saga mentions the sixth-century consecration of a huge Kashmiri brass image of the Buddha, said to weigh nine tons (48,000 *prastha*), which the tenth-century king Kshemagupta destroyed, using the metal to produce another Buddha statue, only to have that one melted down to create an image of the Hindu god Śiva the Auspicious. Other rulers destroyed statues for reasons of anti-iconic austerity, or simply to realise wealth.[22] Some Kashmiri figures, including a surviving eighth-century Śiva,[23] were of relatively pure brass, and most were cast by the lost-wax process, often with chased and polished surface details, and with inlaid material to emphasise and enliven the eyes and lips.[24]

Lost-wax metal-casting technology had been practised in the pottery-producing Middle East since before 4,000 BC.[25] The technique was widely used for brass-casting in the early centuries AD and continued long afterwards, in some places to this day, so it is worth discussing how it was done. First, a rough basic clay shape (inner

---

[19] Biswas 1993: 311
[20] Sharma 2000: 90, figure 28
[21] Reedy 1997: 281-290, metallurgical tables
[22] Pal 1975: 11-12
[23] Biswas 1993: 311
[24] Pal 1975: 13
[25] Davey 2009: 152

core) was made. Onto this, successive wax coatings were poured, then gradually built up and modelled to form the image itself. The finished wax sculpture was coated with clay to form a mould and, once the clay had hardened, the mould was heated until all the wax was drained out, or 'lost'. The inner core was supported by chaplets – short rods protruding into the outer clay mould, which helped to keep the two clay moulds in the same relative position once the wax melted. Molten brass was then poured through tubes into the empty cavity, filling it from the base upwards to allow any gas bubbles to rise and escape through vents. The clay coating was chipped off, leaving the original wax model reproduced in brass, which then needed scraping, plugging and polishing. Before the use of wax, clay moulds formed the core of cast brass figures. It is thought that, from the 4th century AD, lost-wax metal-casting technology, together with Buddhism, diffused from Central Asia to China, where leaded bronze, rather than brass, was used.[26] From the 10th century, cementation brass, called *toushi*, imported into China from Central Asia, prompted small-scale local craftsmen to experiment with making it,[27] but there is scant evidence for brass-making in China before the 15th century.

Figure 13. Padmapani, god of compassion, in sorrowful thought, 16 cm high

Tibet, however, on the high plateau then lying between China and western Central Asia, produced imaginative figures in brass. A brass image from Himachal Pradesh (Himalayan area bordering Tibet) surviving from about AD 700 shows Padmapani, the Bodhisattva of Compassion, seated in a pensive, sorrowing pose, and another figurine depicts the Vairocana – the celestial Buddha embodying the quality of Emptiness.[28] This small figure, found hidden inside a slightly later Buddha statuette, dates to AD 1000-1200. West Tibetan brass images of the 12th to 13th century include Dorje Phurpa, representing deep truth, piercing like a dagger to replace ignorance,

---

[26] Cowell *et al*, 2003: 80-81
[27] Zhou Wenli personal communication
[28] Ashmolean Museum, University of Oxford: EA 1971.14 and EA 2000.43

Figure 14. Talismanic plaques, Tibet: horses, AD 701-900; peacocks, AD 801-1000, 7.8 cm high

and Mañjuśrī, personifying transcendent wisdom.[29] The spire or bell-shaped summit of a Tibetan shrine (*chorten*) was usually of copper alloy,[30] normally a heavily leaded 'brass', the alloy most commonly used in West Tibet.[31] The *chorten* was a domed edifice symbolising the path to enlightenment, normally surrounded by steps rising from a square base, and similar pinnacles appear on many Tibetan sacred buildings.

As trade and movement increased, stylistic influences diffused from one area to another. Later Tibetan imagery, for example, was strongly influenced by the art of the great Buddhist monasteries flourishing from the 8th to 12th centuries in the Pala Empire in today's Bangladesh, West Bengal and Bihar (a ninth-century brass Buddha figure from Nalanda survives). Indian Pala or Pala-Sena metallic art expressed itself in bronze rather than brass, but stylistically it influenced most of South-East Asia as well as Kashmir, Nepal and Tibet. One product of this influence was the more slender, oriental-looking, twelfth-century brass image of Maitreya from the Potala Palace, Lhasa[32] (Maitreya is the Bodhisattva who will represent the state of Enlightenment far in the future, when it might otherwise be forgotten). Brass plaques or talismans from Tibet, dating from the 8th to 12th centuries, illustrate vigorous pairs of horses or peacocks confronting each other,[33] and thirteenth-century leaded-brass images from Tibetan monasteries include several majestic Bodhisattvas,[34] but another culture was now contributing artistic brass-work to the Himalayan area.

---

[29] Reedy 1997: 191-2, figs. W 137 and W 142
[30] Brautigam 2007: 208-210
[31] Reedy 1997: 83
[32] Brautigam 2007: catalogue number 32
[33] Ashmolean Museum, University of Oxford: EA 2000,108 and 2001.154
[34] Museum für Asiatische Kunst: 87, *Tibet: Klöster öffnen ihre Schatzkämmern*, W15, W140, W149

The seventh-century Muslim conquest of the neighbouring Central Asian region of Khurasan introduced a great flowering of metalwork artistry under Ummayyad and Abbasid caliphs, particularly in the later 12th to early 13th centuries. The items produced included brass wares richly inlaid with calligraphy, images and abstract design. These brass objects, probably influenced by earlier Kashmiri inlaid metalwork, formed a vital aspect of Islamic society, and often carried words that faithfully reflected contemporary beliefs. Khurasan's brass wares reflected the region's fascination with astrology and with foretelling future events – actively attempting to correlate movements and events in the cosmos with the ebb and flow of human lives.[35]

When Islamic regions ran short of silver in the 11th century, and tin was not found in brass was an appealing alternative because of its gold-like appearance. The elite, who had previously vaunted their wealth through finely wrought silver, now displayed their status through well-crafted brass. By the 12th to 13th centuries a wider spectrum of Khurasan society could enjoy and afford brass objects, previously enjoyed by only a few.[36] Fairly pure brass was chosen for Islamic objects that were hammered into shape, with about 17% zinc and low percentages of lead and tin.[37] Impurities such as arsenic and antimony made brass-work harder to hammer without cracking. Islamic metalworkers used any grade of brass for casting, but used only purer metals – low in impurities – for their wrought (hammered) work, so they understood the problem.[38]

In parts of Persia (Iran), zinc oxide was prepared for brass-making by the sublimation method mentioned in the previous chapter. Travellers through Persia described the process. In the 10 century, for example, al Muqaddasī visited Kerman province where villagers brought zinc ore down from the mountains to be heated in 'amazing long furnaces'. He saw ore being heated until zinc vapour steamed out – sublimed – to be collected as solid zinc oxide powder or *tūtīyā* (meaning vapour or smoke) onto finger-like baked clay rods.[39] For brass-making, sublimation helped to eliminate most impurities, including excess sulphur from the sulphur-rich zinc ores found in the Iranian mountains.

Sublimation is an alteration occurring in certain forms of matter, including zinc, which can change directly from solid to gas when temperature and pressure reach a point when the compositions of the solid, liquid and gaseous phases exist in equilibrium. When the temperature coincides with the (relatively low) pressure required to reach equilibrium, zinc will sublime to gas (vapour) without first melting to liquid. Once zinc gas forms in air (which causes oxidisation) it evaporates into the atmosphere unless it immediately contacts a cooler surface, onto which it will condense and form

---

[35] Allan 1998: 11-13
[36] Allan 1998: 15
[37] La Niece 2018: 390
[38] Peter Northover, personal communication
[39] Allan 1979: 40

a coating of zinc oxide.⁴⁰ In Persia, zinc oxide was collected in this way on thin ceramic rods above the fire. As well as zinc oxide, small amounts of actual zinc metal (metallic zinc) were sometimes found.

The advantage of this zinc metal was that it could be directly mixed with molten copper to form brass. Al Muqaddasī who witnessed the zinc sublimation process in Kerman province (Iran) did not connect it with brass-making. However, Ibn al-Faqih, a Persian travel writer, noted in about 900 AD that brass manufacture was a Kerman government monopoly, which suggests that the main use for the collected zinc oxide and zinc metal was brass-making.⁴¹ Al-Hamdani from the Yemen, writing of Persia in the mid-10th century, described a powdered form of zinc oxide being spread onto the surface of molten copper. Al-Birūnī (973-1048), a great scholar from Khurasan, wrote in the earlier 11th century that a tin-like substance (zinc), derived from ore, was collected on pegs above a fire as scales, or flakes, that were used to 'turn copper gold' – meaning 'to make brass'.⁴² Analysis of early clay furnace-bars recovered from Kushlik, central Iran, revealed 2% of lead in their zinc oxide coating,⁴³ because sublimation removes fewer impurities than distillation. On his way to Persia in the 1260s, Marco Polo saw ore heated in a furnace at Kuh-banan and vapour collected on a grid above the hearth as zinc oxide (*tūtīyā*), which, he was told, was used for an eye salve.⁴⁴

In about 1222 AD, a wandering scholar, al-Jawbarī, described a square Persian furnace with a fitted lid. The middle shelf was a lattice of earthenware rods coated with sticky clay slip rolled in finely ground white earthenware, with space left between them for the zinc vapour to pass through. The lid was replaced and the ore fired to red heat in the furnace, after which the rods were extracted and quenched in cold water, the process being repeated three times. When the pegs had cooled, they were lightly hammered, and white zinc-oxide scales or flakes were knocked off ready for use.⁴⁵ This kind of ancient clay rod has since been found to be widespread throughout Iran but particularly in the area north of Kirman, eastern Iran⁴⁶. For the actual brass-making process, Persian scientist al-Kāshānī (died 1274) described *tūtīyā* (zinc oxide powder) being sprinkled onto molten copper, the crucible remaining covered for long enough for the zinc to penetrate the copper before it cooled. The lid kept out oxygen, allowing charcoal (carbon) to act more efficiently as a reducing agent. By this period, Islamic culture had spread throughout much of the Middle and Near East, and Al-Kāshānī reported that brass was made in the same way in Syria.⁴⁷ The zinc-oxide-sprinkling

---

⁴⁰ Rehren 1999: 1085-1086
⁴¹ Forbes 1972: 284
⁴² Allan 1979: 39-40, translation of al Hamdani and Al- Birūnī
⁴³ Barnes 1973: 15
⁴⁴ Polo 1298, 1958 translation: 31-32
⁴⁵ Allan 1979: 41
⁴⁶ Barnes 1973: 8
⁴⁷ Allan 1979: 42

method of brass cementation left little time for zinc to penetrate the copper, which may help to explain why Islamic brass was generally quite low in zinc.[48]

Eastern Iran, during the 11th and 12th centuries, saw the production of the most sophisticated and skilfully inlaid brass-work of the contemporary world. The ruling Ghaznavids inherited influences from seventh- to eighth-century Buddhist bronze- and brass-work of the Kashmir area. Buddhism, in fact, continued alongside Islam in Afghanistan and the Swat valley, and flourished in Tibet. From the mid-12 century, in today's eastern Iran, Islamic brass objects were delicately inlaid with silver and copper, combining pictures, designs and inscriptions. The flowing designs of the minutely applied inlays illustrate contemporary Islamic beliefs and ways of life, including zodiac astrology and the aspirations of a wide range of members of Khurasan society, intent on pursuing an ideal and pleasurable life and a rightful destiny. Ibn Sa'id (1213-86), a geographer from Spain, described the Mesopotamian city of Mosul on the River Tigris (northern Iraq) as a source of fine inlaid brass vessels, popular for presenting as prestigious gifts to foreign rulers.

The brass masterpieces created in Herat, within the eastern Persian Empire (now Afghanistan), reflect the local tradition as well as a rich diversity of outside influences derived from silver-work in Central Asia and further west. The quality of the intricate, mid-twelfth-century work from Khurasan and Herat had never before been matched. The brass objects had religious significance of their own, representing aspects of the total Islamic system of belief. The huge output included cast-brass spherical incense-burners, bowls, candlesticks and inkwells, forms that were linked to Islamic religious observance, ranging from cleansing to sacred calligraphy. By the late 11th century, silver, even for coins, became scarcer in Iran and Afghanistan, so silversmiths transferred their skills to brass, which, with its golden glint, was a more appealing alloy than bronze.[49] Brass thus became the main artistic metal worked during this era of Islamic scholarship and prosperity.

Zinc-distillation was practised at Zawar (now Jalor in north-west India) from at least the 9th century. In response to the wider demand for brass in Islamic regions, zinc-distillation at Zawar increased production on an increasingly industrial scale from about the 12th to the 14th centuries.[50] Long before this, in the period when the Gupta family ruled in India (5th-6th century) the production of zinc metal was documented on the Indian subcontinent, and it is thought to have been produced there even earlier. By about AD 990, Persian authors were aware of zinc oxide made from a heated tin-like metal. Abū Dulaf twice wrote that zinc oxide (*tūtīyā*) normally came from the vapour of an ore, but that Indian zinc oxide came from the vapour of tin (*rasās qal'ī*). Zinc metal is several times described as tin in early historic accounts, because tin

---

[48] Ponting, 2003: 96
[49] Allan 1998: 13-17
[50] Craddock, Freestone *et al* 1998: 42-46

and metallic zinc looked very similar. Two other Persian writers also mentioned that Indian *tūtīyā* was different.[51] Two Greek physicians, Dioscorides (*c*.AD 40- 90) and Galen (*c*.129-c.216) mentioned the use of zinc oxide for medicinal purposes, pronouncing Indian zinc better for the purpose than Persian.[52] This was because distilled Indian metallic zinc was pure – free of contaminants – whereas Persian zinc oxide produced by sublimation still contained some other elements, like lead.[53] An earlier Indian cup made from distilled Indian zinc, belonging to the Persian King Darius, was mentioned in the previous chapter.

The huge Zawar zinc-distillation plant in the Aravalli hills of Rajasthan, excavated in the 20th century, was found to be surrounded by ruined temples of the Jain religion, dating to the 14th to 16th centuries. This culture, related to Hinduism, encouraged beautiful art forms, including a graceful eleventh-century brass 'chauri-bearer' (fly-whisk attendant),[54] and a brass group dated 1283, originating from Zawar, depicting Śiva and Parvati.[55]

India, which had been urbanising southwards over the 1 millennium AD, was moving into its great Classical Period as the world's richest empire, probably controlling more than a quarter of global wealth. From the 1st to the 7th centuries, brass was reportedly used in many parts of India.[56] By the 3rd century AD, in central southern India, new dynasties were rising to replace the earlier Mauryan rulers. From the time of the Vakataka dynasty (*c*.AD 200-550), brass images of the standing Buddha survive from the ancient settlement of Phophnar, in Madyar Pradesh.[57] In June 1961, a local man ploughing his field uncovered seven leaded-brass Buddha images dating to the 5th century, all cast by the lost-wax method.[58] Although made under the Vakataka dynasty, their style is reminiscent of earlier priestly images. The Buddha grew up as a prince and heir-apparent, but later renounced worldly wealth for the contemplative life. Images of the Buddha therefore often included both aspects, with the ascetic's bare right shoulder and hair pulled back into a turban-like twist on the head, but also showing hints of regal pleated robes and jewellery. The Phophnar brass images were found 40 kilometres from the painted caves of the Buddhist monastery of Ajanta (founded in the 2 century), which likewise reflect the rich, sensuous early life of the Buddha and his simpler compassionate life after conversion.[59]

---

[51] Allan 1979: 43-44
[52] Gunther 1934: 623-625
[53] Craddock, La Niece and Hook 1998: 74
[54] Pal 1994: 192, figure 74B
[55] Ashmolean Museum, University of Oxford: EA 1965.5
[56] Biswas 1996: 132
[57] National Museum, New Delhi: L556, L657, L660, L661; Sharma 2000: 77, 83, 85, figures 20-25
[58] Sharma, 2000: 88
[59] Sharma 2000: 77, 87

Medieval Europe and far beyond

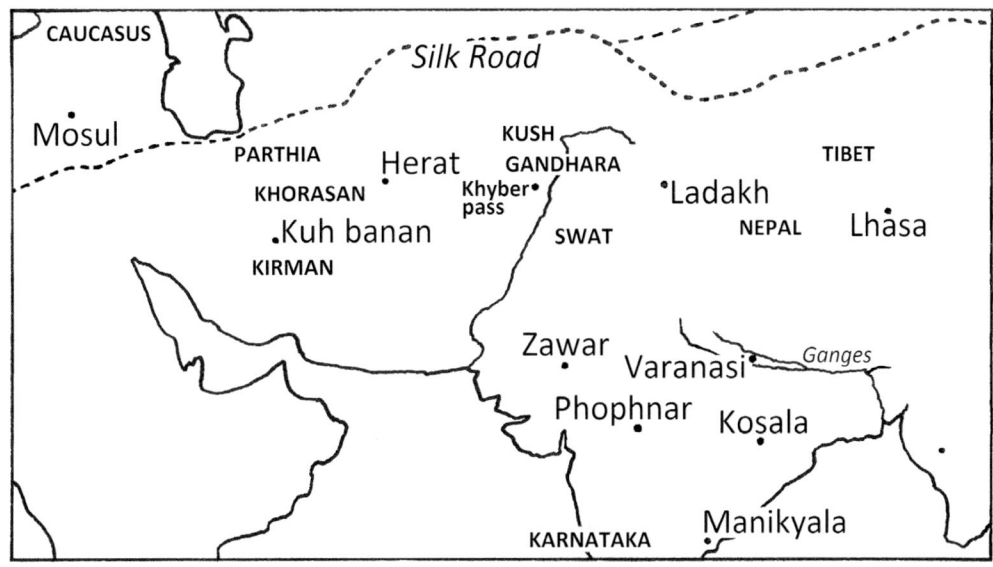

Figure 15. Middle East and Asia, map showing some sites mentioned in the text

When the slender, graceful, statues of the Buddha were created at Phophnar, the Vakataka dynasty, patrons of the arts, ruled the central Deccan and became allies of the rising Gupta family (5th to 6th century), during whose rule Hindu metal figures rose in importance, whilst Buddhist images were also still produced. In southern India both hollow and solid casting was used – a surviving fifth-century hollow-cast cup-shaped octagonal bowl from Andhra Pradesh is mounted on a base of elephants and cobras, standing on a flattened tortoise.[60] This octagonal bowl was made from a quaternary alloy, relatively high in zinc (tin, probably sourced from Karnataka, had been used for centuries in southern India).[61]

Was there another Indian zinc-ore source besides the mines at Zawar? To try to find out the origin or source of a particular metal or alloy, metallurgists study traces of lead isotopes and the presence of specific elements, comparing a wide range of in-situ ores and finished objects. As patterns emerge, an expert scientist can begin to say where the metal in a certain object came from. Patterns for southern India are starting to emerge. One very early metallic zinc ingot carries a Deccan Brahmi inscription dating it to about the 4th century. This ingot therefore shows fourth-century evidence for the distillation of metallic zinc. Its lead isotope and trace signature suggest that both the ingot and the zinc contained in an associated fifth-century octagonal bowl were sourced from the Andhra area, well south of Zawar.[62] By the 12th century, the states of southern India had long worked in brass, though no zinc distillation plant has yet been

---

[60] Victoria and Albert Museum, London: IM-9-1924,
[61] Srinavasan 2008: 382-384
[62] Srinavasan 1999: 204-5; 2008: 387-389

unearthed in the Deccan region. Excavation of a ruined Buddhist *stûpa* (constructed sacred mound) at Ghantasâlâ, south India, revealed an eleventh-century Hindu brass image of Śiva, about 35 cm high, and brass utensils for *pûja* ritual.[63]

Across the Bay of Bengal, in today's Myanmar, the Burmese brought particular enthusiasm to brass-making, investing an almost religious fervour into producing a 'living' metal that would enable its owner to fly freely through land, water and air. From the 5th to the 11th centuries they practised alchemy, with the aim of changing lead into silver and brass into gold. Burmese artefacts dating to the Pyu period, the 5th to 8th centuries, included bronze and brass caskets, miniature stupas and images of the Buddha. A few Hindu and other deities were also portrayed, influenced by contemporary Indian styles. In the 11th century, the earlier obsession with alchemy was discouraged in Burma on religious grounds, but the practice quietly continued.[64] The Burmese kingdom based on the city of Pagan, was founded in the period 1044-77, accompanied by invasion of the Shan and other neighbouring states and a revival of Buddhism and its culture.

The empires of Central Asia and northern India were trading with the Roman Empire at a time when much of the West, including the eastern Mediterranean lands, still lay under Roman rule. It is useful at this point to return to the history of brass in the Near East and Europe in the later Roman period. A Jewish revolt (AD 63-74) against garrisons installed by the Romans in their Near-Eastern provinces left two abandoned Palestinian siege sites, Masada in the south, and Gamla in the Golan Heights.[65] Roman-type military brasses from these two sites are similar in metal content to first-century Roman military brass excavated on European sites. This suggests that the Roman imperial army in Europe delivered ready-alloyed brass or brass equipment to military craftsmen in their remoter provinces where no good local source of zinc ore ($ZnCO_3$) was known.[66]

The Eastern Roman Empire provided the next major scenario for brass production in the eastern Mediterranean. In the 3 century, the Balkans, Turkey, the Near East, Egypt and the North African coastal strip were all ruled from Rome, but Emperor Diocletian (r. 284-305 AD) divided his one vast empire into two areas, one to the east of a line running through today's Balkans and the other to its west. In AD 330, his successor, Emperor Constantine (r. 306-337), moved his capital city from Rome to Constantinople. One of his successors, Theodosius I (died 395), the last Roman emperor over both Eastern and Western Empires, bequeathed his Eastern empire to one son and the Western empire to another but, in the 450s, Rome was attacked by Huns, Vandals and a mixed army under Odovacer, so the Western Emperor and senate fled to Ravenna,

---

[63] Rea 1969 42: plates 33-34
[64] Fraser-Liu 2012: 123
[65] Ponting 1999: 112
[66] Ponting and Segal 1998: 109, 118

where the final emperor was deposed in AD 476. In 489 the Ostragoths moved west from north-west of the Black Sea, and took Rome in 493[67]. Ostragothic copper-silver brooches and belt-buckles found at Spoleto (Umbria) contain significant percentages of zinc.[68] Theodoric, leader of the Ostragoths, ruled over both Romans and Goths. The civilisation of the Eastern Empire grew slowly, flowering from the 7 to 11 centuries.

A swing from bronze to brass was precipitated because the Eastern Empire had lost its tin sources in Britain, Spain and Pannonia (today's Hungary), on which both east and west Romans had depended for making bronze.[69] In the Byzantine Near East, Palestine's neighbour, Syria, lay on the main trade route from the Middle East to Egypt, so it shared a rich culture with the entire surrounding area. From Antioch (then in Syria, now in Turkey), an engraved brass bucket dating to the 6 century portrays men stoning a griffin, and a swordsman and archer attacking a leopard.[70] In the 6th to 7th centuries, the Coptic Christians in Egypt created a metalworking tradition south-east of the Mediterranean, using, for casting, a mixed (quaternary) copper alloy containing variable amounts of zinc (averaging around 12%), and roughly equal amounts of lead. For sheet or raised metal-working they used a relatively pure brass with more zinc and little or no tin or lead.[71]

The Eastern Roman, or Byzantine, empire, meanwhile, still held sway in parts of the eastern Mediterranean, evidenced for instance by low-zinc-alloy archaeological finds dating to the late antique period, from Sardis, a Byzantine trading city in the province of Lydia (today's eastern Turkey). Excavated Sardis finds include early Byzantine brass spoons (perhaps for cosmetics), strigyls (skin-scrapers for bathers), buckles, hooks and lamps.[72] The zinc-rich Anatolian mountains, with their long tradition of zinc-ore mining, were readily accessed by land or sea from Sardis. Early objects dating from AD 135-640, excavated from northern Jordan, hold similar zinc percentages to those in other contemporary Roman-controlled areas, suggesting some degree of centralised control.[73]

In the Near East, after the death of the prophet Mohammed in AD 632, Islamic brass technology and artistry slowly emerged. The nomadic tribes of the Arabian Peninsula organised themselves into an efficient military force to appropriate much of the Middle East and parts of Europe, Asia, and northern Africa, causing the Byzantine Empire to retreat slowly before it. Within a decade, Palestine, Syria, Egypt and the

---

[67] Talbot-Rice 1965: 158
[68] Riederer 2002: 131
[69] Campbell 2006: volume 1, 142
[70] Ashmolean Museum, University of Oxford: AN 1975.308
[71] Craddock, La Niece *et al* 1998: 77, 89
[72] Waldbaum 1983: catalogue figures 227: 619, 628, 641, 694, 698, 701, 728
[73] Cooper, Harold 2000: PhD dissertation, University of Arkansas, Fayetteville

Persian Sāsānian Empire had fallen to Muslim troops, who invaded southern Spain in 711, reaching Khurasan and India at about the same time.

In Palestine, there seems to have been a swing from bronze to (mainly recycled) brass, using the lost-wax process and (for simple objects) sand-casting. From the 9th century, Islamic metalworkers used both techniques to produce brass artefacts, and some were cast in bronze casting-flasks (small metal frames).[74] Sand, being naturally porous, allows gas to escape during casting, reducing the risk of bubbles accumulating in the metal. Fine sand, mixed with a little binding material such as clay, loam or oil, is pushed into two frames. A wood or clay model of the object to be cast is pressed into the sand of the two moulds, to give the shape, and is then removed. The moulds are clamped together and molten metal is poured into the void left by the mould. Cast items include ninth-century brass ewers with low-relief decoration, and numbers of candlesticks and incense-burners.[75] Fragments of a twelfth-century decorated brass hinge, plate and bracelet were excavated at Tiberias[76]. Early Islamic brass tended to be fairly low in zinc, with oil lamps and stands containing more lead, and everyday cauldrons, buckets and basins less zinc.[77]

Many Near- and Middle-Eastern countries produced brass for astrolabes, which were widely used for religious purposes. A Syrian bishop, Severus Sabokt, in about AD 650, described astrolabes as flat, circular instruments made from brass.[78] By the later ninth century, Arab contact throughout the Near and Middle East meant that astrolabes were produced not only in Syria – for example by Khafif, apprentice of Alī ibn ʾIsā – but also in Persia, where, in Isfahan, 'Ahmed and Muhammad, the sons of Ibrahim', made and signed astrolabes.[79] Elegant, decorative brass astrolabes were used by astronomers and astrologers to calculate the user's position from the time of day; to find the direction of Mecca, and to use the astrolabe's current position to work out the hour of prayer. They were also used to predict the movements of the stars and planets. These advances led to their widespread use for making horoscope predictions and for navigation. Brass astrolabes survive from late eighth- to ninth-century Isfahan and from Arab-dominated ninth-century Syria and Egypt.[80] An astrolabe dated 1081-2, from Guadalajara in Spain,[81] was made of relatively pure brass, and Islamic astrolabes were still being made in the 13 century in Sicily and Spain.[82] Initially, copper for Islamic brass came from south-east Asia, and later from Europe, but zinc ore sources existed

---

[74] Shalev and Freund 2002: 22, 24
[75] British Museum: 1959,10-.3.1; La Niece 2003: 91-92
[76] Lester 2004: 64-5, figure 5.3, numbers 7, 9 and 12
[77] Riederer 2002: 141-146
[78] Gunther 1932, vol. 1: 82
[79] Museum of the History of Science, Oxford University: 47632 and 48470; 33767
[80] Museum of the History of Science, Oxford: 33769, 48470, 47632
[81] Museum of the History of Science, Oxford: 52473
[82] Museum of the History of Science, Oxford: 50769; 45307

within the Islamic world in Iran, Afghanistan and Anatolia.[83]

Brass-work flourished in Syria and Egypt under successive Islamic dynasties of the Fatimids (909-1171) and Ayyubids (1171-1260). A surviving celestial brass instrument, dating to AD 1241-2, and probably from Damascus, relates to the heavenly bodies and was used by a 'geomancer' to foretell his client's future. It was made by Muhammad ibn Khuthilh al Mawsili, of Persian origin.[84] Eighth-century brass objects of the Ummayad dynasty (661-750) from Jordan, and late tenth- to eleventh-century objects of the Fatimid

Figure 16. Fatimid domestic vessels, early 11 century

dynasty were made from brass alloyed in Caesarea by the cementation process, using zinc oxide produced from calcined (slow-roasted) zinc carbonate ore, and cast by the lost-wax method.[85] About 1.5% lead was added to improve casting, but brass intended for hammering into sheets contained far less, so makers understood the technology.[86] In 1204-6, al-Jazarī, chief engineer at the palace of a Mesopotamian official in eastern Anatolia, wrote that both the sand-casting and lost-wax casting methods were used in creating a brass door for the imperial palace. Al-Jazarī invented several unusual devices involving brass, including an elaborate mechanical clock mounted on a model elephant.[87]

The influence of the Eastern Roman, or Byzantine, Empire was meanwhile felt further west. Since the end of the third century, new influences from the Black Sea area had been introduced to northern Europe by westward migrations of traditional metal-working tribes[88]. Decorative Byzantine belt or girdle parts were often made of gold, but the components of one very splendid Byzantine girdle dating to the 780s AD were of brass.[89] The girdle came from an Avar grave at the Hohenberg, near Graz in the

---

[83] La Niece 2003: 90
[84] British Museum: ME OA 1888 5.26.1
[85] La Niece 2003: 91
[86] Shalev and Freund 2002: 23
[87] Hill 1974: chapter 1 and 194-5, translation of Al Jazarī
[88] Sas 2004: 343
[89] Daim 2010: 67

eastern Austrian Alps.[90] The Avars were a group of horsemen of steppe origin, by then settled in Eastern Europe. Their leaders had longstanding links with Byzantium and a penchant for prestigious Byzantine goods and fashion, stimulated by generous yearly payments from Byzantium in recognition of historic Avar cavalry support.

The Avar girdle from the Hohenberg is understood to have arrived in the Alps from Byzantium, through Italy. Girdles were officially approved as diplomatic gifts to influential potential friends or foes, so it seems that the belt was buried with a prominent horse-riding Avar. A belt with similar brass fittings has since been found in a plundered grave in the church of Saint Vigilius, Bolzano, on the trade route north from Italy into the Tyrol. A fresco in the church of Santa Maria Antiqua in Rome, dated AD 741-52, illustrates a similar belt indicating high rank, being worn, it is thought, by the young step-son of Theodotus, donor of the church and supreme commander of imperial troops in Rome.[91] Ornate Byzantine belt and buckle parts and strap-ends, dating from the 5th to the 8th centuries, have been discovered as far afield as today's Crimea, Asia Minor, Spain, Sicily, Italy, Greece, Albania, Hungary, Czech Republic, Turkey, Egypt and Morocco, but most of those which survive are made of gold or of 'copper-alloy' that has not been analysed.[92]

The Western Roman Empire, following the death of Constantine the Great (306-337) was meanwhile weakened by a succession of attempts to seize leadership. The sparsely populated extreme western part of the Roman Empire – including Britain, Gaul and Spain – was for a while governed from Trier, and its mainland European region was eyed enviously by adventurous tribes pressing in from the east, ahead of others from even further east. Local and incoming peoples sought a share in late Roman prosperity and territory. The Romans at Trier, for their part, welcomed incomers to farm the land and boost their dwindling defending army,[93] so ethnic immigration contributed materially to both the Roman rulers and the indigenous inhabitants of the Rhine-Meuse area (Roman *Germania Inferiore*).

From AD 351, the Western Roman Empire suffered a period of poor administration, while intruders continued to arrive from further east. Europe was invaded by that roving Asiatic steppe leader, Attila the Hun (died AD 453), who ruled over parts of Scythia (a vast area north and east of the Black Sea) and, ultimately, several Germanic tribal lands, controlling them on the hoof and dominating by terrorising, exacting tribute and distributing rewards. His highly organised nomadic band introduced to metalwork a new decorative animal symbolism rooted in shamanism. The Huns worked mainly in gold, spreading their Scythian-influenced animal design motifs from Central Asia to the northern Black Sea area and onwards to the

---

[90] 47°54N, 15°37E
[91] Daim 2010: 62, 67-68
[92] Entwhistle 2010: 21-25
[93] Alcock 1998: 27

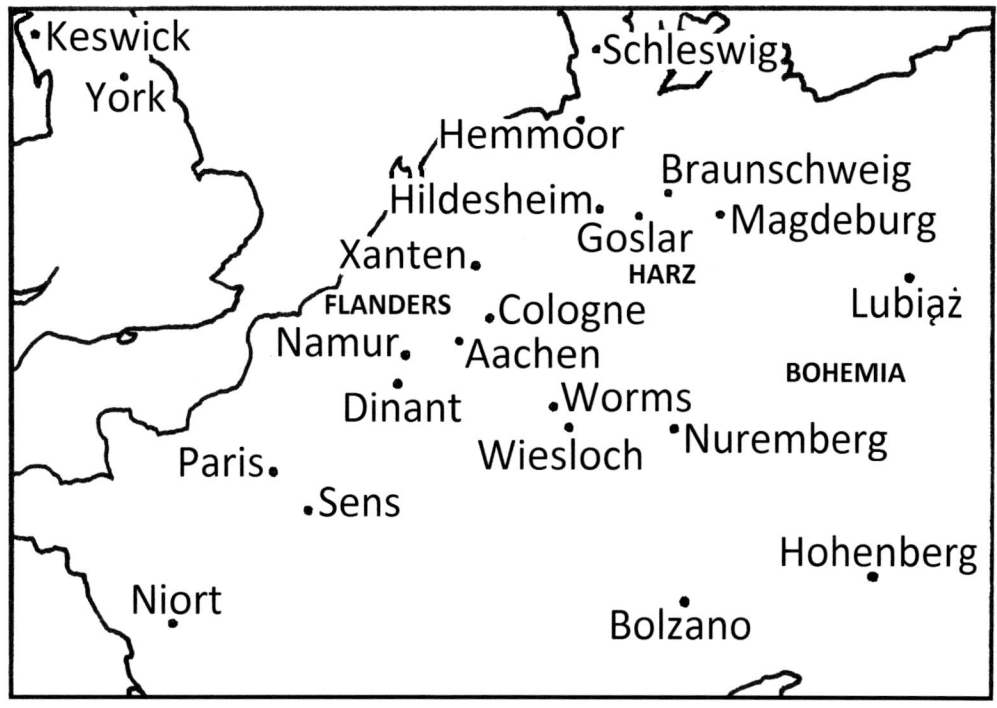

Figure 17. Europe, map showing some sites mentioned in the text

Baltic, influencing the decorative metal-work of Hungarian, central Germanic and Scandinavian tribes.[94] By the 6th century AD, another roving tribe, the Visigoths, after sacking Rome in 410, had settled in southern Gaul. However, after defeat by the Franks in 507, they were forced to retreat and to consolidate in Roman Hispania (Spain)[95]. A surviving sixth-century leaded-brass Visigothic harness pendant portrays two animals facing each other.[96]

The Franks, another incoming tribe, occupied former Roman territory near the richest source of exceptionally pure calamine at Moresnet in the Ardennes forest, 20 kilometres south-west of Aachen. Roman miners, who dug zinc carbonate ore from this source, had named the place 'Kelmis', from *calamis*, meaning calamine ($ZnCO_3$, zinc carbonate). The Kelmis zinc source would later be called the Altenberg or Vieille Montagne. Gressenich in the Rhine-Meuse region, eight kilometres east of Aachen, has evidence for Roman calamine mining, and zinc-rich lead ore was also extracted

---

[94] Hedeager 2007: 43, 47-49
[95] Talbot-Rice 1965: 168
[96] Metropolitan Museum of Art, New York: 1990.52

around the northern edge of the Sauerland slightly further east.[97] Calamine was also mined in the 2nd and 3rd centuries AD at Wiesloch, five kilometres south of Heidelberg.

Roman industrial settlements in the Stolberg neighbourhood, five kilometres east of Aachen, used local Stolberg and Gressenich calamine to make brass. Nineteenth-century Roman copper-alloy finds from Stolberg Roman villa reportedly included buckets, horse harness and figurines.[98] A substantial collection of fourth-century cylindrical copper-alloy buckets, high in tin, was excavated in 1892-3 at Hemmoor, then a key north-south trading port on the River Bentwisch, some sixty kilometres north-west of Hamburg and an equal distance north-east of Bremerhaven.[99] Some of these buckets were constructed of fine brass sheet, one millimetre thick or less (which would have needed almost lead-free high-zinc brass).[100] Crucible evidence shows that, in the third- to fourth-century, brass was made slightly further south at Xanten in the Roman *colonia* of Ulpia Traiana, for brooches containing an average of 18 per cent of zinc.[101] Mining and brass-working in the Stolberg area must have declined once the Roman army left, but pockets of expertise perhaps survived in the hands of Roman-trained local artisans. In the 6 century they would have melded with the invading Franks who (like the Huns) had come from the Black Sea area, where brass was known.

Many fragments of copper-alloy bowls dating mostly to the 4 or 5 century, each decorated with a rim of beading, have been found in graves in Norway and Sweden. One Norwegian example from Blindheim had been turned over to cover a Westland cauldron, a vessel containing cremated bones, a gold button and pieces of bronze, and topped with shells of mussel and a great scallop.[102] Others have been found in the Namur area, and some near Worms in the Rhine Valley. Half the samples analysed were low in zinc, but 36 of the bowls contained considerably more. Although these bowls were of quaternary alloy (composed of four elements, copper, zinc, tin and lead), their medium zinc content brings them closer to true brass. Tin would have been obtainable from Cornwall, Spain or the Erzgebirge in Roman Pannonia (today's eastern Czech border). The beaded-rim brass bowls from Norwegian graves were the purest, containing more zinc and less tin, while six samples from the Meuse Valley held very little or no zinc but more lead and tin.[103] The Byzantine-Italian brass belt fittings from an Avar grave at the Hohenberg (780s AD) had similarly beaded edges.

---

[97] Lammers 2009: 60, 69
[98] Roderburg 1927: 9
[99] Fundorte der Hemmermoorer Eimer: www.geschichtsatlas/de/-gr36/fundort.html
[100] Deutsches Kupferinstitut 2011: brochure, *Messing - ein moderner Werkstoff mit langer Tradition*, paragraph 6
[101] Rehren 2002: 151
[102] Kulturhistorisk Museum, Oldsaksamlingen, University of Oslo; Blindheim is at 66.4462 N; 6.3825 E
[103] Bollingberg 2005: 488-489, 491-92

Figure 18. Beaded-rim bowl from Norway, 5 to 7 century AD, 140 cm wide

Not far from the Meuse valley, certain high-status late Roman burials in the Cologne area were accompanied by wooden caskets, with brass lock-plates and closure-bars. Several were excavated from a Roman burial-ground. Analysis of 37 brass lock-plates from caskets dating from the mid-2nd to the 5th century show that the zinc content remained relatively constant (12-24%). This was in slight contrast to brass found in late Roman Britain and most of mainland Europe, which showed a gradual drop in zinc content over the same period. Because these lock furnishings were of true brass, but not of the typical northern European late Roman type, their analysts suggest that they may not have been made in the area, but further south.[104]

One surviving Cologne (Köln) casket lock-plate and mechanism, believed to date from the late 4th to 6th centuries, is decorated with spirited engraved animal images, full of movement, portraying deer-hunting.[105] The precise woodwork, metal technology and zinc content point to a Mediterranean origin, suggesting that the casket had arrived by trade.[106] The decorative style of this casket is possibly closer to known fourth- to sixth-century Byzantine designs from the Mediterranean, particularly in view of the frieze or border of leaf designs and beading.[107] The huge, lean, hairy hunting-dogs chasing the deer on the Cologne lock-plate resemble Irish wolf-hounds, highly prized by the Romans, who discovered them after invading Britain in AD 43-44. The Roman consul Quintus Aurelius Symmachus (345-402), wrote to thank his brother Flavian for sending him seven Irish hunting dogs, which won awe and admiration at the Games in Rome in AD 391.[108] Hunting dogs were prestige exchange gifts, perhaps reflected in the writing of Saint Patrick of Ireland who mentions a shipment of hunting dogs to Rome around 400. Their reputation would have inspired the depiction of these shaggy Irish hounds on a casket lock-plate.

In the Meuse valley, brass-workers under the Merovingian dynasty of the early Frankish Empire evidently exploited the same zinc ore sources that the Romans had used. The fourth to fifth century cemetery at Samson – the name has since disappeared – was a spectacular site known today as the Rochers de Frène. It overlooks the River Meuse (Maas) between Namur and Dinant. It yielded objects that included

---

[104] Schmauder and Willer 2004: 147-148, 155
[105] Römisch Germanisches Museum, Cologne: 2005.14
[106] Schmauder and Willer 2010: 681-683
[107] Schmauder and Willer 2010: 675-681
[108] Rouse 2015: 27

Figure 19. Deer hunt engraved on the lock-plate of a late Roman casket, Bonn, after photo by Dr Michael Schmauder, with his kind permission

Merovingian Frankish iron belt-fastening plaques inlaid with wrap-around designs in fine brass wire.[109] Brass wire was also used to provide a zigzag decoration set into a delicate Franco-Roman glass drinking horn, excavated in fragments from Samson and now reconstructed.[110] Good brass is more ductile than bronze, so it would have been the better metal for drawing out into fine wire. Furnaces, crucibles buckets, pans and many other objects were found there, including small fibulae and enamelled jewellery. When first found,

---

[109] Balon 1863: plate opposite page 16v.
[110] Musée Archéologique de Namur, Franco-Roman section: (no accession numbers)

some fourth- to fifth-century early Merovingian objects from the Samson excavation reportedly had the yellowish sheen typical of brass, but they were described in the 19th century, before modern examination techniques existed.[111] The Franks also used brass and silver for decorative inlay work on a seventh-century iron belt plate.[112]

Four or five analysed Anglo-Saxon 'great square-headed' brooches, dating from the early 6th century, contain small amounts of zinc. This suggests that Roman or Franco-Roman brass was re-melted for casting in Britain (re-melting causes significant loss of volatile zinc).[113] This was not true brass but a quaternary alloy, and no Anglo-Saxon jewellery items made in England were of true brass, since they contained more lead or tin than zinc. However, three analysed Scandinavian brooches of the same period were of brass as defined here (more than 8% zinc and significantly lower percentages of lead or tin).[114] Scandinavian brass perhaps profited from strong Baltic contacts with the Black Sea through the Russian river routes. Anglo-Scandinavian scrap copper-alloy excavated at Coppergate, York, was all intended to be recycled, using crucibles later excavated there. The very few actual brass pieces (including eighth-century helmet parts) were more numerous from the 10 century onwards, both at York and at the Flaxengate site, Lincoln.[115] The *styca* coinage of Æthelred, king of the English from 978-1016, was consistently made of brass, averaging about 20% zinc to just over 70% copper.[116]

By the 7th century, metal-workers in the Rhine-Meuse area had apparently been producing brass chalices for religious use, because a canon (church law) was issued by the Council of Rheims in c.AD 625-630, advising that chalices should no longer be made of brass or copper. This was because it was alleged that the wine would combine with corrosion on these metals to provoke vomiting.[117] Pure brass may oxidise, due mainly to the effects of ammonia and mineral acids, which form tarnish (cuprous oxide), especially on brass that contains little zinc, but neither alcohol nor organic acids (like those in wine) normally react with brass in this way. Tarnish can, over time, impart a slight metallic taste to food, but is not toxic. The copper content in brass even has a slow-acting germicidal effect which could perhaps have inhibited infectious vomiting. Contact with the copper in brass destroys bacteria within minutes, especially if the surface is dry, but the process is not yet fully understood.[118] The truth may have been

---

[111] Marmol 1859: 345-391 and plates I-VIII
[112] Unprovenanced material in the Metropolitan Museum of Art, New York: 17.193,412
[113] Brownsword and Hines 1993: 3
[114] Kershaw 2013: Appendix B, tables 1, 2 and 3
[115] Bayley 2014: 803, 808-810
[116] Metcalf and Northover 1987: 4-5, 14-21
[117] Mansi 1759, canon XXII: 603; *De aere aut Orichalco non fiat Calix; quia ob vini vitutem aeruginem parit, quae vomitum provocat*. 'Do not make chalices of copper or brass, because the wine and the rust together will cause vomiting'.
[118] Copper Development Association 2005: Brasses, Preparation and Application, revised by Peter Webster, http://admin.copperalliance.eu/docs/librariesprovider3/pub.117-the-brasses_whole_web-pdf, 48

that the goldsmiths had the ear of those in power, because the canon clearly specified that brass chalices should be replaced by others made of gold or silver.

Conversely, an Islamic canon or Hadith recorded by Imam Muhammad al Bukhari (c. 810-870) decreed that no man or woman should use or drink from vessels made of gold or silver, as these metals were too ostentatious for use in this life[119]. In another Hadith, it was reported that the Prophet himself (570-632) had been brought water in a brass vessel, and had used it for his ablutions[120]. Brass was therefore acceptable for ablution and other uses, and became the main alloy employed by medieval Islamic societies. For Christian ecclesiastical ritual, Isidore of Seville, writing in the 7 century, mentions the use of a water-filled aquamanile (ewer) and liturgical cloth, for washing the chalice or the priest's hands, but the metal from which this ewer was made is not mentioned.[121] Later European medieval aquamanilia and cauldrons were generally made from either brass or quaternary alloys (containing copper, zinc, lead and tin). With more artefacts produced and greater contact and trade around Europe, markets for such brass items were becoming increasingly important.

In AD 744, the Frankish king Pepin (reigned 751-68) decreed that every town should have a market, introduced a new standard for currency and granted exemption from certain tolls. Within 120 years the number of yearly markets and trade fairs had risen steeply and communication networks had improved, greatly facilitating trade in Frankish metal goods. By the 9 century there were over 200 markets in the area between the Rhine and the Loire, where trade in brass flourished from the 10 century. The greatest markets and workshops developed around the centres of wealth and power – the courts, palaces and abbeys – and at the border-posts and coastal harbours of the Frankish domain. Some harbours and border-settlements traded merchandise to Scandinavia, Britain and the Slavic East. Coins were used in the barter of goods such as metals, and brass and bronze weights were used for measurement.[122] In antiquity, brass makers are not known to have sought out refractory (heat-resistant) clay for crucibles, but in the early medieval period they did so at Huy and at nearby Frankish sites between the Rhine and Meuse, where fragments of casting-moulds for seventh-century Frankish (Merovingian) brooches have been found.[123] A lively Frankish trade existed in transporting Slavic slaves to Cordoba (then predominantly Muslim), from where Frankish merchants returned with silver and Islamic-inspired goods that influenced metalwork design in their home area.

Brass was produced in the time of the Frankish emperor Charlemagne (r.768-814), because from 794, he standardised weights, measures and prices for commodities,

---

[119] Sahih al Bukhari, book 024, Hadith 5134, hadithcollection.com/sahihbukhari.html
[120] Sahih al Bukhari, book 001, Hadith 196, hadithcollection.com/sahihbukhari.html
[121] Barnet and Dandridge 2006: 4
[122] Steuer 1999: 406-407, 413-414
[123] Willems 1973: 57-58

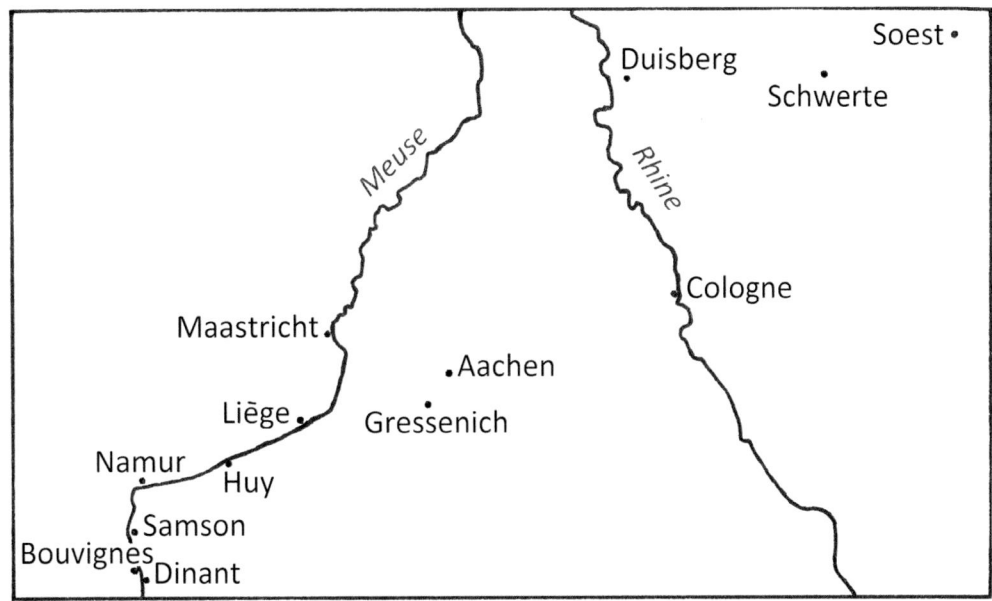

Figure 20. Rhine/Meuse area, map showing some sites mentioned in the text

listing prices in *denarii* for a pound of copper ore (*aeris cypri*) and for a pound of calamine (*caerae auricalci*).[124] Alcuin of York, adviser and teacher at Charlemagne's court, indicated that brass was familiar to those around the emperor, by posing his students a mathematical problem: a dish weighing 30 pounds is made of gold, silver, brass (*aurichalcum*) and tin. By weight, the dish contains three times more silver than gold, three times as much brass as silver and three times as much tin as brass. What, he asked, is the weight of each type of metal?[125] Charlemagne's palace site at Aachen (Aix-la-Chapelle), formerly occupied by first-century Roman military thermal baths had now become a market and centre for craftsmen and merchants.[126] The late 8 century saw increased hostilities between Saxons and Franks, culminating, in 772, in Charlemagne's Saxon War, which isolated parts of his empire, extending from Duisberg to the Sauerland.

In the same area, at Soest, Dortmund and Schwerte, along a contemporary east-west trade route – the *Hellweg* – crucible evidence has been found for brass-making and casting from the late 8th century onwards. The three settlements were relatively close, and seams of calamine existed in the area, particularly at Iserlohn. At the

---

[124] Borst (ed.) 2006: 1063, 1309-1310
[125] Singmaster 1995, translating Alcuin of York (804), *Propositiones Alcuini Doctoris Caroli Magni Imperatoris ad acuendos Juvenes* 'Est discus, qui pensat libras XXX sive solides DC, habens in se aurum, argentum, aurichalcum et stannum. Quantum habet auri, ter tantum habet argent, Quantum argenti, ter tantum aurichalci. Quantum aurichalci ter tantum stanni. Dicat qui potest, quantum in una quaque specie pensat?'
[126] Giertz and Ristow 2013: 59 (figure 1), 61

Plettenberg site at Soest, Westphalia, vestiges of rectangular loam casting-moulds were excavated. They were let into the ground, and were used for casting leaded brass sheets. Metal debris on crucibles showed that they contained not much less lead than zinc.[127] Excavations of sites dating to around AD 900, beside a town-fortification tower at Dortmund,[128] and at Kückshausen site at nearby Schwerte (a few centuries later), also revealed crucible fragments used in producing brass.[129] At Soest, where the main metals worked were iron and lead, the brass finds included a twisted piece of bung (used to stop up a conical casting-vessel), a disc-brooch, and a brooch fragment encircled by brass wire and decorated with an engraved cross.[130] Certain lively early ninth-century manuscript illustrations show clothing secured by disc brooches, some of them probably brass.[131] Craftsmen such as brass makers did not normally work alone, but formed part of the entourage of influential kings and bishops.[132]

At Charlemagne's capital, Aachen, three recently excavated areas close to the walls of Aachen cathedral contained evidence for glass-making workshops dating to around 800, including crucibles of good heat-resistant clay. The clay was of the type excavated from a context dating to about AD 800 at Huy in the central Meuse (Maas) region, site of a court of the Frankish Empire. Charlemagne's creative energies evidently injected new life into craftsmanship. However, in 774 he granted his zinc-rich mining area at Gorno and the surrounding Italian alpine valleys east of Bergamo, to the Abbey of St Martin of Tours, which largely abandoned the mining of calamine for several centuries, in favour of silver.[133]

The spread of the religious use of brass in Europe came at a time when prestigious relics in Aachen cathedral, assembled for Charlemagne in the early 9th century,[134] attracted contemporary and later pilgrims from around the Baltic and from as far afield as the Balkans. The lively contact maintained between Aachen and Scandinavia may have increased the widespread popularity of religious brass in Scandinavian lands. Byzantine Christian culture, evidenced by 9th to 10th-century pectoral crosses[135] influenced Sweden by way of long-distance Eastern European river trade routes, and may have affected Aachen through its northern trade networks.

Four hundred kilometres east of Aachen were the Harz Mountains, a mineral-rich outcrop emerging dramatically from today's central northern German plain. At Düna, the court of a contemporary overlord near Osterode in the Harz, metal was

---

[127] Lammers 2009: 33, 72-73, 180
[128] Rehren and Martinon-Torres 2008: 173
[129] Lammers 2009: 73
[130] Lammers 2009: 174, plate 19, figures 280-281, 287, catalogue numbers 217, 102, 114
[131] Stuttgart Psalter (Wurttemburg Library): 24v, 25v, 67v, 112r, 135v, 145r
[132] Lammers 2009: 73; Steuer 1999: 413
[133] Furia 2012: 37, 39
[134] Butzer *et al* 1997: 259
[135] Musée d'Art et de l'Histoire, Brussels ACO 75 1.1

processed and ninth-century enamelled brass brooches and the base of a twelfth-century chandelier have been excavated.[136] At the Rammelsberg, a hill near Goslar in the Harz, a copper-mining town was established in the 10th century around a stone-built chieftain's court, at a time when Harz miners processed both silver and copper.[137] A chronicler wrote that the mines were opened under Otto I (912-973).[138] In 1009, Holy Roman Emperor Henry II (972-1024) is said to have granted the Rammelsberg to a Frank named Gundelcarl, from the Rhine-Meuse area of the Frankish Empire, which included Aachen (Aix-la-Chapelle), close to the rich zinc-ore source of the Altenberg (Vieille Montagne). Gundelcarl is said to have travelled home to fetch experienced metal-workers and returned to set up mining and metalworking at Goslar.[139] The lofty, Romanesque, 'Frankenberger' church, built in Goslar in 1108 on the profits of mining, is still standing. In 1150, miners dug a channel to drain water from the local Rammelsberg mine, and in 1250 a 10-ft water-wheel was installed to pump out water.[140]

Excavations provide interesting evidence of a contemporary colony of brass-makers, possibly of Frankish descent, with a tradition of burial with their crucibles. Sites at Mell and nearby Niort (Deux-Sèvres, central western France) have revealed the stone burial cists of metal-workers, each interred with one or more crucibles. All the smaller crucibles showed evidence of contamination by zinc, as well as copper and a little silver, and proved to have been heated to over 1100ºC. These small crucibles dating from the 12th to the mid-14th centuries, were luted (edges sealed) with soft clay.[141] This site, with possible links to the Franks, lay almost midway between Aachen and Cordoba.

Charlemagne's cathedral at Aachen had been contemporary with the great Cordoba mosque, which had established lively new connections between Western Europe and the Islamic Near East. The eastern Mediterranean ports had traditionally traded mainly with one another but, by the 10 century, merchants from Genoa and Venice had set up trading centres at Constantinople and along the Syrian coast down to Egypt, stimulated in part by the first crusade (1095-9). European leaders who took part in crusades or pilgrimages to Jerusalem introduced Islamic-inspired design and craftsmanship to other parts of Europe. Large 'Veneto-Saracenic' cast bronze figures of creatures like griffins, stags or lions found their way from the Syrian coast to Italy and Spain.[142]

---

[136] Segers-Glocke *et al* 2000: 21
[137] Segers-Glocke *et al* 2000: 4
[138] Asmus 2018: 27
[139] Boyce 1920: 15, 20
[140] Website, 2007: wikipedia.org/wiki/Mines_of_Rammelsberg
[141] Thomas 2006: 53-56
[142] Barnet and Dandridge 2006: 10

The 12th and 13th centuries saw the flowering of elegant aquamanilia – ewers with pouring-spouts in the shape of a sometimes-fabled animal or bird, in the same style as the large-scale Veneto-Saracenic sculptures. Aquamanilia, which were used mainly in religious hand-washing rituals, were found in Islamic, Indian and European contexts. One copper-alloy aquamanile made in the Meuse valley in about 1120 represents a dragon, and another in the form of a wyvern (a two-legged dragon), was made around 1200, probably also in northern Europe.[143]

In about 1125, a German monk and metal-worker called Theophilus, or Roger von Helmarshausen (*c.*1070-1130), recorded the brass-making process that he evidently practised. He probably worked at Helmarshausen monastery near the junction of the Diemel and Weser rivers, thirty kilometres west of Göttingen, with previous experience in the Meuse Valley and at Cologne. When the crucibles are red-hot, wrote Theophilus, the brass-maker should take calcined calamine (zinc carbonate ore), that had been slowly roasted with charcoal, sometimes for three weeks, to convert it to zinc oxide and rid it of most impurities. He should then pound it up very fine with charcoal, and place it in each crucible until about one-sixth full. The crucibles should be filled up with copper and covered with a reducing layer of charcoal. This last layer sealed the zinc oxide layer below it from oxygen by melting a layer of hot copper over both. Once alloyed, the brass was cast in hollows dug in the ground, and used for making cauldrons, kettles and basins.[144] This method may have been more like the co-melting of zinc oxide and copper than cementation (where zinc vapour penetrates warm but still-solid copper).[145] Small crucibles excavated close to the early medieval city walls of Dortmund bore evidence for brass being produced in this way.[146]

About 260 kilometres north-east of Dortmund, Braunschweig (Brunswick) was patronised as a metal-casting centre around AD 1000 by Bernward, bishop of the town of Hildesheim (about 40 km further east).[147] In the 12th century, Bernward's successor, Duke Henry the Lion, of Saxony and Bavaria, had a great lion made from quaternary alloy (gunmetal) and erected at Braunschweig to remind people of the power of lions such as himself.[148] Such complex hollow-cast shapes needed an inner framework of armatures to support them. In 1226, a richly decorated baptismal font for Hildesheim cathedral was created from a quaternary copper alloy rich in zinc but with rather too much tin and lead to be true brass.[149] Hildesheim was only about 35 kilometres north of the copper-rich Harz Mountains and, from around 1100, copper-delivery documents relating both to brass production at Huy and Dinant in the Meuse Valley

---

[143] Barnet and Dandridge 2006: 88-89
[144] Hawthorne and Smith 1963: 143-144
[145] Craddock and La Niece 1998: 75
[146] Rehren 1993: 312-313
[147] Beddies 1996: 92-93
[148] Riederer 2000: analysis, 200-201
[149] Dandridge 2018: 213

and to brass-working at Braunschweig and Hildesheim, mention copper from Goslar in the Harz Mountains.[150]

Harz copper has been detected in brooches, styli, knife-hilts and coins, found over a widespread area and thought to have been crafted from the mid-tenth century at the Düna court workshops near Osterode in the Harz.[151] Although metal-casting workshops have been excavated at Hildesheim and Magdeburg,[152] and brass enamelled brooches have been found at Düna, there is so far little evidence that brass was actually alloyed in that area in the 10 century. The eleventh-century doors and richly decorated font (c.1200) of Hildesheim Cathedral were cast from gunmetal (possibly using recycled brass, plus significant amounts of lead and tin). Harz Mountain silver-rich lead ores also contained zinc sulphide (the silver on the Rammelsberg is said to have been exploited from 968 AD, when a horse named Ram is fabled to have stumbled upon it during a stag-hunt).[153] From the 11th to 12th centuries these ores were heated to obtain silver by separating it from the lead. During this process, powdery white zinc oxide called *Öfengalmei* (with some solidified droplets of metallic zinc) collected as thick layers of accretion in crevices or recesses in the chimney-flues over Harz furnaces. Although initially ignored, workers eventually noticed the potential for using zinc-oxide accretions for brass-making.[154] At first, Harz furnace zinc oxide was exploited locally for brass-making, but in due course it would become more widely recovered and distributed.

Along the broad River Meuse (Maas), brass was made and worked into objects at Huy and Dinant. Brass was produced at Huy from the 11th to early 13th centuries. Johan Reinart (pen-name of Hugues de Pierrepont, Bishop of Liège, 1190-1240), author of an early version of the *Roman de la Rose*, mentions 'Huy, where the cauldrons are made'.[155] The Huy metalworkers competed with those in Dinant during the 11th and 12th centuries, but in the 13th century they turned over to cloth-working and were no longer competitors. The Meuse Valley was well placed for local timber, and its brass-producers could obtain zinc ore from the Vieille Montagne (Altenberg) in the Duchy of Limbourg and from another smaller calamine source alongside a five-kilometer stretch of the River Meuse between Huy and Andenne.

There was no copper in the province of Liège, in which Dinant lay, but its excellent river transport and trade route connections allowed copper deliveries, via Cologne,

---

[150] Beddies 1996: 92
[151] Segers-Glocke, *et al* 2000: 1, 20, 62
[152] Barnet 2006: 14-15
[153] Segers-Glocke *et al* 2000: 54-5; ZnS, and occasionally chalcopyrite (CuFeS)
[154] Hoover and Hoover 408-10: with footnotes
[155] '*Huy ou l'on fet les chaudières*'; Reinart, Johan, *Roman de la Rose*: verse 5528, www./persee.fr/web/revues/home/prescript/article,rbph_0035-0818_199_num_77_2_4358

from Goslar in the Harz Mountains.[156] Metalworking at Huy and Dinant was mentioned in a tenth-century tariff list drawn up at Visé customs house on the River Meuse (Maas).[157] Excavated Dinant workshops date from the 12th to the early 16th century, with the earliest twelfth-century workshops lying beneath today's town-centre car park. The remains of a dozen medieval houses have been excavated in Dinant, together with 15 back-yard brass workshops, each with one or more furnaces set into the ground. This thriving industrial settlement lay outside the Dinant town walls, alongside the River Meuse.[158] Excavations uncovered many furnaces, crucibles and fragments of foundry-moulds for casting brooches, ornaments and belt-buckles. In the industrial quarter established in the 13th century, most of the houses – in which the craftsmen both lived and worked – fronted the street. In an earth courtyard at the rear stood the furnace where the alloy was prepared. Excavated furnace debris included crucible and metal fragments, remains of furnace brick, and slags, glassy from the furnace heat.[159] Verdun, too, had several brass workshops from the 13th to the 16th centuries, and archaeologists have also excavated foundry evidence at Maastricht and two further Meuse settlements.[160]

Dinant products became celebrated far and wide. The earliest French-speaking medieval Dinant craftsmen to shape and emboss brass were described as *batteurs*, or beaters, because their sheets of cast brass were beaten into shape.[161] They used techniques like repoussé, where decorative shapes are beaten out from the reverse, or 'chasing', where the design is adjusted by hammering from the front. In the 10th century, the Flemish Bishop of Liège (Lüttich) had granted Dinant the right to hold its own market. The more influential ecclesiastical patrons began to pay for brass incense-burners, baptismal fonts, paschal candle-sticks and beautifully crafted eagles, pelicans and griffins mounted on lectern pedestals. By the early 13th century, the artistic Meuse Valley craftsmen were selling their brass wares far and wide.

By the 12th century, besides decorating brass by the repoussé technique, Dinant craftsmen were casting it.[162] A celebrated gunmetal baptismal font, with three-dimensional animals and figures, was cast, perhaps in the Meuse Valley, for the baptistery of Liège cathedral.[163] Fine Dinant aquamanilia – animal-shaped pouring jugs – dating to the 12th to 14th centuries, were used mainly for washing rituals in churches and cathedrals.[164] This was the basis for the development, around 1450, of cast monumental

---

[156] Douxchamps-Lefèvre 2005: 9
[157] Haedecke 197: 30
[158] Verbeek 2014: section IV, 35, 37
[159] Thomas *et al* 2013: 173-174
[160] Nicolas Thomas, personal communication
[161] Beddies 1996: 123
[162] Douxchamps-Lefèvre 2005: 10
[163] Now in the church of S. Barthelmy, Liège
[164] Douxchamps-Lefèvre 2005: 13

brass images.[165] The Meuse Valley masterpieces included crucifixion figures,[166] and a shrine to St Servatius, Maastricht,[167] still carried in procession through the streets once in seven years. Plain, thin rectangular plates of Meuse Valley brass were cast in Flanders, Cologne or Paris for memorials to prominent people, including Archbishop Pierre de Corbeil of Sens, who died in 1225. Bishop Nikolaus of Schleswig (1208-33) had a fine brass seal made for himself in about 1210,[168] perhaps cast in Schleswig from brass alloyed in the Meuse Valley. Most analysed Flemish and English monumental brasses of the mid-14th century contain from 11 to 19 per cent of zinc. Three monumental brasses made around 1300 and 1305 for Lubiąż in Upper Silesia (today's Poland) contain a slightly higher percentage of zinc than Flemish ones, and significantly more tin, so their metal was probably alloyed in Upper Silesia or Bohemia, where both zinc and tin were available. For large thin plates, brass was more stable to work than bronze.[169]

Because Britain produced little or no brass of its own, the rectangular plates for memorial brasses to be engraved in London[170] are thought to have been sent over from the Meuse Valley. However, Henry III of England (1216-72) did cause copper to be dug from the Keswick area of Cumberland (now Cumbria), possibly with a view to brass-making,[171] and in 1319, Edward II (1284-1327) sent a surveyor to Caldbeck, Cumberland, after hearing of silver and copper mining there.[172] Keswick lead-copper sulphide ores yielded about 25% copper, but processing copper from this sulphide ore, which included repeated slow roasting and melting, took at least 18 weeks and burned up literally tons of peat and wood.[173]

Back in the Meuse Valley, Bouvignes was a rival brass-working centre adjacent to Dinant, lying on the west bank of the river. It lay in Namur Province and had an enviable supply of excellent plastic and refractory (heat-resistant) white crucible clay. Refractory clay was less prone to crack under high heat than the clay used for domestic pottery. Heat transfer into the crucible increased, the greater the difference between the heat outside and inside it, so a strong refractory clay was needed to withstand these conditions, whilst also allowing the temperature within to be raised faster and maintained longer.[174] Bouvignes was in stiff, sometimes aggressive, competition with Dinant, which lay two kilometres upriver in the separate Liège Province, controlled by the Bishop of Liège. At *La Porte Chevalier*, beside Bouvignes, several thirteenth-century workshops were excavated at the base of a spur jutting from the steep wooded hillside

---

[165] Oosterwijk and Badham 2018, 379
[166] Musée Diocésain, Namur: 455; Musée d'Art et de l'Histoire, Brussels, 3672 and 6649
[167] Gérard 1958: 12
[168] Archäologisches Landesmuseum, Schleswig: 1950/1
[169] Norris 1978: 30-31 and 34-35
[170] Each measuring about 76 cm x 61 cm
[171] *Calendar of State Papers Domestic* 1547-1580: 287
[172] *Calendar of Fine Rolls* Edward II, vol. 2: 179
[173] Smith 1994: 116-122
[174] Bayley and Rehren 2007: 49-50

west of the river. Brass workshops dating from the 13th to the 14th centuries lay within the city walls.[175] They were among many such workshops around Bouvignes, some of which continued until the 16th century. A document dated 1375 made a distinction between the respective Bouvignes makers of cauldrons, basins or pans.[176]

By the 13th century, about 500 kilometres to the south-east, Nuremberg craftsmen initially cast and hammered Meuse Valley brass alloyed at Dinant and Bouvignes. The brass sheets or ingots were transported to Nuremberg along a north-south trade route. Ores, brass ingots, wire and semi-prepared brass-wares were normally packed in barrels, and carried on trains of horse-drawn wagons, each loaded with up to half a ton.[177] In Nuremberg, a craftsman who worked brass by beating it into shape was called a *Rothsmiedt*. For centuries, European scholars believed brass to be copper that had been caused to grow, gain weight and turn yellow by the mysterious power of calamine ($ZnCO_3$), an ore that was only later understood to contain the element zinc, so German and French writers consistently (but confusingly) used the words *Kupfer* or *cuivre* for both brass and copper, distinguishing them by colour (*cuivre jaune* or *rouge*) rather than by metal type. However, at times, for brass, they used the German term *Messing* (in contemporary English *maslen* – a mixture) or the French term *laiton* (in contemporary English *latten* or *laten*; Arabic *latun*, Italian *ottone*). In the medieval period latten is understood to have meant copper intentionally alloyed with zinc but together with other elements.[178]

From about the 8th century, brass could potentially enter Western Europe from the Black Sea area by routes up the Russian rivers. By the 9th century, metal and metal objects from the Black Sea region were energetically traded between Baltic Sea settlements. Trading links were simultaneously established across the north-western European lands of the Franks and along the Danish and Baltic coastlines. The sheltered harbours of Frankish Haitabu, Danish Ribe, and Polish Wolin and Menzlin all took part in the Baltic metal trade network.[179] The southern forested areas of today's Russia formed the empire of semi-nomadic Khazars from the northern Caucasus region along the southern shores of the Black Sea. From the mid-8th century, the Khazars followed the Russian river systems northwards, to set up a fur-trading post, which became a settlement, at Staraya Ladoga (Old Ladoga, then called simply Ladoga) south of Lake Ladoga in today's northern Russia. By 750 the settlement had increasing hectares of grave-mound cemeteries. It lay near the mouth of the River Volkhov, a small river which connected southwards, via a short portage section, to the River Dniepr, along which goods were transported to Gorodishche (Novgorod), Kiev and, ultimately, the Black Sea and Byzantium. During the 8th century, Norse Scandinavians, the original

---

[175] Thomas *et al* 2013: 172-173
[176] Gérard, 1958: 5
[177] Stahlschmidt, 1970: 131
[178] Blair and Blair 1991: 84
[179] Tolocko 1998: 213-215

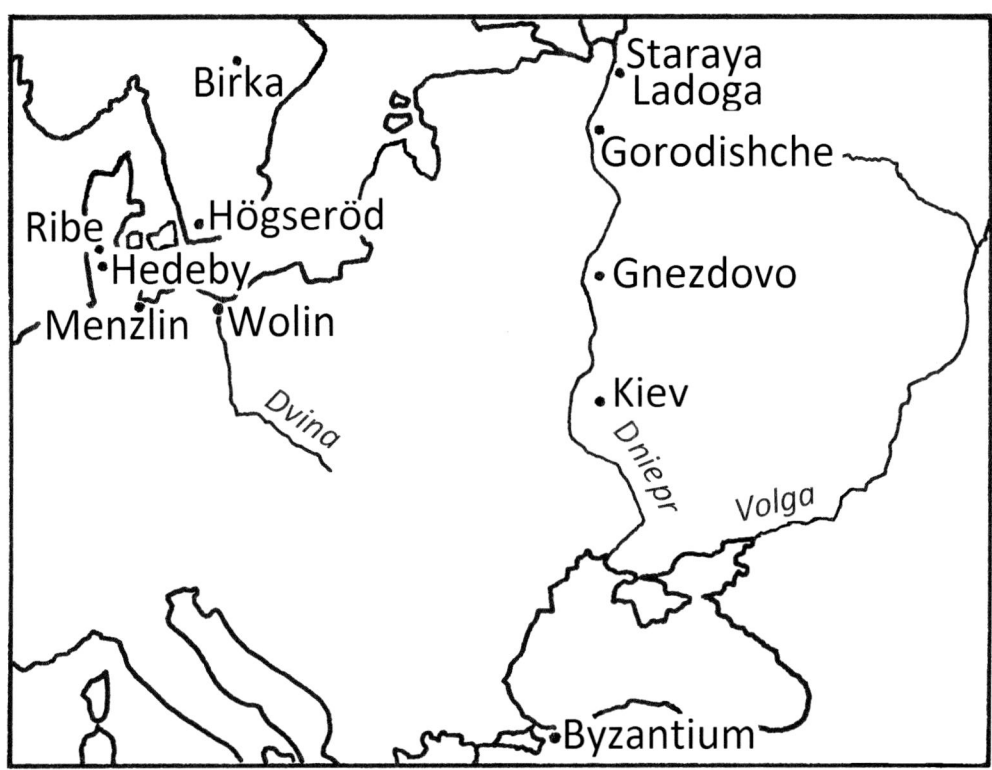

Figure 21. Russian rivers, map showing some sites mentioned in the text

Rus or Rhos,[180] moved south from the Baltic, and began to dominate the river trade.[181] Early northern European brass imports began to show signs of Byzantine influence.[182]

The Rus became closely associated with Byzantium,[183] which provided a strong link between the Black Sea and the Baltic civilisations. Scandinavians infiltrated the Dniepr river route in the early to mid-9th century, rapidly developing their trade. During the mid-10th and late 11th centuries, Byzantine coins were circulated by Scandinavian and Russian mercenaries serving in the Byzantine army. The late tenth-century Danish settlement sites of Sigtuna and Lund have both yielded numbers of Byzantine coins, amphorae and other finds. Scandinavian trade with Turkestan and Middle-Eastern Islam was also active in the 10th century through Kiev and the Dniepr. Both the Khazars and the Scandinavians bartered goods for silver dirham coins circulated by the Abbasid caliphs of Baghdad, and craft workshops have been excavated at trading sites along the banks of the Volga and Dnieper river highways. Excavations at contemporary Birka, central Sweden have revealed numerous Middle

---

[180] Jones, 1968: 246, note 3, derived from a Finnish word for boatmen from Sweden
[181] Duczko 2004: 14, 61
[182] Shepard 2008: 496-497
[183] Oldeberg 1942: 500-501

Eastern items.[184] Evidence from one Russian river settlement shows that trade, over time, changed its inhabitants from an indigenous population to an ethnically mixed one. The Norse cemeteries of a tenth-century settlement at Gnezdovo, a vital tenth-century exchange point on the Baltic-to-Black-Sea trade route (by today's Smolensk) even revealed Abbasid Arab coins in its merchant hoards.[185] Traders could only complete the dangerous journey from Kiev to Byzantium and the Black Sea when the waters were high in the spring but, even then, they faced hazards from hostile nomads, sandbanks, rapids and storm-force winds.

At a twelfth-century exchange and boat-repair centre at Gorodishche on Lake Ilmen (by today's west Novgorod), crews and cargoes from the River Volkhov could prepare for their long journey down the Dnieper. At cosmopolitan Gorodishche, eleventh-century Byzantine, English and German coins have been excavated, showing its cosmopolitan character.[186] From the 10th to the 13th centuries, metals were worked at the Troitsky site (on the banks of the river Volkhov at the Lyudin end of Novgorod). Excavations of medieval Novgorod have produced a wire-drawing plate and a coil of brass wire.[187] From c.1230, after the Teutonic knights conquered Baltic territories, Germanic merchants regularly traded to Russia through the Viking island of Gotland in the Baltic Sea,[188] and in 1235 they gained a market in London, setting up an office there two years later.[189] The eastern Scandinavian trade routes are thought to have inspired certain brass ringed pins introduced to Scotland by Norse settlers, and found in their Scottish graves. Locally-made belt-fittings excavated from Norse Scottish graves were insular in style, rather than Scandinavian, and had a different metal content, suggesting that the zinc in their alloys was recycled from melted-down true binary brass. Norse objects such as oval brooches were also found in the graves.[190]

Around the Baltic in the 9th to 11th centuries, one ubiquitous trade commodity was the brass bar or rod, which served both as a brass ingot, to be melted down and cast, and as an item of exchange or barter. Viking exchange trade in consistently good-quality brass bars (averaging about 20% zinc) had developed from the 9th century around the shores of the Baltic Sea, with the Swedish trading island of Gotland at its hub. In 1847, a ploughman uncovered a hoard of brass trade bars at Myrvälde on Gotland, where about 20 rod-shaped ingots (some broken) emerged among other finds dating to the 10th century when Myrvälde trading centre was at its height. Ingots found elsewhere were mostly triangular or square in section, but those at Myrvälde were longer and hemispherical in section. Contamination of the copper and zinc by

[184] Mikkelsen 1998: 39, 41
[185] Duczco 2004: 170, 174
[186] Janin and Gadjukov 1998: 345
[187] Eniosova *et al* 2014: abstract, 5
[188] Dollinger 1963: 24-25, 29
[189] Davies 1997: 340
[190] Paterson 1998: 125-126

other metals was so slight that the brass bars seem likely to have been made of fresh, not re-cycled, brass.[191]

Hedeby (Haitabu) was a major Frankish town and harbour just south of today's Danish-German border. Excavations at both Haitabu and Ribe, on the Atlantic coast, have revealed brass workshops dating to the 8th to 11th centuries, where a series of brass bars, 24 cm long and square in section, were excavated, together with crucibles, casting moulds and needles.[192] While other Haitabu brass objects were probably made from Rhine-Meuse ores, recent lead-isotope analysis suggests that the most likely metal source for the freshly-cast brass bars was the Bor region, in eastern Serbia, close to the River Danube trade route.[193] Eleven freshly-cast brass bars or fragments, mid-late Saxon in date (c.650-1060), have been recovered from Kingsway, London. Similar to the Frankish ones but longer (35cm), and averaging about 23% zinc to 74% copper – analysis suggests that their copper may have derived from the Harz mountains.[194] Early Baltic sites, despite using fresh brass, may not have produced it. In that case, fresh brass trade bars must have been imported from an alloying site further afield, to be cast in local moulds. Haitabu was located on an isthmus with easy access to both the Baltic and the North Sea – a huge trading advantage. This harbour was a border trading-post at the northern extremity of the Frankish Empire, lying on the ninth-century trade route to Viking Scandinavia. In the early Viking period – the late 8th to early 9th centuries – coinage finds show that Frankish Haitabu was under strong Carolingian influence. Imported goods excavated at Haitabu show the extent of its trade – carnelians arrived from the Carpathians, jet from the English East coast, walrus ivory from Iceland and Greenland, mercury (used in gilding) from Central Asia, and furs from Russia.[195] Western connection to the Russian river trade is evidenced by a belt fastening of the Volga Bulgars, found at Haitabu.

Haitabu harbour had direct access to the Baltic Sea by waterways, and a 16-kilometre defended overland portage route connected it with river transport to the North Sea, so the brass trade bars supplied to Haitabu and Ribe could have arrived by sea from Meuse valley brass-making centres. When ingots or rods of freshly made brass are re-cast into artefacts, they lose some zinc (and a smaller amount of other less volatile elements). Brass objects found on Scandinavian sites include dishes and cauldrons of the type made in the Meuse Valley. A site in Västmanland, Central Sweden, has yielded a Viking brass double spiral ornament,[196] but Byzantine Orthodox pectoral crosses have been found near Lake Mälaren, central Sweden.[197] Other Viking-age

---

[191] Sindbaek 2001: 51-55
[192] Drescher 1983: 174-178
[193] Merkel 2018: electronic source
[194] Bayley et al, 2014: 122, 128
[195] Wikinger Museum, Schleswig
[196] Oldeberg 1942: 58, 145
[197] Roslund 1997: 249

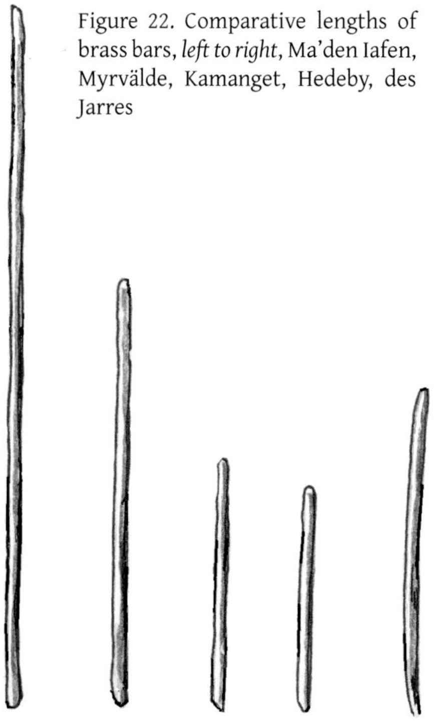

Figure 22. Comparative lengths of brass bars, *left to right*, Ma'den Iafen, Myrvälde, Kamanget, Hedeby, des Jarres

finds include a tenth-century heavily-leaded brass key from Fole, Gotland; and from southern Sweden an eleventh- to twelfth-century brass candlestick from Sovestäd, and an aquamanile from Högseröd, Skania. Heggen church, Buskerund, Norway had an early eleventh-century brass and gilt copper weathervane.[198]

Viking-age hoards made up of brass rods in bundles of consistent length have been excavated from three ninth-century sites on the narrow Swedish island of Gotland in the Baltic, which was something of an entrepôt, attracting and trading Byzantine, Viking and other goods. Brass bars from Viking sites at Myrvälde and Kamänget, on Gotland, contained a level of zinc then only achievable with good-quality raw materials. The tenth-century Myrvälde brass bars were slightly longer and heavier than Kamänget bars, which were probably ninth-century. Other Baltic sites have yielded similar brass-rod fragments in portable bundles of precise, consistent length. A tenth-century brass bar was excavated at Wolin on the north-west Polish coast and 84 early ninth-century bars were found at Birka, northern Sweden, a major port and trading centre.[199] Goods were traded from Haitabu and Old Lübeck, on to Wolin (called the home of 'Slavs, Greeks and Barbarians'), then to Gorodishche (west Novgorod) and through to Kiev.[200] A whole brass bar and a fragment were excavated from the Gorodishche settlement and nine brass bars or fragments from old Ladoga,[201] where, in the early 13 century, German merchants regularly called in on their way up the River Neva.[202] Another 50 brass bar ingots were dredged up from the River Rhine, a few kilometres east of Mainz, presumably lost in transit.[203]

In the Mediterranean Sea, two brass bars have been salvaged off the coast of Provence from the wreck of the mid-tenth-century vessel *Des Jarres*. With them in the wreck was the skeleton thought to be of a man of north-African type, bearing a sword. Pottery

[198] Williams *et al* 2014: 219
[199] Sindbaek 2001: 51, 55
[200] Toločko 1998: 215
[201] Sindbaek 2001: 51, 56-57
[202] Dollinger 1968: 27
[203] Eiwanger 1996: 222

evidence suggests that his ship had sailed from Iberia and was probably bound for the island of Lerins, off Cannes, where an influential monastery had stood since the early 5th century. Lerins was a vital key to the transfer and spread of eastern thinking and knowledge to the west, especially since it was only 40 kilometres west of a harbour base at Sainte Marguerite at the mouth of the River Var. The Var provided a navigable entry from the Mediterranean to a major route connecting the Mediterranean to inland alpine Europe.[204]

So, did the brass for the traded bars come from the Middle East, the eastern Mediterranean, Western Europe or Morocco? In Europe, brass produced in the Rhine-Meuse area probably supplied Haitabu. Byzantine brass, though, cannot be ruled out for Gotland and Baltic Scandinavia, arriving by river northwards to Riga in the eastern Baltic.[205] Long-distance Germanic merchants sailed down the river Dvina to Vitebsk and on to Smolensk (Gdnezdovo), a great market with contacts to both Novgorod and the Black Sea.

The Saharan Berbers were simultaneously making their own active long-distance trading contacts, using camel transport across the deserts of northern Africa. From the 10th to the 12th centuries, the Saharan and eastern coasts of the Mediterranean Sea were dominated by the stable and successful Fatimid Empire.[206] When their silversmiths ran short of silver in the later 11th century, they readily adapted to working in brass, a brighter alternative to bronze.[207] Excavations at sites in Caesarea (Israel) and Denia (southeast Spain) show that Fatimid craftsmen hammered out very fine early eleventh-century brass braziers, bowls, lampstands, buckets, jugs, ewers, boxes and candlesticks, often with rich embossed or inlaid decoration. Besides hammering, they cast objects in metal-framed sand moulds.[208] Saharan desert traders would have been familiar with this work.

From the 11th century onwards, Almoravid Berbers, who had earlier defeated Cordoba in southern Spain, started to trade brass southwards in exchange for slaves and gold. Gold from richly forested West Africa was a metal coveted further north. The Andalusian geographer, Al-Bakri (d. 1074), mentioned Kumbi, the capital of the Ghanaian empire (a state that covered today's northern Senegal and southern Mauretania from about 800-1200), where gold was exchanged for objects of worked copper, cast in a local desert settlement and sent to a trading centre at Awdaghust (Audghost).[209] Berbers from the land of the Sus (eastern Morocco) traded brass to

---

[204] Eiwanger 1996: 220-221
[205] Shepard 2008: 507, map
[206] Wickham 2009: 336
[207] Allan 1979: 45
[208] Shalev and Freund 2002: 22
[209] Levtzion and Hopkins 1981: 69, translation of al-Bakri

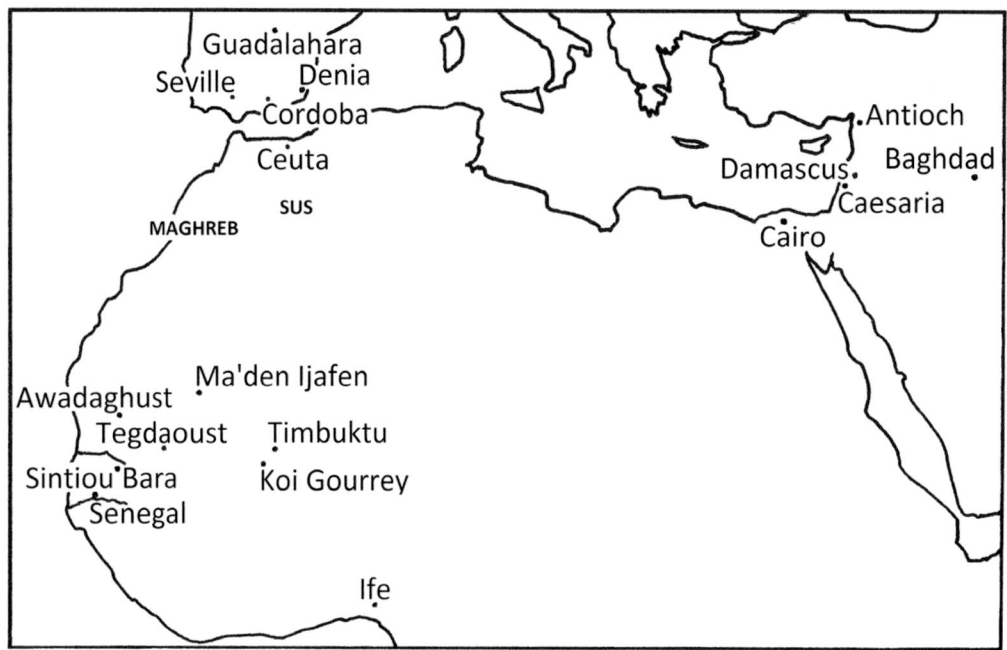

Figure 23. Northern Africa, map showing some sites mentioned in the text

the Maghreb (Morocco) and Andalucia.[210] In the 1190s, an anonymous traveller visiting Fez, capital of the Maghreb, described merchants' caravans leaving to trade brass to the Sudan and other regions further south.[211] The traveller's account is supported by a chance desert discovery in 1964 by French explorer, Theodore Monod, at the Ma'den Ijâfen (now in southern Mauretania). Monod chanced upon the wreckage, dated to about 1165, of a southbound camel caravan, carrying brass bars, longer, but somewhat similar in shape and zinc content to contemporary bars found around the Baltic Sea area.

The camel caravan found by Monod was travelling southwards from the Maghreb (Morocco) towards Timbuktu, carrying more than 2000 cast brass trade bars, tied in six bundles of consistent portable length. On average 72 cm long, and 50 g in weight, the bars were far longer and heavier than the Myrvälde Swedish brass bars because they were designed for camel transport. They were slightly thicker at one end, where molten brass had entered its mould,[212] (a narrow furrow in the sand)[213] and have, like those from the Baltic, been described as bread-loaf-like in section, due to being cast in shallow open moulds, leaving molten brass bulging just over the brim, like baked loaves from a bread-tin.[214]

---

[210] Levtzion and Hopkins 1981: 95, translation of al Zuhri, Geography, 1130s-80s
[211] Levtzion and Hopkins 1981: 139, translation of al-Bakri
[212] Monod 1969: 303-306
[213] Hawthorn and Smith 1963: 144
[214] Sindboek 2001: 54

The brass bars or rods of the camel caravan disaster were found alongside sacksful of cowrie shells from the Maldives (Indian Ocean), which were then highly valued as currency along the West African coast. Brass craftsmen already existed not only in Isfahan and Mesopotamia, but in North Africa and Islamic Spain – by AD 718 a Berber and Arab army had captured southern Spain for the Ummayad caliphate (661-750) based in Damascus.[215] The empire of the succeeding Abbasid caliphate of Baghdad (750-1171), that now included Spain and the Maghreb, traded regularly with India and China.[216] However, at the time of the caravan disaster, the rest of North Africa was ruled from Cairo by the Fatimid caliphate (909-1171), which had overthrown the Abbasids in Egypt and northern Africa, but had failed to wrest Spain or the Maghreb (Morocco) from them. In 1081-2, Muhammad ibn Sa'id as-Sabban made a brass astrolabe at the Spanish town of Guadalajara,[217] and by the 12th to 13th centuries, complex brass astrolabes were produced throughout the extensive Islamic territories.[218] There were strong links southwards from Islamic Spain and Morocco via the trans-Saharan camel caravan routes.

Berber and Arab merchants operated a metal-working centre at Tegdaoust (next to Awdaghust, Audghost) in the southern Sahara.[219] Archaeological evidence suggests that Tegdaoust metal-workers received the long brass bars brought south by camel from Morocco. In their furnaces they melted down the brass, adding a regular amount of lead and some local copper, enough to increase its bulk and weight but not to lose the golden colour of the alloy. This increased its trade value, so it was re-cast into moulds to form shorter brass bars, 25-26 cm long, suitable to trade south into the forested area on donkeys, to exchange for gold and slaves. The brass was suitable for making the bracelets, rings and harness that were popular in the vast southern forested regions of the Ghanaian Empire. It was a high point for trans-Saharan trade, when empires ruled the southern Sahara and sub-Saharan Africa, and north-south contacts were strengthened by a northern desire for southern gold, slaves and timber, and a southern yearning for salt and brass. The Tegdaoust metal workshops, were, however, destroyed during Almoravid raids in 1054-5.[220]

In the ancient, extensive Ghanaian Empire, brass was known from the 9th century. Excavations at Sintiou Bara, ancient capital of the Silla kingdom (in today's Senegal) revealed brass bells and end-looped bars dating from the 10th to the 13th centuries.[221] Other finds included iron horse trappings decorated with brass, and the mid-eleventh-century geographer, Al Bakri, wrote of horses in harness at the Ghanaian

---

[215] Wickham 2009: 338
[216] Mikkelsen 1998: 40
[217] Museum of the History of Science, Oxford: 52473
[218] Museum of the History of Science, Oxford: 44141, 49033, 57878, 49861, 41122, 45307
[219] Garonne-Marot and Mille 2007: 165
[220] Garonne Marot 2009: 6
[221] Garnne-Marot and Wayman, 1994: 45, 47-49

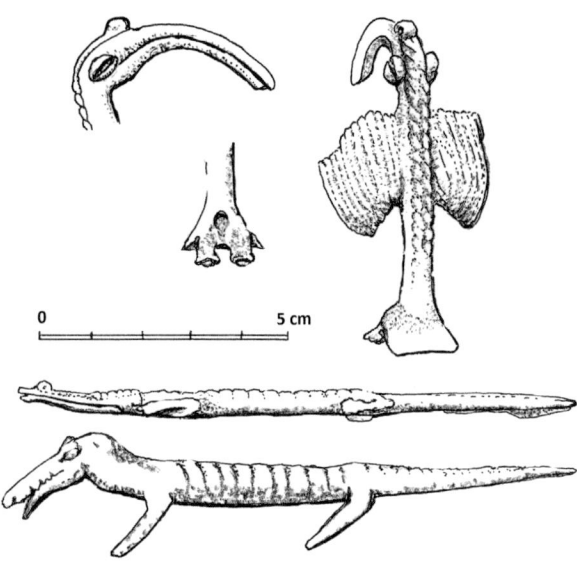

Figure 24. Koi Gourrey figurines: *above*, hornbill; *centre*, female crocodile; *below*, male crocodile, after illustration by Dr Laurence Garenne-Marot, with her generous permission

court. In 1904, when an infantry lieutenant, Louis Desplagnes, excavated a burial mound at El Oualadji, containing two male burials, he noticed that it corresponded well with Al Bakri's description of the burial-mound of a Ghanaian ruler. Koi Gourrey monumental burial mound in the ancient Ghanaian inland Niger Delta (in today's Mali) produced three of the most exciting finds – three brass animal figurines, perhaps part of a cult of religious animism.[222] One, a male crocodile, has since been lost but the other two very north-African-looking creatures, a hornbill and a female crocodile,[223] were made of what was, for their time, very high-zinc brass.[224] The percentage of zinc they contained (27-31%) was higher than any produced in the contemporary Islamic Middle East, Europe or other parts of Africa, and the figurines were cast by the lost-wax technique. Four brass bracelets and a ring from Koi Gourrey contained considerably less zinc than the animal figurines.[225]

The human remains recovered from the two Koi Gourrey tumuli wore jewellery made either from copper or from low-zinc brass, but never both worn by the same person, so the contrast between wearing reddish copper and yellower brass is thought to have symbolised different social relationships.[226] Copper is known, from Arabic sources, to have been mined later at sites in today's West Mali and Niger, along the western desert trade routes. Zinc ore from the Sus area of the Atlas Mountains (south-west Morocco) was exploited for making brass to be transported further south on camel caravans.[227] In gold-mining regions of the Ghanaian Empire, brass was especially highly esteemed by local working families, though copper was valued for use in rituals. As in the earlier Middle-Eastern city-states, brass may have been viewed as exotic, new, and produced in a mysterious way, whereas gold, so highly valued elsewhere, held little attraction

[222] Garonne-Marot and Mille 2007: 160
[223] Musée de l'homme, Paris: 03 8 2 and 03 8 3
[224] Garonne-Marot and Mille 2007: 162
[225] Garonne-Marot *et al* 2003: 78
[226] Garonne-Marot and Mille 2007: 165
[227] Monod 1969: 31

for Ghanaian miners. In fact, details from Arab texts suggest that local gold-mining families feared and shunned real gold. Even today, local people link gold with genii representing the historic dangers suffered by miners who dug gold to satisfy the demands of the local elite and their Arab merchant contacts, taking copper and salt in return. The merchants, who had no fear of gold, saw it only as a valuable exchange commodity.[228]

During the period 1250 to 1350, when Timbuctu was the major trading centre of the Mali Empire of the Mande group of peoples, Berbers and Arabs traded brass arm-rings (manillas) southwards along Sahara caravan routes from Ceuta (Morocco), where brass may have been produced. After 1259, when Damascus became a celebrated centre for making and trading brass, Berbers may have imported brass through merchants from Venice.[229] By around AD 1200, further west along the Guinea Coast, the beautiful Ife 'bronzes' were being cast. Two-thirds of the heads from Ife (in today's Nigeria), dating to the 12th to early 15th centuries, are in fact cast by the lost-wax process from leaded brass that was similar to the brass of contemporary Viking and eastern Mediterranean brass artefacts. The original belief system of the people of Ife concentrated upon one god, Olukun, creator of the earth – with Ife the sacred place at the centre of creation– but the legendary settlement became overgrown by a great sacred grove – the *Igbo*. By tradition, the treasure of the culture was buried beneath the *Igbo*.[230]

The Middle Ages saw more and more brass casting, using both lost-wax and sand-casting methods, practised from northern and western Africa to the Himalayas and from the Baltic to the Mediterranean and the Black Sea. It was an impressive era of empire building in some areas and territorial shrinkage in others. These changes, coupled with increased trading, brought far-reaching, but as yet tentative, cultural interchange between Europe, India, Central Asia, China and Africa. The growth and spread of cultures inspired craftsmen to use brass in new ways for images, holy vessels and sacred objects. The succeeding centuries would see an increasing variety of both religious and secular brass objects produced.

---

[228] Garonne-Marot and Mille 2007: 166-167
[229] Allan 1998: 18
[230] Drewal and Schildkrout 2010: 22

Ninth-century brass astrolabe cast in (today's) Syria by Khalīf, apprentice to ʿAlī ibn ʿĪsā. Its plates cover latitudes 33° to 41°. Added inscriptions show that it was used in Armenia. Astrolabe Inv. 47632, © History of Science Museum, University of Oxford

Chapter 3

The Sacred and the Salesmen
Later medieval years *c*.1250-1500

From the later 13th century, as demand increased, markets gradually became more centralised and better organised, though overland transport remained arduous. Widening trade contacts increased the demand for brass and broadened its uses, which in turn accelerated production, leading to the need for more centralised markets, or entrepôts, for buying and selling a diversity of goods from further afield. The great religions – Buddhism, Hinduism, Islam and Christianity – made ever more confident statements in brass, producing religious objects which those at subsistence level could venerate rather than possess. In this later medieval period, however, domestic brass objects, though still mainly for the wealthy few, began to be used by a rather wider range of society.

The history of anything, brass included, is supported by the evidence left behind, which varies greatly from one period to the next and therefore affects the narrative. The pre-Roman period, for example, relies on very sparse artefact evidence, the study of fragmentary texts and the excavation of vestiges of ancient archaeological sites. The later medieval period is richer in evidence from surviving religious images, illuminated manuscripts and the vessels and tools used in religious ritual, which may have survived due to being protected by veneration. Excavation of later medieval sites has also provided technological evidence about how contemporary brass was made, particularly at Dinant on the River Meuse.

Moving forward to the 16th century, the story becomes enriched by the survival of fine artefacts that were still mainly religious, and by more detailed knowledge of brass works and technology. Earlier seventeenth-century Europe, at the outset of the era when the written word became all-important, was severely influenced by wars, which affected brass-working, though the later part of the century saw the Age of Enlightenment, with its increased demand for scientific and other secular brass items. The 17th century includes evidence for brass as part of the international Slave Trade, for world-wide trade in metallic zinc, and for fine artefacts surviving from India, Persia and West Africa. The 18th century is dominated by brass as a component of fashion, and by more detailed industrial accounts of individual brass works, the people working in them, and their innovations and intrigues. The history of brass therefore weaves closely into the background history of each period but the emphasis constantly changes, depending on the evidence available.

For the later medieval period considered here – from about 1200 to 1500 – it was still important to look east. Himalayan Central Asia continued to make inspired use of brass for creating or gilding sacred Buddhist figures, including a thirteenth-century Tibetan stone figure, gilded with brass, representing Jambhala, who personified the elimination of poverty.[1] The meanings embodied in such images indicate the daily philosophy of these resilient mountain people. For example, surviving thirteenth- to fourteenth-century Tibetan figures made of true brass include Amoghasiddhi who showed the Buddhist way;[2] the Bodhisattva Mañjuśrī; personifying transcendent wisdom;[3] Mahāsiddha Virūpa, expert in yoga, embodying the way to psychic perfection,[4] and a dancing Vajravārāhī joyfully personifying triumph over ignorance.[5] In the 14th century, craftsmen sometimes cast brass portraits of respected Tibetan monks. Copper-rich Nepal could supply Tibet with the copper that it lacked, and trees for charcoal then grew in some Tibetan valleys. Surviving fifteenth-century Tibetan Buddhist brass images include the richly-adorned, primordial Buddha Vajradhara, representing full enlightenment,[6] and a Tibetan Lama named Tsong-Kha-pa, whose teachings showed the graduated way to enlightenment.[7] Three more examples are the shrine of the trinity of Amitāyus, the Buddha of infinite light attained through performing good deeds in past lives,[8] and images of Padmasanbhava, seen in Tibet as a second Buddha,[9] and Vajrayoginī, representing the transformation of everyday experiences, including death, into paths of spiritual enlightenment.[10] The ten Tibetan analysed brass figures described here contained, on average, 78% copper to 20% zinc, with less than 2% lead or tin.

In Persia (encompassing today's Iran and Iraq) in the 12th to 13th centuries, most objects containing zinc held more lead than Tibetan objects, and a certain amount of tin. Middle Eastern zinc ores were probably first sourced from the Anatolian mountains of today's north-east Turkey and by Marco Polo's day from the mountains near Isfahan, southern Persia (Iran). Surviving true brass examples from this twelfth- to thirteenth-century period of Islamic Persia include table-tops, pen-boxes, and ewers,[11] intricately decorated and inlaid with silver and a black composition.[12] Some feature calligraphy and bear roundels illustrating musicians.[13] One surviving brass bowl, dating to 1350-1400, carries inlaid silver designs with depictions of horsemen,

---

[1] British Museum 1983,1109.1
[2] Reedy 1997: figure W 150, 136
[3] Reedy 1997: figure W 149, 195
[4] Reedy 1997: figure C 171, 203
[5] Reedy 1997: figure C 173, 204
[6] Reedy 1997: figure C 182, 207
[7] Reedy 1997: figure C 183, 206
[8] Reedy 1997: figure C 185, 207
[9] Reedy 1997: figure C 186, 207
[10] Reedy 1997: figure C 189, 209
[11] Allan 1979: 143-45, analyses, table 21
[12] Melikian-Chirvani 1982: 170
[13] Victoria and Albert Museum, London: 381-1897

Figure 25. Ewer, Western Iran, c.1220-40; *right*, detail showing a musician

flowing calligraphy and thick foliage.[14] Beautiful boxes, inlaid with silver and gold,[15] were made of brass in Persia in the 13th and 14th centuries.[16] The drum-shaped base of one Persian candlestick, dating to the 1340s, bears a roundel inlaid with gold and silver,[17] thought to depict Tashi Khatun, the poetess and powerful Mongolian mother of Sheik Abu Ishaq of Shiraz (1343-1354). She had fine Q'urans produced, and built a shrine, a religious school and the dome to a mausoleum,[18] and is said to have appeared, unveiled, to address the people of Shiraz, inciting them to oppose invaders who had captured her husband and sons.[19]

A gradual diffusion of ordered Islamic culture and design into Hindu Indian metalwork occurred following Islamic invasion of northern India, which started with incursions and then conquests by the Afghan leader, Mahmud of Ghazna, from c.1000 to 1027, accompanied in 1017 by chronicler Al Biruni.[20] This fusion of two cultures, Islamic and Hindu, appears in a few surviving vessels, including a symmetrical cast brass pilgrim flask (*c*.1200) probably made in Khorasan but found in the Punjab, with Arabic script inlaid in silver, and with life-like figures of curly-horned goats protruding from each rotund side, their horns forming handles.[21] By the later 13th century, the Islamic invaders held a huge area stretching from Gujarat to Bengal and stretching south to include the Deccan.[22] During the sultanate period (1192-1526), three of the major religions were present over much of India. The ancient Indian Jain religion, embodying non-violence and self-control, was

---

[14] Ashmolean Museum, University of Oxford: 1974.9; 1967.122; X 3206 and EA X 1202
[15] Musée du Louvre: 3355
[16] Zebrowski 1997: 264
[17] Museum of Islamic Art, Doha: MW 122.1999
[18] Brookshaw 2019: ebook
[19] Allan 2002: 37
[20] Sachau 1989: ix
[21] Melikian-Chirvani 1982: 170
[22] Zebrowski 1997: 19 and 200

also practised. More brass objects created at this period were functional, for example bowls or utensils and sometimes scientific instruments.

Sacred images, however, continued and flourished. From western India, a twelfth-century seated image of Santinatha,[23] the revered teacher, saviour of the Jain faith – invoked at times of disaster,[24] proves to have been made of brass. This image and a thirteenth-century brass Buddha from Kanchipuram, Tamil Nadu,[25] prove (on lead-isotope analysis) to contain zinc probably sourced from a western Indian zinc-lead mine at Ambaji.[26] Another Jain brass image, a shrine to the Tirthankatra Kuntunatha (the liberator who shows the way), inlaid with silver and copper and dating to 1476, is known to have been commissioned by a merchant of Vasatagash.[27] A leaded brass figure of the benign Jain goddess Ambikā, dated 1350, was found in Gujarat, and another from Bengal survives from 1521, before the Mughals reached that area.[28] A number of Jain brass images from Gujarat, dating to the 15th century, are models of the spiritual Nandisvara Island, representing a continent of rejoicing and pleasure, the outermost island of Jain cosmography, filled with lush gardens, lotus-filled lakes and an ordered number of concentrically arranged figures.[29] From Gujarat, one brass model dates to 1416 and another four-doored example, measuring about 10 x 24 cm, dates to 1480. A figure of Vishnu-Narayana from Gujarat, dating to 1485, evidently contained some distilled zinc, probably from Zawar.[30]

Bateshwar on the River Yamuna, a tributary of the Ganges, seventy kilometres south of Agra, has lain beside a major highway since the 4th century BC and was a major cattle, horse and camel fair for both the Hindu and Jain religions. Horses played an important role, and one surviving brass image from Rajasthan shows a Rajput on horseback.[31] Rajput literally meant 'prince', but became applied to a member of any substantial northern or central Indian landowning family. 101 temples to the god Śiva lined the banks of the River Yamuna at Bateshwar. Their great brass and copper bells (*ghantas*), were rung to focus the minds of the entering worshippers and to alert the deity of their arrival. At one time the local bandits (*dacoits*) are said to have offered brass bells to the temples before setting out to rob a wealthy traveller. Some Hindu temples were surmounted by a pitcher-shaped brass crown (*kalasha*) on each pinnacle.

Cleanliness in relation to food is an important aspect of Hindu belief, so the wealthy used easily-cleaned brass and copper utensils. Where possible, Hindus used brass for

---

[23] Victoria and Albert Museum: 930-IS
[24] Srinavasan 1999: 205
[25] Victoria and Albert Museum: IS-44-1966
[26] Srinavasan 1999: 205-206
[27] Ashmolean Museum, University of Oxford: EA 05.108;
[28] Biswas 1993: 325; Ashmolean Museum, University of Oxford: EA 2005.17
[29] Pal 1995: 121
[30] Kharakwal and Gurjar 1966, vol. 1: 7
[31] Biswas 1993: 325

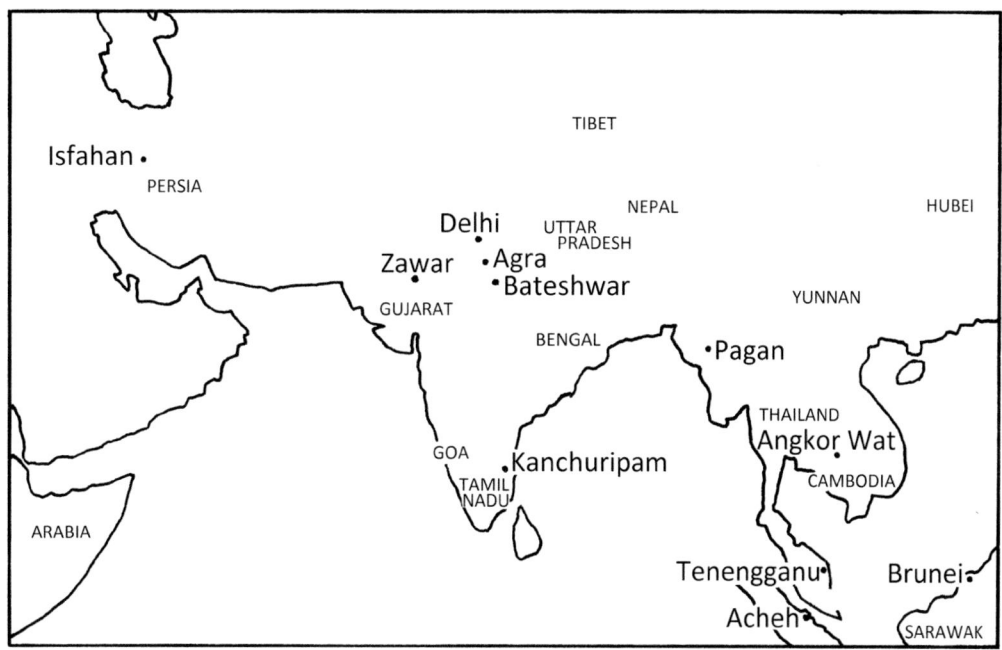

Figure 26. South-East Asia and India, map showing some sites mentioned in the text

large water-vessels, cooking pots, trays, bowls, lamps and spoons, but poorer people might carry just a brass *loti*, or drinking-cup. Large brass pots were used for cooking at religious festivals, marriages and feasts, and spare ones were kept by the wealthy to be lent to poorer people for such occasions. An Indian dance called the Tarangam of Kuchipudi, was danced with the feet on a brass plate, and certain Indian dancers wore small brass ankle-bells. Excavated sites in Gujarat, Uttar Pradesh and Madhya Pradesh have yielded brass bangles, caskets, household utensils, finger rings and religious objects.[32]

During the later medieval period, zinc distillation flourished at Zawar (now Jalor) along the Tiri River valley in the Aravalli hills of Rajasthan. A nineteenth-century writer describes how 'in the rains, a foaming river roars past a ruined town' with 'temples of hoar antiquity', surrounded by 'hills clothed with verdure', which 'rise to a great height on every side'.[33] Evidence suggests that some zinc ore was processed at Zawar by 900 BC,[34] probably initially being roasted to produce zinc oxide, but the main period for producing distilled zinc metal for use in brass-making in northern India and the Middle East lasted from the late 14th to early 16th centuries, eventually

---

[32] Kharakwal, J. S. and L. K. Gurjar 2006, Zinc and Brass in archaeological perspective, *Ancient Asia* 1, 139-159 http://doi.org/10.5334/aa.06112

[33] *The India Antiquary* 1: 13-14, reprint of an anonymous article from the *Times of India*, January 1892

[34] Kharakwal, J. S. and L. K. Gurjar 2006, Zinc and brass in archaeological perspective, *Ancient Asia* 1, 139-159 http://doi.org/10.5334/aa.06112

ceasing in the 18th century.[35] Distilled zinc metal could improve brass production by allowing better control of the proportion of zinc to copper and enabling a higher zinc percentage than was possible with the cementation process, rising to a point where the brass could be cold-worked, without need for annealing.

The Zawar zinc-distillation process was carried out very skilfully and on an industrial scale, in a great multi-furnace distillation plant. By the 1500s, there were several banks of seven square furnaces, each furnace holding 36 upright retorts, fired simultaneously all along the row.[36] Precise heats had to be maintained for every single retort and condenser in every furnace – from the furnace-centre to the outer corners – an impressive feat which by the 16 century produced an estimated output of 60-80 tonnes of metallic zinc a year.[37] At Zawar, the distillers crushed zinc sulphide ore, roasted it with powdered charcoal to dispel sulphur and produce (slightly impure) zinc oxide. They mixed it with a flux of salt, borax, cow dung and water, rolled into tiny balls, dried in the sun, and placed in clay retorts (10-15 cm in diameter and 30-35cm long), and thrust a stick down the middle. They then inverted the retorts and placed them vertically in a chamber on a perforated brick plate. They completely surrounded the retorts with charcoal, and inserted clay wedges to prevent the tightly-packed retorts from touching, and allow heat to circulate. The charcoal was ignited, fanned by draughts through small holes in the perforated base. At about 600°C zinc vapour started to distil up to the top of the retort, condensing to a liquid as it descended through a central tubular gap (left after the charred stick burnt). From there it dropped through an outer condenser tube, leading below the brick shelf, into a cooler second retort below, to solidify as zinc metal. The condenser tube was maintained at the temperature at which zinc just starts to vaporise (between 415°C and 550°C), but the temperature inside the retort reached a maximum of 1200°C (zinc oxide reduces most efficiently from about 1100°C).[38]

The Zawar method collected the distilled zinc downwards. The Chinese, however, collected distilled zinc upwards. The rising zinc vapour emerged from the top of the retort to collect in an upper chamber over a pocket or saucer known as the 'bird's nest' and there condensed into the saucer as solid metallic zinc.[39] Although the Chinese perhaps distilled zinc equally early, unlike the Indian distillers they seem not have made brass from it until the 15th century.

In 1271, when Mongol leaders captured China's Yunnan Province, refugees from the Shan mountains of western China were driven westwards into today's Myanmar (Burma), whose ruby-bearing Shan mountains had long been exploited for zinc ores.

---

[35] Craddock 2000: 43
[36] Craddock 1984: 23
[37] Craddock *et al* 1998: 42-43
[38] Craddock, Freestone *et al* 1998: 30, 37-38, 41
[39] Bell 2017: www.thebalance.com/zinc-history-pt-ancient-chinese-zinc-production-2339710

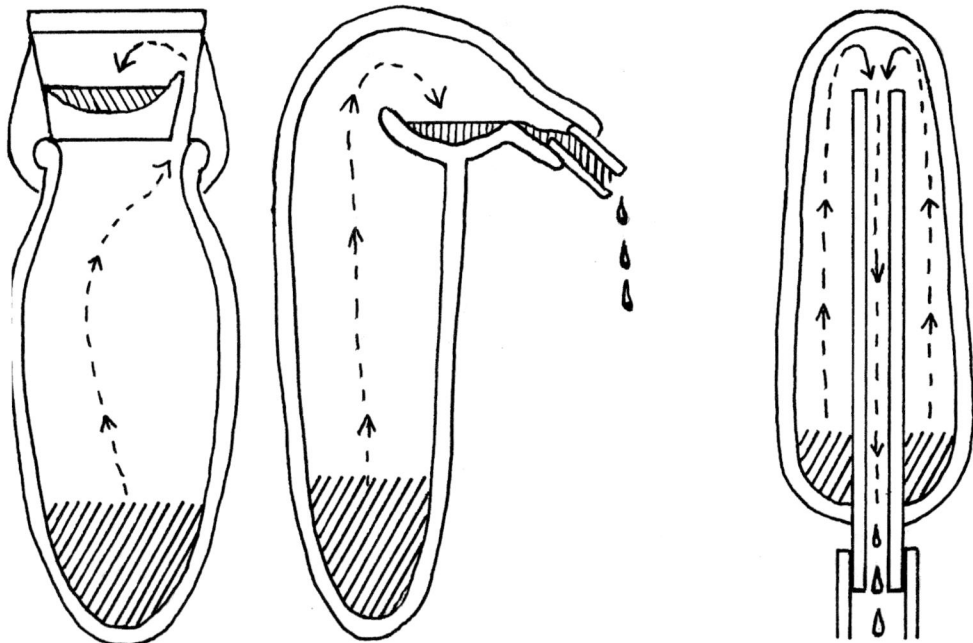

Figure 27. Diagram showing *left*, Chinese method, *right* Indian method distillation retorts

The Chinese refugees arrived at the end of Burma's period of Buddhist revival (1044-1287), based around the city of Pagan (now Bagan). The relic chambers of pagodas dating to the post-Pagan period contained bronze statues of the Buddha, sometimes attended by small figurines of worshippers making lost-wax votive offerings, in copper, bronze and brass.[40] The mountain-born Chinese Shan settlers set up open-sided bamboo and thatch metal-workshops in Burma or worked in the open air, using shallow hearths dug in the ground. Their clay crucibles were handled in bamboo cradles, and were heated by charcoal kept at an even temperature by air forced through a pair of cylindrical bellows made of bamboo, with feather-covered pistons.[41]

In Cambodia, by the 12th to 13th centuries (the Angkorian period) the Khmer metal-workers had received no tradition of brass-making, although from the 9th century, zinc crept into bronzes or tin-bronzes in small amounts, resulting in quaternary alloys, rather than true brass. Only one analysed example, an adorned Buddha from Angkor Wat, protected by a *nāga* (great serpent), was of leaded brass, probably alloyed elsewhere. As in other areas, the zinc content could have arisen from adding re-melted brass objects or using zinc-rich copper ores. In Thailand, brass (some containing over

---

[40] Fraser-Liu 2010: 123, 128
[41] Fraser-Liu 2010: 128

10% zinc) was first produced in the 15th century,[42] when it seems that brass technology was finally introduced to Thailand and Cambodia, under Chinese influence.[43]

Already, by 1260, the Mongols had conquered south China and founded the Chinese Yuan dynasty (1279-1368) which actively encouraged trade and united China with most of Asia by reviving the Silk Road. A low-zinc twelfth- to thirteenth-century figure of Guanyin and an inscribed Buddha dating to 1403-24 were probably made from cementation brass imported into China.[44] After Hongwu (1368-93), first Emperor of the Ming dynasty, defeated the Mongols in 1386, he decreed that all this Chinese overseas trade should cease. His son, Emperor Yongle (1403-24), participated instead in diplomatic gift exchange with foreign leaders. Between 1405 and 1426, his admiral, Zheng He, was dispatched with seven imperial naval tribute fleets to negotiate in South-east Asia, India, Arabia, and along the East African coast, distributing and accepting valuable gifts.[45] The gifts are not known to have included brass, though astrolabes from Arabia could have been involved, but the exchanges established cultural east-west contacts which would facilitate future brass trade.

Other fifteenth-century analysed brass objects from China include a bearded man dating to 1436 and a Tibetan-style figure of the Bodhisattva Avalokitesvara.[46] The early figures may have been produced in Central Asia and brought to China as gifts for the emperor. On the other hand, imperial court foundries are understood to have cast gilded brass Buddhist figurines as gifts for monks arriving from Tibet to honour the emperor, since thirty such gilded Buddhist statues bearing Yongle-Xuande (1403-35) reign marks have proved to be made of brass, using the cementation method.[47] Brass was also used for certain architectural features which would traditionally have been carved from wood. Recent analysis shows a base layer of brass lying beneath the worn gilding of ornamental features decorating the attractive 'golden hall' constructed in 1416 during the Yongle period (1403-24) on a peak of the Wudang Mountain chain, Hubei Province. The brass layer, or substrate, was of leaded cementation brass (10-17% zinc), and a Chinese armillary sphere built in 1437 also had measurement-taking components made of leaded cementation brass.[48] Some incense-burners, with reign-marks of the Xuande period (1426-35), are made of 29-35%-zinc brass.[49] China was, and still is, very rich in zinc carbonate ores, and its copper supplies were good, but

---

[42] Bourgarit *et al* 2003: 116
[43] Vincent *et al* 2012: 136, 141-142
[44] Cowell *et al* 2003: table 1
[45] Vainker 1991: 140
[46] Cowell *et al* 2003: table 1a
[47] Quanya Wangan and Sascha Priewe 2013: 67
[48] Zhou Wenli 2012: 1 and 46-48
[49] Zhou Wenli 2016: 11-12

The Sacred and the Salesmen

Figure 28. Golden Hall temple, Wudang Mountain, China, 1416

the vital timber used elsewhere for fuel was scarce in northern China and Tibet in the medieval period, so coal often replaced it.[50]

Shortage of timber, however, was so far not a problem in the West, where extensive forests of suitable trees existed to provide fuel and charcoal for brass-making. Sweden, particularly rich in forests, was increasingly exploiting copper from the *Kopparberget* mine at Falun in sparsely-populated Dalarna, north central Sweden, where it had been mined locally since the 9th century. In the late Middle Ages, therefore, Falun mines started overtaking the Harz mountains as Europe's principal copper source. This rich Swedish copper supply raised the possibility of producing far more brass in Europe, but Sweden itself lacked suitable calamine (zinc carbonate ore) for the cementation process. A plentiful source of very pure calamine ($ZnCO_3$) existed at the Vieille Montagne, or Altenberg (near today's Belgian hamlet of Moresnet, which in 1439 became part of the Duchy of Limbourg). In 1308, a cargo of this calamine was sent by Dinant merchants to Hull, England.[51] Copper is said to have been mined at Newlands, Keswick, Cumbria until 1270,[52] so brass-making might have been planned, but perhaps failed. From the 15th century, calamine was also transported from the Vieille Montagne to the Cologne, Frankfurt, and Bruges

---

[50] Zhou Weirong 2007: 18
[51] Suttor 2014: 23
[52] Camden 1637 edition, 822

trade fairs and to Braunschweig (Brunswick). The Tyrol provided a calamine source further south, and smaller amounts were available from the Teutoburger Forest (in today's Central Germany) and from Iserlohn (Rhineland), where lapsed calamine-mining rights were revived in 1474.[53] Iserlohn traded calamine through Cologne and the nearby Hanseatic town of Soest.[54]

Very few medieval European sites actually alloyed brass fresh from the ores, but those few centres delivered brass to numerous forges and foundries to be worked into objects. The work was therefore already becoming industrialised, with different workshops performing different stages of production. In some, brass was simply alloyed, cast into ingots and distributed to foundries, whilst others cast both ingots and sheet metal. Some mills cast both brass sheet to slit and draw into wire, and slightly thicker plate brass to be hammered into 'semi-prepared' goods – usually part-finished cauldrons and basins – hollow but unfinished. Semi-prepared brass wares were sent on to be worked further, sometimes far from the original mill, by specialists, who shaped, scraped, filed, etched, polished and added legs, handles or spouts. Mills which installed wire-drawing equipment carried the alloyed brass through to finished or part-finished coils of marketable wire of varying thickness.

By the 12th century, beside the cauldrons and pans hammered out at Dinant, Bouvignes and throughout Flanders and northern France,[55] Dinant founders now produced figurative castings. These included a twelfth- to thirteenth-century candlestick in the form of a horned stag, and an aquamanile portraying an alert animal somewhere between a lion and a dog.[56] A fourteenth-century Dinant brass-worker, Jean Josès, cast a fine eagle lectern and an elaborate paschal candle, over 250 cm tall, with fierce brass lions as feet, both now in the church of Notre-Dame de Tongres (Tongeren), eastern Belgium.[57] It was becoming important to find overseas markets.

Markets for brass products needed to expand in the later medieval period, some growing in size and scope until they became central entrepôts for saleable goods which arrived from near and far, to be displayed on stalls, sold and distributed all over the continent and beyond. From the early 13th century, merchants from Lübeck and other Baltic ports traded to the English east-coast ports of Lynn (later Kings Lynn), Yarmouth, Hull and Boston and in the later 13th century the numbers of ships visiting these ports rapidly increased.[58] In 1157, the brass-founders of Cologne, who made objects from Aachen-produced brass, gained British royal permission to market their wares in London. In 1281 the Hanseatic League cities of Lübeck

---

[53] Klostermann, Rolf, 2013, 'Historische Einordnung', unpublished presentation at Iserlohn Museum, 6
[54] Klostermann 1996: 30
[55] Yante 2014: 17-18
[56] Thomas et al 2014: catalogue numbers 197 and 189,
[57] Saint-Amand 2014: 77-78
[58] Dollinger 1968: 39

and Hamburg gained the right to trade in London,[59] and in 1329 Edward III (1327-77) extended London Hanseatic trading rights to include Dinant merchants. This allowed them to export their brass bowls, cauldrons, drinking vessels, cooking-pots, candlesticks, scissors and sickles to the Thames-side wharves of the German steelyard (*Stalhof*).[60] A fine thirteenth- to fourteenth-century tripod ewer with a curved animal-head spout, found at Gilsland near Carlisle, northern England,[61] was probably imported from Dinant in this way. The London market halls charged no border tax, which made selling in London a profitable option. Almost every state in mainland Europe charged stringent tolls or taxes on any goods crossing its borders, which made overland carriage of brass or its raw materials costly compared to direct sea transport to a flourishing market on the River Thames.

Figure 29. Brass market stall, Champeaux market, Paris, 1403-4

Cumbrian lead-copper mines at Keswick, northern England, were worked in the 13th century and again mentioned in the 15th century,[62] but Cumbrian copper did not satisfy demand – because customs records for the first of July 1384 show Britain importing large quantities of copper, twelve barrels of 'calamys' (calamine) and about fifty kilos of brass (probably scrap brass to add to the crucible).[63] This indicates that fourteenth-century English craftsmen were already making brass.

A legal dispute further suggests that fresh brass was produced in England. The dispute was between two Oxford braziers active in the 1390s, involved in making brass from good imported zinc carbonate ore, probably from Stolberg (*stelebeke*), alloyed with '*graycober*' (probably part-smelted copper sulphide ore or *matte*), and scrap brass.[64] Graycober contained significant amounts of arsenic and antimony, which lowered the melting temperature and rendered the brass alloy less viscous, making a good casting

---

[59] Dollinger 1968: 40
[60] Beddies 1996: 123-124
[61] Bryan Stich, personal communication; artefact in private hands
[62] *Calendar of Patent Rolls*, Edward IV/Henry VI: 506-7
[63] The National Archive: E 122/71/8
[64] Blair, Blair and Brownsword 1986: 82-87

brass. Early sites in Cornwall produced *graycober*, sometimes up to 80% copper.[65] Graycober and six cartsful of clay for casting-moulds were due to be delivered to Richard Gyles, the brazier or brass-maker working beyond Oxford's North Gate, who passed on ingot brass to dependent craftsman for casting pots, stop-cocks and small bells.[66] On the south side of the Bristol Channel, five late fourteenth-century flagons have been found, of which two, higher in zinc and iron, were probably cast from imported brass. The other three, probably re-cast from recycled metals, held less zinc and iron but were surprisingly high in both arsenic and antimony, so they are likely to have contained Cornish 'grey copper'.

In areas like Cornwall, pure 'native' copper sometimes appears in delicate dendritic, or tree-like, veins in rock fissures, and in much larger masses in certain areas, particularly in North America.[67] Normally, though, copper occurs combined with other elements, as copper sulphides or carbonates. The most useful ores now lie well below the surface because rainwater, over time, causes reactions that produce enriched copper ores at deeper levels. Rainwater containing weak carbonic acid (CO_2) converts copper sulphide to water-soluble copper sulphate, releasing sulphuric acid and iron sulphate into the solution. This sets up chemical reactions that leach copper into a zone beneath, causing copper-enriched ores such as malachite and cuprite to form. Deeper, below the water table, lie enriched copper sulphide ores such as chalcocite,[68] or chalcopyrite (34 per cent copper),[69] found, for example, in parts of the Alps.

Figure 30. Wool cards in use

Brass was widely used in medieval England, where imports included semi-prepared hollow wares to turn into cauldrons, basins or jugs; sheet brass for making things like memorial slabs, spoons and other small objects, and brass wire for wool cards. Wool cards were pairs of wooden rectangles, each with one wooden handle, with a leather layer fitted into the underside of the rectangle, through which were threaded thousands of densely packed fine wire bristles, pointing outwards. The wool cards were held one in each hand and used to card – comb out straight – the fibres of sheep's wool, laying them in one direction ready for spinning. The commonest available wire was probably made from iron, but brass wire was preferable to iron wire for wool-

---

[65] Chalcocite, tetrahedrite or tennantite
[66] Blair, Blair and Brownsword 1986: 82-87
[67] Craddock 1995: 94
[68] Cu_2S
[69] CuFeS

carding, being rigid and rust-free, and for this reason brass wire soon became an important European product.

For non-luxury cast objects, cheaper metals like lead were added to the expensive imported brass in the melting-pot. Too much added lead would make the metal brittle, and if too little zinc remained in the alloy the golden colour would be lost. If zinc was diluted by adding other metals, the ratio of zinc to iron remained constant – less zinc meant correspondingly less iron. After analysis of the Bristol-area flagons, this knowledge helped to estimate the composition of the original brass and the amount of dilution that occurred when casting and re-casting it.[70]

High-status European artefacts included twelfth[71] to fourteenth-century[72] Irish harps, described in contemporary latin texts as being strung with brass wire.[73] From the fifteenth to the early seventeenth century, the lute-like cittern had brass lower strings and steel upper strings.[74] Brass book-clasps of varied design have also been excavated from twelfth to thirteenth-century Winchester. They were originally attached to leather thongs that held precious bound library books together. Further protection was sometimes afforded by adding brass bindings along the sides and at the corners of book covers; and the reader was helped to keep the pages under control by using devices such as brass page-holders or clips.[75] A surviving calendar binding dated 1486-1506 has a decorative brass clasp and corners.[76]

Across the North Sea, in central Paris, copper and brass workshops excavated at the Hotel de Mongelas site were active between 1325 and 1350. In this roughly 750-square-metre industrial unit, just outside the inner city walls of Paris, the finds included belt and harness buckles and small sequin-like castings thought to have been sewn onto clothing. Casting-moulds, together with fragments of larger vessels like cauldrons and bowls, were also excavated. Small castings from the workshop and furnace area (averaging 9% zinc, by weight) were mostly of a consistent type of sheet brass, so they derived from one source, sometimes with added lead to facilitate casting. Buckles and

---

[70] Brownsword, 1997: 14-16
[71] MS. Topographia Hibernica, translation dated c. 1200, of the 1185-1188 original by Gerald of Wales. National Library of Ireland digitised collection, MS. 700, folio 36r. *Aeneis quoque magis utuntur chordis quam de corio factis*
[72] Ranulph Higden (died 1364), Polychronicon. *Hibernici tamen in duobus musici generia instrumentis, (cithara scilicet et tympano aereis chordis armato)*. Translated by John Trevisa (1342-c, 1402) as 'in harpe and tymbre [th]at is i-armed wi[th] wire and wi[th] strenges of bras.'; or 'an harpe and a tympan stryngede and armede with cordes of brasse'.
[73] http://www.wirestrungharp.com/material/pre-1700_string_references.html 1-3/8 and 7/8 (contemporary translators gave *aes* or *aeneis* as brass [strings], whereas later translators wrote 'bronze', a more brittle alloy and duller in tone)
[74] Arnold, 1983:400
[75] Biddle 1990: 755-756
[76] British Museum: Add MSS 18852, binding

small decorative elements for personal or harness ornament were both hammered and cast.[77] A manuscript dated 1350-1400 shows an array of such objects deposited with a pawnbroker.[78] Wire produced in the Paris workshop in the early 1300s, averaged ten per cent zinc, quite a bright, warm gold-coloured brass.[79] For objects to be shaped by hammer or drawn to wire, it was better for the brass to contain only minimal amounts of lead, otherwise the metal might be too brittle, and crack under impact or tension. The excavators of this site found evidence that individual tasks involved in the technical side of production were carefully planned for economy, especially necessary for large vessels, which used up more raw materials. For example, experts created matrices (original models) from which to form multiple moulds for the body, legs and handles of a vessel, ready for casting. The expertise therefore lay in designing the matrix, not in casting from the mould, which would normally be discarded after use.[80]

At the neighbouring Meuse Valley sites of Dinant and Bouvignes, the excavated remains of brass workshops and furnaces date from the 13th to the 16th century. These two towns had successions of up to 20 workshops, each with a deep circular furnace, lined with refractory (heat-resistant) bricks, in which stood a perforated platform for the crucibles. The crucibles were set above a hearth or combustion chamber equipped with an air supply leading in below ground level.[81] Evidence has been found for stone structures, crucibles, hammered metal and workshop debris. Experiments simulating Dinant materials and conditions suggest that the brass-making process took several workers around three hours, producing a maximum of about 50 kilos of brass a day, or roughly two and a half tons a year. The Dinant crucibles could hold more than two litres, but the thermal inertia of such large-capacity crucibles was such that the process took longer. Once heated through, the furnace temperature had to be maintained constant for three hours at between 1,200 and 1,400°C, which took a considerable amount of fuel. The process is quicker with smaller crucibles but, of course, less brass is produced. It was a combination of these factors that led the brass-makers to turn, for their cheaper products like cauldrons, to more economical and rapidly made alloys like quaternary (four-metal) alloys that included more lead and tin.[82]

A Hanseatic League wreck report shows that, in 1345, workshops at Dinant and Bouvignes imported copper from Sweden.[83] From 1300 onwards, Dinant brass wares penetrated a wider, more secular, household market, with hair and dress ornament,

---

[77] Thomas & Bourgarit 2006: 56-57
[78] British Library: Add MS 27695 (1350-1400), folio 7v
[79] Bourgarit and Thomas 2012: 3055
[80] Thomas 2010: 34
[81] Plumier 2014: 31-34
[82] Bourgarit and Thomas 2011: 12
[83] Gadd 2007: 3, translating Lübeck Archive, Hansisches Urkundsbuch III

including the tiny, shiny sequin-like metal pieces that were sewn onto clothing.[84] By the 15 to 16 centuries these workshops produced many candlesticks, tripod ewers with long spouts, and tripod cauldrons with carrying-handles. Such cauldrons, featured in illuminated manuscripts, were found in most kitchens by the late 15th century, suspended over the fire on chains, or standing on the hearth on their three legs, and were used to carry and pour liquids.[85] By the 15th century, Dinant wares became known as *dinanderies*. Dinant yellow brass, glowing like gold, was highly prized, but analysis has shown that, for its less prosperous customers, Dinant metalworkers also produced cheaper cauldrons from copper, bronze and less costly quaternary alloys of copper, lead, tin and zinc.[86]

Dinant brass-workers sold their wares from the stalls at the important Leipzig and Frankfurt trade fairs, as well as marketing them in France, England, Antwerp, Bruges and Brussels. During the 15th century, to local indignation, the Dukes of Burgundy gained control of the territory around the River Meuse. Around 1450, Jacques de Gérines, a founder from Brussels, was casting brass sculptures or effigies of angelic mourners for Burgundian ducal monuments.[87] Since about 1450, in the Low Countries, brass monumental sculptures had been cast, especially in the workshops of Tournai, Brussels and today's Holland.[88] In 1466, during the Liège Wars (1465-68), Charles the Bold (1433-77), heir to the Duke of Burgundy, was sent to put down a rebellion in Dinant, which he besieged and destroyed, throwing 800 burghers into the river and torching the town. The surviving Dinant brass craftsmen scattered in all directions, to nearby Flemish cities and further afield. Examples of their craft can be glimpsed in Flemish paintings, including the spherical-frame, baluster-shaft chandelier in the 1434 Arnolfini painting by Jan van Eyck.[89]

The Duke of Burgundy enticed some Dinant brass-workers to Namur (25 kilometres north of Dinant), where, over the succeeding century and a half, their products included shapely incense-burners, holy-water buckets, tripod jugs and double-spouted hanging cauldrons, each spout tipped with an animal head through which water could pour.[90] Following his 1466 conquest of Dinant, Duke Philip of Burgundy had gained Aachen (Aix-la-Chapelle), as well as the very pure and prolific calamine source at the nearby Altenberg (Vieille Montagne) mine, near Moresnet. Several Dinant brass-workers settled in Aachen, which had abundant water power, and forest timber for charcoal. In 1450, brass-maker Daniel von der Chamen had, with a partner, founded the Aachen brass industry, assisted by skilled workers, labourers and domestic servants.

---

[84] Thomas and Urban 2014: 62-63
[85] Thomas and Urban 2014: figures 65, 67, 69
[86] Thomas and Bourgarit 2014: 51
[87] Wiersma, 2014: abstract 30
[88] Wiersma 2018: 377-380
[89] National Gallery, London, 186, van Eyck painting, 1434, Arnolfini and his wife,
[90] Namur, Musée provincial des Arts anciens du Namurois: coll SAN, 63

Figure 31. A batteur at work, after a woodcut by Jost Amman, 1568

The city council, hoping that he would bring wealth to the city, allowed him to start a brass craftsmen's guild and provided him with a site within the city walls.

A more industrial brass-making picture emerged when certain Aachen brass guild members became capitalist entrepreneurs prepared to advance money. They funded the time-consuming processes involved in producing and marketing semi-prepared brass, which meant sheet metal cut into shapes and hollowed-out by hand with hammers into preliminary bowl or cauldron forms.[91] Heavy capital outlay was required for raw materials, plant, equipment and wages, but the entrepreneurs bore the costs until such time as the raw materials could be converted into profit, often a matter of months. The brass quality at Aachen rose noticeably with the influx of Dinant brass-makers who partnered themselves with local aristocratic patron families. From the 1400s onwards the number of Aachen brass-works steadily increased, but the artisans became over-dependent on the entrepreneurs to provide them with workshops and semi-prepared brass, and to purchase and distribute their products around Europe via the established trade routes and markets.

The *batteurs* from Dinant shaped and embossed their later decorative brass work – brass being more malleable than bronze for hammering out embossed designs.[92] Examples of their work include a bucket-shaped vessel for holy water, made for the chapel of St Jacob, Aachen, as well as a hanging cauldron with two spouts, and a brass memorial plate to Abbot Herbert von Lülsdorf (died 1481), for nearby Cornelimunster Abbey.[93] The former international trade pre-eminence of Dinant began to transfer

---

[91] Gerard 1958: 6
[92] Haedecke 1970: 37
[93] Gérard 1958: plate opposite 15

itself to Aachen.[94] In 1471, the Bishop of Liège mediated an agreement with King Edward IV of England (1461-83) to allow refugee Dinant craftsmen now working in the Flemish cities of Huy and Middelburg to trade in London. Flemish craftsmen were already casting monumental brasses, to be fitted onto the lids of tombs or grave-slabs, at first engraved mainly in Ghent and Tournai, and later in Antwerp. The engraving of monumental brasses also took place as far afield as Paris, Rouen, Cologne, Nuremberg, Goslar, Lübeck, London, Norwich, York and Wroclaw.[95]

Monumental brasses, in the later 14th to early 15th centuries, were designed to commemorate the deceased and to encourage the living to pray for their souls. In mainland Europe, the surviving monumental brasses commemorate royalty, nobles and archbishops. English brasses survive in greater numbers, having escaped the subsequent destructive European wars. They therefore represent a wider spectrum of society including merchants, wool traders, parish priests, canons, knights, esquires, craftsmen and tradesmen.[96] Different brasses were used in different regions, and lead contents varied. Silesian brasses contained more tin (then available in Bohemia).[97] Analysis shows that fifteenth-century English monumental brasses contained, on average, similar amounts of tin, 1% more zinc and 3% more lead than those of the fourteenth century.[98] By that time, designs were more standardised – the figures often reclining with their hands in prayerful pose and their feet resting on a heraldic dog or lion.

Brass astrolabes were also current, known in Europe since the 'crusades' to the Near East, which finished in the 1290s. Astrolabes consist of many components, so analysis often shows components to be made from different grades of brass (some are even revealed to be modern replacements). Average zinc contents in astrolabes gradually rose Europe-wide, so an early fourteenth-century English-made astrolabe embellished with animals and mythical beasts contained a similar amount of zinc[99] as a Hebrew instrument made in Moorish Spain.[100] One astrolabe dated 1326 shows that lower-zinc brass was used for its cast components, whereas higher-zinc alloy was used for a hammered sheet-brass component called the rete, a thin flat plate with cut-out designs.[101] Surviving instruments include a quadrant dated 1350,[102] a French astrolabe of 1400 and an English astrolabe made in 1342.[103] European brass technology seems to have greatly improved during the 14th century.

---

[94] Roderburg 1927: 13
[95] Norris 1978: 48-49
[96] Norris 1978: 53-57
[97] Norris 1978: 34-35
[98] Clare. Calver, 1990. Magno cum artificio: medieval metallurgy and the monumental brass industry. MSc dissertation, Newnham College, Cambridge, Appendix 1, metallurgical tables, folio 68 ff
[99] British Museum: M & ME Sloane 64 (c.1320s) Sloane astrolabe
[100] British Museum: 1893,6 16 3 (c.1350 ?), Spanish-Moorish (Hebrew) astrolabe
[101] British Museum: 1909 0617 (1326), astrolabe
[102] Turner 1987: 12, 15, 19
[103] British Museum: 1853 1104 (1342), Blakene astrolabe

Continuing Continental trade to East Anglian ports (Hull, Kings Lynn and Ipswich) may explain recent Norfolk finds of brass scientific instruments dating from the 14th to 15th centuries,[104] which include a *horologium* or early type of sundial, a quadrant and a nocturnal.[105] These instruments contained around 11-12% of zinc but enough lead and tin to class them as quaternary alloys (appendix 3.9). Burial in soil containing chlorine ions could have leached out some of the zinc from these long-buried instruments, so the original zinc content may have been greater. In other more drastic situations, exposure to sea water, ammonia, damp chlorine or strong mineral acids gradually causes the zinc to leach out of brass – a process called 'dezincification'. However, brasses are resistant to the effects of many alkalis, organic acids and most foods, and dezincification can be slowed down by including a little extra arsenic, or 1% of tin, to the brass mix, though this was perhaps not understood at the time.[106]

In 1392, Geoffrey Chaucer (*c*.1343-1400) wrote a treatise on the astrolabe, and his Squire's Tale recounts that the 'King of Araby and of Inde' had sent the squire's knight a 'steede of bras' [brass horse] that could tell him where he was, day or night.[107] An astrolabe[108] dating to about 1330-1340, made in England or northern Europe, probably for Edward III, appears to have been destined for use by the Medici or Strozzi families in Florence.[109] One surviving astrolabe of *c*.1450 is made of typical low-zinc English brass,[110] but two of its components are high in zinc, implying the use of distilled metallic zinc. This suggests that the components were modern, although there is the outside chance that they were imported from Chaucer's 'Araby or Inde', that is, from an eastern area where high-zinc brass astrolabes were made. Certainly zinc-rich Anatolia, Iran and the Golconda-Hyderabad region of India were all ruled in 1450 by zealous Shi'a Turkmanic leaders who would have used astrolabes.[111]

By the 14th century, Nuremberg had gained rights to trade in brass. It lay on important trade routes from Venice to Hamburg, Breslau to Ghent and Vienna to Brussels, and rapidly gained a reputation as a brass-trading centre.[112] In May, 1398, King Wenzel of Bavaria granted a privilege to certain burghers of Nuremberg to operate hammerworks with furnaces and ores, and to employ smiths, foresters and other workers.[113] Many early brass products were exported through the port of Bruges but, by the later 1400s, most export trade transferred to Antwerp. Frankfurt and Leipzig, These fairs, rather than producing brass, were the most important mainland European metal

---

[104] I am grateful to Dr John Davis for drawing my attention to these artefacts
[105] Norfolk Museum Services: NMS-82C940, quadrant and horologium; NMS-D40DC2, nocturnal
[106] Copper Development Association 2005: 48
[107] Chaucer 1957 edition: 129
[108] Adler Planetarium Menzel-26
[109] Davis 2017: 27
[110] British Museum: 1914 0219.1
[111] Reedy 1997: 321
[112] *Mitteillung des Vereins für Geschichte der Stadt Nürnberg* 54, 1966: part IV, 65
[113] Staatsarchiv Nürnberg: MS Rep 1a, Reichstadt Nurnberg, Kaiserliche Privileghien, number 228

Figure 32. Europe, map showing some sites mentioned in the text

markets and trade fairs for buying and selling raw materials, including copper, brass, finished brass articles and zinc ore from the Altenberg (Vieille Montagne). From the mid-1400s until about 1630, the wealthy Fugger family of Augsburg were the most influential metal merchants in Europe, already transporting copper from the Balkans to Venice. Nuremberg and Aachen both produced and sold brass plate, semi-prepared wares and finished objects in brass.

In 1388, on a rise above the south bank of the River Pegnitz, just inside Nuremberg's eastern city walls, a merchant named Konrad Mendel established a charitable trust – the 'Twelve-Brothers' House'. The illustrated records of this house give a lively glimpse into small-scale medieval brass workshops. The city's skilled workers (for trade secrecy reasons) were not allowed to move away from Nuremberg,[114] but successive communities of twelve disabled or destitute elderly city craftsmen could live in brotherhood at the house and ply their various crafts. Over the centuries, a painted manuscript portrait and short description was made of each brother practising his trade. In 1425, a wire-drawer is shown sitting on a swing so that his legs, combined with the impetus of the swing, could give extra force to his strength of arm, enabling him to draw wire through graded sizes of holes in a securely fixed template. In 1458, a brass-worker member of the community is shown filing a two-branched candlestick. He is surrounded by a delicate gothic monstrance (for displaying the holy bread to the faithful), candlesticks, circular brass platters in various sizes and a small box. In 1471,

---

[114] Hachenberg and Ullwer 2013: 48

another member of the community, a maker of moulds for cast brass candlesticks, is depicted modelling clay candlesticks from which to form his moulds.[115]

The brass-workers who retired to the Twelve-Brother's House came from Nuremberg foundries that had been established when they were younger men. By about 1400, Nuremberg foundries produced beautiful aquamanilia, some of them representing lions with what look like tufts of flame rising along their curling brass tails. Nuremberg aquamanilia in the shapes of lions, unicorns, dogs and horses had stop-cock-style taps protruding from their chests for pouring out water. Nuremberg fittings in similar style include lion door-handles and chandeliers.[116] One Nuremberg treasure made in 1424 is a reliquary casket, heavily embellished with a brass diamond lattice pattern and with a lid in the shape of a pitched roof.[117] Nuremberg's craftsmen had great ability at crafting semi-prepared brass vessels, sheet-brass and wire into desirable finished articles, which included decorative, ornamental and artistic wares. There were even Nuremberg specialists who made trumpets, trombones,[118] small tinkling bells or clashing cymbals.[119]

In 1453, master brass-founder Hermann Vischer came to Nuremberg, where he set up a workshop and family brass-foundry, which probably stood inside the eastern city walls south of the River Pegnitz, near today's Vischergasse (Vischer Lane). Nuremberg expertise became known elsewhere. From 1470 to 1485, Kunz Mülich, a Lübeck merchant with Nuremberg origins, imported Nuremberg brass-wares to Lübeck on the Baltic Sea.[120] In 1485, a Nuremberg brass trader and producer named Holzschumacher (clog-maker) set up a brass works at Neubrunn (now Sachsenbrunn), on a trade route 140 kilometres north of Nuremberg, where he is said to have trained Erasmus Ebener, later known for brass-making in the Braunschweig (Brunswick) area, as well as members of the Kanler family, who in 1492 converted the Nuremberg mill on the River Pegnitz at Hammer, Laufamholz, to brass-making.[121]

Nuremberg was full of brass craftsmen, but lacked copper and calamine resources. However, the low-lying Heroldsberg area north-east of Nuremberg produced abundant refractory (heat-resistant) clay, ideal for making crucibles, and local forests provided timber for charcoal and furnace-fuel, only later to be depleted by over-exploitation. By 1474, Nuremberg was making its own brass,[122] and six brass 'battery' trip hammers were at work along the River Pegnitz which runs through the centre of the city.

---

[115] Stadtarchiv Nürnberg: MSS Mendel I, Amb.317.2°, folios 79-80 and 89-90
[116] Mende 2006: 1, 20, 24
[117] Römisch-Germanishes Museum, Nürnberg: KG.187
[118] Hachenberg and Ullwer 2013: 47-48
[119] Westermann 1971: 38
[120] Meyer, 2006: 321
[121] Dietz 1921: 198
[122] Stahlschmidt 1970: 134 and footnote

Copper came from Saxony, and in 1476 Hans Mayenstetter from Augsburg secured rights to supply Nuremberg brass producers with calamine (zinc carbonate ore) from the Tyrol.[123] On 30 April, 1492, a Nuremberg trader named Amelreich had a hundred tons of calamine delivered to him by river, probably for Hammer Mill, Laufamholz.[124] Long-distance delivery of zinc ores was expensive, but the Nuremberg craftsmen had established very strong markets throughout central and southern Germany, so they could afford the costs. At the Frankfurt trade fairs, Nuremberg craftsmen like the famed brass founder Hermann Vischer, could buy brass made in Aachen from purer high-quality calamine sourced from the Vieille Montagne (Altenberg). Through the same Frankfurt trade fairs, Nuremberg merchants sold Bohemian or Tyrolean copper north to the Aachen brass-makers.

By the fifteenth century, European brass-making no longer depended solely on the manual strength of the worker operating the bellows or wielding the hammer. Water-power came into its own to drive water-wheels, which, through geared axles, powered bellows to heat furnaces or drove grinding-mills to pulverise zinc carbonate ore. Once craftsmen had seen the advantages of trip hammers, or 'battery' hammers, used in this way, they harnessed water energy to turn wheels to power rotating horizontal shafts.

Trip hammers had a long history. Operated by foot treadle in China and Greece they had been used for corn-crushing by the start of the first millennium AD, and in Roman Italy for ore crushing. In the mid-1400s, the hammers, with different-shaped hammer-heads, each on the end of a long beam, were designed to beat sheet metal into basic hollow shapes or flat sheets. A water-driven wheel turned a shaft, from which protruded tooth-like lugs, acting as cams, each of which, in turn, tripped the end of a long beam, hinged on a fulcrum. As the shaft turned, the cam depressed the base of the beam, which acted as a pivoted lever raising the heavy hammer-head, so that, once the base of the beam passed the lug, the weight of the raised hammer-head dropped it onto the metal, only to be raised again by the next rotating lug – resulting in rapidly repeated up-and-down hammer movements. The same shaft often had enough lugs to operate several battery hammers, in sequence. Each hammer-man sat astride a bench before his anvil, guiding the brass with his hands, so as to take advantage of each successive hammer-strike.[125] The recurring problem with water-powered machinery was of course that it was affected by droughts in summer and by deluges of water in winter.

In Nuremberg, the number of brass battery-workers and hand beaters evidently rose rapidly, and there were probably three brass furnaces.[126] Nuremberg had at first obtained copper from the extensive copper and silver mines of the Krušné Hory

---
[123] Wittek 1984: 115,
[124] Dietz 1921: 199
[125] Wilson 2002: 16, 22, 31
[126] Hachenberg and Ullwer 2013: 49

(Erzgebirge) mountain range between Saxony and Bohemia (now Czech Republic), and by 1463 Nuremberg operated its own silver and copper refinery, where impurities were refined out of the copper, making it possible to produce purer brass.[127]

By the later 15th century (first mentioned in Nuremberg in 1453), German silver-workers operated '*Saigerhütte*', or liquation plants, where they separated out the silver from silver-containing copper ore. This copper was first melted together with a greater amount of lead, and cast into cakes. As it cooled, the copper and lead separated into two phases, with the silver passing from the copper to dissolve in the lead, for which it has an affinity.[128] When the ingot was heated a second time, enough for the lead to melt but not hot enough to melt the copper, the silver-rich lead flowed out of the pores in the copper 'sponge'. This process was liquation. The silver was then separated from the lead by cupellation - oxidising the lead in a furnace formed of an open bowl lined with bone ash. This calcium-rich powder absorbed the lead oxide, leaving the relatively pure silver metal (which did not oxidise) on the surface of the hearth.[129] Copper, the other desirable product obtained, had to be further refined to remove the remainder of the lead and other impurities before it could be any use for brass-making.[130]

During the later 15th century Nuremberg merchants started financing copper mines and smelting works around Mansfeld (southern central Germany). From the 1490s, the Fugger merchants of Augsburg supplied Nuremberg with copper from Schwaz (Inn valley, now Austria), and from Prettau (now Predoi, near Bolzano, Italy), southern Hungary and Bohemia.[131] In the 14th and 15th centuries, at Steyr on the River Enns in Carinthia (Kärnten, eastern Austria), iron knife production flourished and the Steyr cutlers' statutes had, in 1407, specified that their knife-handles should be decorated with openwork (*gefensterten*) brass. In 1470, craftsmen at a brass foundry at Steyr used Nuremberg brass to cast ornamental rosettes for the knives, a craft which continued for centuries.[132] From 1489, a Nuremberg merchant named Köberer, who lived in Steyr, traded its decorative knives.

In 1488, in Nuremberg, Peter Vischer 'the Elder', son of brass-founder Hermann Vischer, drew parchment designs for the celebrated pinnacled brass shrine to Saint Sebald, a masterpiece of late medieval artistic and technological achievement.[133] In 1518-19, Peter Vischer and his sons produced brass castings for this shrine, whose tallest pinnacles reach to a height of twelve metres. A cast figure of Peter Vischer

---

[127] Palme 2000: 137-138
[128] L'Heritier and Tereygoel 2010: 136
[129] Bayley and Eckstein 2004: 145-147
[130] Dr Jean-Marie Welter: personal communication
[131] Palme 2000: 19, 137
[132] Stadtarchiv Steyr: MS. Hack, Irmgard, Steyrer Messerhandwerk
[133] Meller 1925: 28; in Sankt Sebaldskirche, Nuremberg

the Elder wearing his brass-founder's apron (and formerly holding a small hammer), can be discovered at one corner of the tomb.

In Nuremberg, two generations of Peter Vischer went on to cast many fine brass figures. Brass monuments by Peter Vischer the Elder were made of a good casting-brass that could also be forged.[134] The Italian renaissance had been leading to a revival of classical design styles, inspiring the Vischer family to cast some outstanding human and animal figures around the base of Saint Sebald's shrine in Nuremberg.[135] In northern Italy, from 1428, the Venetian republic held power, profiting from the mineral-rich alpine valleys of the Bergamo and Brescia regions, including Gorno, whose zinc-ore miners had surrendered themselves to Venice the previous year, tired of years of power struggles between the powerful Visconti and Malatesta clans. The Venetians granted the miners rights to build houses and workshops by the streams and to collect timber for fuel and charcoal.[136] At Bienno, at the head of the nearby alpine Camonica valley, copper sulphide appears to have been mined and refined by roasting with charcoal well before iron was mined there.[137] In the later Middle Ages, however, Milan's craftsmen, received supplies of sheet brass and brass wire through the Ravensburg market, to produce a variety of brass-wares, including parts for armour and firearms for export to the Levant, Spain or north Africa.[138] One example of early northern Italian renaissance dress ornament is a very sumptuous belt-buckle, featuring stamped brass, crafted in Lucca in about 1450.[139] Calamine ($ZnCO_3$), then available from the nearby Massa Marittima, was one of the few common metal ores that did not belong to the monarch.

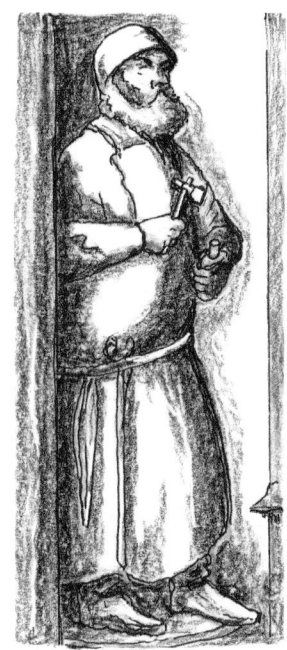

Figure 33. Peter Vischer the elder with his hammer, St Sebald tomb, Nuremberg

In 1381, a decree prohibited Nuremberg-produced brass from being worked by craftsmen beyond the city. This meant that brass produced within the city boundaries at Hammer, Laufamholz, had to be worked into objects by Nuremberg hand craftsmen (a rule that seems later to have lapsed). In the 15th century, talented Nuremberg craftsmen overtook Dinant in supplying central and southern Germany with such

---

[134] Riederer 2000: 170
[135] Slafski 1962: plates 11 and 41
[136] Furia 2012: 42
[137] Tizzoni 2001: 195
[138] Westermann 1971: 40
[139] Victoria and Albert Museum, London: 4278-1857

wares as brass gun barrels, candlesticks, keys, scales and weights.¹⁴⁰ By the later 15th century, city craftsmen in Nuremberg were using hand-powered winding-gear to draw brass wire through ever-narrower holes in a draw plate or former (*Zieheisen* or *Locheisen*) attached to the work-bench.¹⁴¹ Wire was mentioned in a document from Iserlohn, dated 1394, as being used for wool cards and mail armour-rings but it was iron wire, not brass.¹⁴²

The Hammer brass-mill, Laufamholz, Nuremberg, obtained its calamine from the same Altenberg (Vieille Montagne) source as the Aachen brass-houses, so artefacts made by Peter Vischer the Elder (c.1455-1529) contain similar average zinc ratios to the brass made by his father Hermann, who also used Altenberg zinc ore.¹⁴³ However, two samples, dated 1494-5, taken from a memorial brass made by Peter Vischer the Elder for Magdeburg cathedral, contain over thirty per cent of zinc. Goslar, one of Magdeburg's main suppliers, lay eighty kilometres further west, so, were sublimed zinc metal furnace accretions from Goslar perhaps added to the brass when casting, to raise zinc content? Brasses containing thirty per cent of zinc could be hammered cold or engraved – useful for memorials.

Figure 34. Brass figure of Theodoric by Peter Vischer the younger, c.1519

Brass memorials created by Peter Vischer the Elder included one to Johanns von Heringen in the cathedral at Erfurt,¹⁴⁴ and numerous rectangular brass plates in Meissen Cathedral commemorating members of the Saxon royal family (dating 1463-1476),¹⁴⁵ one of which commemorates Prince Ernst of Saxony with a curly-maned lion at his feet. Three brass memorials to bishops, dating to the 15th to 16th centuries, are in the same cathedral. The Innsbrück court church (*Hofkirche*) holds a great brass statue of Theodoric the Ostrogoth (454-526) in a flexible pose, also understood to be the work of Peter Vischer the Elder.¹⁴⁶ This renaissance-style statue was designed to flank the tomb of Emperor Maximilian I (1459-1519).

---

¹⁴⁰ Palme 2000: 13-14
¹⁴¹ Palme 2000: 126-127
¹⁴² Hildenbrand 1983: 14-16
¹⁴³ Riederer 2000: 170-173
¹⁴⁴ Meller 1925, 87
¹⁴⁵ Custenson 2011: 316-317, (memorials: Schönberg 1463, von Schönberg 1476, von Wurzburg 1472)
¹⁴⁶ Meller 1925: figs. 34 and 60,

Peter Vischer's sons, Hermann (1486-1528), Peter (1487-1528) and Hans (1490-1549), used brasses with a slightly lower zinc average than the previous generation.[147] Their brass was probably produced locally at Hammer, Laufamholz, still exploiting the Altenberg calamine source but possibly economising slightly on zinc, since the Vischer brothers used brass for casting, rather than forging. In 1505, Philip I of Spain and Burgundy renewed an agreement for Antwerp and Brabant to trade freely through the city of Nuremberg.[148] This was significant for the Nuremberg brass industry, since it ensured the continued supply of pure Altenberg calamine from Aachen.

The existence of a brass-mill at Hammer, Laufamholz, alongside the fast-flowing River Pegnitz in outer Nuremberg, meant that the Vischer workshop could work with the plentiful locally-produced brass. From 1490 to the 17th century, those in charge of the Nuremberg *Halle* – the great market halls – leased out rights called 'privileges' to proprietors who operated the business. Battery hammers for brass were set up at the Laufamholz mill by 1492, and a weir held up a head of water to drive them. Three gaps were constructed in the wall of the mill channel or leat, through which raw materials, transported by water, could be delivered to the mill and brass could be brought out.[149] Although the mill lay within outer Nuremberg, it was seven kilometres east of the city walls, which gave it the advantage of being an industry within part of the city but beyond its walls, so it was not bound by guild restrictions. The disadvantage was the lack of walled protection, which resulted in repeated destruction of the settlement during subsequent wars, closely followed by reconstruction.

Brass-making was also alive in northern Europe, where the free Baltic port of Lübeck was the leading member of the mercantile Hanseatic League, trading all around the Baltic, and inland to cities like Braunschweig (Brunswick). Lübeck, founded in 1159, became an independent city-state in 1226. The great walled city had a well-sheltered Baltic port on the River Trave,[150] and was in a good position to import copper from the *Kopparberget* (copper mine) at Falun, Sweden, where organised copper mining had started before 1100.[151] A brass-smith (*Beckenschläger*) is first mentioned in Braunschweig in 1302, and the Lübeck city's workers' regulation of about 1325 specifies that unrefined brass (*erkol*) and sea-brass (*zemissingh*), meaning brass imported by sea from Upper Hungary or Sweden, had to be refined, hammered or cast in or near Lübeck. The cathedral at Lübeck contains a cast bronze figure of its bishop, Heinrich von Bocholt (1315-71), lying on a brass base-plate, probably locally-made.[152] At Schleswig, 90 kilometres north of Lübeck, a broad, brass hand-washing basin survives, dated 1210 and stamped with the

---
[147] Riederer 1983: 88-90
[148] Staatsarchiv Nürnberg: Rep 205-0, Ritterordnen Urkunden, number 3806
[149] Hussennether 2010: 5
[150] Beddies 1996: 133
[151] Forss, T. 2006: Falu Mine, geonord.org/shows/falueng.html, 2
[152] Custenson 2011: 316

seal of Bishop Nikolaus I of Schleswig (r.1208-33). Another survivor is a brass altar pillar-candlestick, dated 1491, from Leusahn, just 50 kilometres north of Lübeck.[153]

In 1400, a restrictive regulation decreed that copper shipped in to Lübeck from Sweden or Upper Hungary (today's Slovakia) might be used for brass-making only in Lübeck itself, and then only by employees of a Lübeck brass-works owner, under his supervision. Brass craftsmen in Braunschweig, Magdeburg and Hildesheim (now in northern Germany) therefore had to content themselves with forging, casting or engraving brass produced in the Lübeck area.[154] Dating to around 1400, two surviving aquamanilia containing some zinc, were probably created in that area. They represent horses and riders – one a falconer and the other a man dressed for battle.[155]

In the Middle Ages, the great trading port of Lübeck not only imported Swedish copper but exported good utilitarian brass articles back to Sweden. They included clasps, needles, hooks, wire, basins, bowls, candlesticks, mortars, buckets, baptismal fonts, pots and syringes.[156] Sweden still imported all its brass, and immigrant brass-workers called *bältare* – girdlers or belters – worked brass into strap-ends, belts, buckles and harness fittings. In Sweden, girdlers were the only guild craftsmen allowed to work brass, and their trade was profitable enough, outside the bigger cities, to raise their social standing. A girdler was working brass in Arboga, central Sweden, in 1465.[157] A Nuremberg manuscript dating to about 1425 depicts a girdler sitting at his anvil, hammering eyelets into belt-holes after attaching the buckles.[158]

Besides the Swedish mine at Falun, a rich source of copper existed in the Harz Mountains rising out of the northern German plain. From as early as the 11th century, the artistic craftsmen of Dinant imported copper from Goslar in the Harz, 400 kilometres away, because the prolific output of Dinant brass goods far outstripped any more local copper supplies. From about 1200, Harz copper-smelters used water-wheel-driven bellows to raise the furnace heat,[159] and from about 1450 brass-making was started in the Harz Mountains. Initially, the Harz brass-makers exploited local copper from the Rammelsberg, but when accessible seams of copper pure enough for brass-making were running out they turned to Upper Hungary (Slovakia), whose copper industry was currently in crisis for lack of capital. However, in 1473, the local king decreed that no copper should be exported beyond Krakow, so Harz traders were temporarily left importing only Swedish copper. The decree also hit the Upper Hungarian copper trade to Zwickau and Silesia. In the late 15th century, Matthius Corvinus, King of Hungary and of Bohemia, had enough capital

---

[153] Archäologisches Landesmuseum, Schleswig: 1950/c1and 1904/214
[154] Beddies 1996: 133
[155] Barnet and Dandridge 2006: 101-102, 128-129
[156] Ekström 1985: 177
[157] Erikson 1978: 18-19
[158] Nürnberg Stadtarchiv: MS, Mendel I, Amb.317.2°, folios 26-27
[159] Segers-Glocke et al 2000: 17

to re-open certain Upper Hungarian mines and he by-passed Krakow by opening a direct copper-trade route west over the Jablonkov pass through the Carpathian Mountains to Zwickau in Saxony. Taking a further bold step, János Thurzó, a merchant from Upper Hungary, resolved to divert copper deliveries from Upper Hungary northwards towards the Baltic Sea through the Hanseatic ports of Lübeck, Danzig (Gdansk) and Toruń (on the River Vistula), which greatly helped the Harz brass-makers.[160]

European cementation brass-alloying processes used calamine (zinc carbonate ore), which brass-smiths from Goslar in the Harz imported by sea from the Altenberg (Vieille Montagne). Goslar's other calamine source was Ilmenau, deep in the forests seventy kilometres to the south. In the later 15th century, at Goslar, where cementation brass was already made, accretions of zinc oxide and zinc metal were discovered in furnace chimney crevices, where they collected after subliming during the heating of zinc-containing lead and silver ores. However, the zinc accretions seem not to have been universally adopted for use in brass-making, because analysis of late fifteenth-century lidded brass-making crucibles from Zwickau near the Erzgebirge (Krušné Hory) showed traces of natural calamine ($ZnCO_3$) but not of zinc collected from silver and lead-smelting furnaces.[161]

Transport of heavy material such as the ores of copper and zinc was, in the Middle Ages, easier by water than overland. Although brass contains a greater weight of copper than of zinc, the bulk of zinc ore required is greater. Calamine (zinc carbonate ore) normally contains less than fifty per cent of zinc and has a lower specific gravity than copper ore, so it is more bulky to transport. Therefore, when practicable, the brass mill was built near a zinc ore source and the copper was brought to it. Copper and calamine were both moved around western Europe by middlemen based in the free city of Nuremberg and in Hanseatic League cities like Lübeck.

János Thurzó, from Upper Hungary, claimed to have stolen from the Venetians the secrets of producing brass. In 1493, the Polish king, impressed by Thurzó's claims, provided him with capital to buy three Upper Hungarian copper mines. By 1490, however, the mighty Fugger merchant family of Augsburg was already sniffing out trading opportunities in Silesia and Upper Hungary, and in 1495 they formed a partnership with Thurzó to found a 'Hungarian' copper-exporting company.[162]

Trade from Venice to the Levant was thriving, and Thurzó claimed to have learned brass-making secrets from Venice, which traded copper and bought and sold brass objects through Damascus. From 1260 to 1517, when the eastern Mediterranean and North African coastal strip were under Mamluk rule, brass working flourished, particularly during the reign of Mamluk Sultan Qāytbāy (1468-96). Through Cairo

---
[160] Backowski 2002: 17, 21
[161] Martinón-Torres and Rehren 2002: 108.
[162] Backowski 2002: 21-23

and the Red Sea, the Mamluks controlled trade with Arabia and the Far East, managing Middle Eastern merchandise through Damascus. Mamluk contact with Europe was through Venice, whose merchants, by the 12th to 13th centuries, supplied huge quantities of copper to Damascus and Cairo, returning with Mamluk incense-burners with perforated lids, tall-necked brass ewers, bowls and candlesticks intricately inlaid with silver designs, and sometimes with gems, which adorned Venetian villas and churches.[163] In the 12 to 13 centuries, some Venetian merchants took the practical measure of starting a small calamine mining business at Iserlohn (central western Germany),[164] probably to supply their brass-maker contacts in the Near East.

Figure 35. Veneto-Saracenic style ewer, c.1500

Fourteenth- to fifteenth-century 'Veneto-Saracenic' brass objects prove, on analysis, to have been made from freshly mined copper and sublimed zinc sulphide ore (sphalerite, the commonest ore in the Middle East). The iron content in the brass was low because the zinc sulphide ore was prepared by subliming it (not distilling it), which purged the ore of much, but not all, its iron and sulphur and created a relatively contamination-free zinc oxide, suitable for brass-making. Venetian and Mamluk documents show that between the 14th and 16th centuries Venice still supplied European copper to the eastern Mediterranean. Metallurgists therefore suggest that most medieval 'Veneto-Saracenic' objects were made from brass alloyed in the Near East from a combination of European copper and Middle Eastern zinc oxide. The bowls, ewers and incense-burners were then sent back to Venice to be hammered into final shape and finished on a lathe, and their brass contained 10-25% zinc, which exhibited a good golden colour.[165]

In the 15th century, traders to Venice included the seafaring Portuguese, who started exploring the African west coast to find out what the Berbers and Arabs were bartering with people in the forested sub-Saharan areas in exchange for gold and slaves, and

---

[163] Serrao, Judith, 2007: 'A walk into Islamic History', Abu Dhabi exhibition, www.mangalorean.com/browsearticles.php?arttypefeatures&articleid=
[164] Hildenbrand 1983: 37
[165] Ward et al 1995: 237-239 and 248

how they went about it. In 1415, the Portuguese captured Ceuta, northern Morocco, and so discovered how the Berbers profited by moving brass and copper southwards over the Sahara, and gold and slaves northwards.[166] Written orders dated November 1439 and September 1441 commanded the Portuguese seafarer Rodriquez to go to Venice to obtain manillas (bracelets used in West Africa both as arm-bands and as currency), and bring them to Lisbon for trading to the Arab port of Ceuta (in today's Morocco).[167]

In 1454 a 22-year-old Venetian adventurer, Alvise da Ca' da Mosto, took a ship to Lisbon and in 1455 sailed in a Portuguese caravel bound for Senegal and Gambia. He mentions Arab camel caravans that plied constantly between Barbary (Morocco) and Timbuctu, carrying copper and silver south and returning north with gold, salt and ivory. Gold from Mali passed through Timbuctu to be transported by desert routes to Syria, Cairo and Morocco. Alvise da Ca' da Mosto also commented on the huge quantities of cowrie shells (from the Maldive islands in the Indian Ocean) used as currency along the West African coast.[168]

From their base at Ceuta, the Portuguese pitched into this complex Arab-dominated market of shifting kingdoms, by trading brass for slaves and gold at the established desert markets of the Mali Empire.[169] By the late 15th century, however, Portuguese mariners had set up trading posts on the sparsely populated Guinea Coast, and in the late 1480s the Portuguese adventurer, Vasco da Gama, continued southwards to round the Cape of Good Hope. Portuguese coastal trading stations gradually gave them entry into the markets of West Africa and an advantage over the Berber desert traders. As already mentioned, Berber camel caravans carried brass manillas, rods, cauldrons and basins across the Sahara from Ceuta to south-Saharan centres from which trade routes led south into the populated forests. The observant Portuguese acquired similar brass wares through Venice, initially from the Frankfurt trade fairs or direct from Nuremberg. By the year 1494-5, however, Antwerp became their main port of supply for brass bowls, pots and manillas. The Portuguese agent in Flanders supplied them with 71,000 brass manillas (bracelets) for a single cargo.[170]

The seafaring Portuguese were ushering in the Age of Discovery – sponsoring Christopher Columbus' discovery of America in 1492. Imported goods were growing more exotic – in 1495, Lübeck merchant Paul Mulich sent his buyer to acquire luxury goods at the Frankfurt fair, and he returned with brass, pearls, Brazilian timber and Italian fabrics.[171] The Age of Discovery was opening European and Asian eyes to

---

[166] Klein 1999: 50-51
[167] Strieder 1933: 254
[168] Crone 1937: 3-4, 25-26
[169] Liverpool Maritime Museum, 2006: liverpoolmuseums.org.uk/maritime/slavery
[170] Petruszka, Andrew 2011: 252
[171] Meyer 2006: 321

further horizons. In the 15th century, Emperor Yongle's Chinese fleets to east Africa, Columbus' discovery of America and Vasco da Gama's sea-route around Africa to India all heralded new challenges in overseas politics and trade. Not only ports but internal fairs and markets were now improving, as was the quality of brass that passed through them. Late medieval artistic achievements in brass included the Himalayan religious figures showing the way to enlightenment; the fine intricate inlays of Persia; the blend of influences in Indian statuary and shapely vessels; the pure simple work of the Dinant *batteurs*, and the medieval- and renaissance-inspired castings of the Vischer family's Nuremberg workshop. Instruments for recording the movements of the heavenly bodies, whether in Nuremberg, the Chinese imperial palace or rural Norfolk, were becoming more intricate. Brass making was now in a good technological and trade position to advance confidently into the 16th century.

Brass candlestick inlaid with gold, silver and a black substance, 1340s. The side shown is thought to depict Tashi Khatun, Mongolian regent and mother of Sheik Abu Ishak of Shiraz. She is being offered fruit and a book. Inv. 47632. © Museum of Islamic Art, Doha, Qatar.

Detail of brass candlestick inlaid with gold, silver and a black substance, 1340s. The side shown is thought to depict Tashi Khatun, Mongolian regent and mother of Sheik Abu Ishak of Shiraz. She is being offered fruit and a book. Inv. 47632. © Museum of Islamic Art, Doha, Qatar.

Chapter 4

Age of Discovery

c.1500-1600

Moving forward to the 16th century, the story is enriched by the survival of some fine artefacts, still mainly religious, and by more detailed knowledge of brass works and brass technology. By 1500, the Americas were known to a wider world, and the Portuguese traded around the Cape of Good Hope, northwards up the East African coast and across to India. Arab merchants already traded southwards down the East African and Indian coasts and had resident merchants in China. Portuguese merchants now began to trade silks and precious stones between India and China.[1] Therefore, by the mid-16th century, more eye-witness accounts of far-flung cultures and trading opportunities circulated, and exotic man-made objects and materials began to influence taste and fashion. In the early 16th century, a yearly fleet left Lisbon for India and, by its last two decades, merchants from Florence, Nuremberg and Augsburg were resident in Goa for part of the year, showing that long-distance trade had adopted a more settled pattern.

By the late 16 century, most India-bound European merchant shipping from Venice and Marseilles passed through the Red Sea, but seven per cent of cargoes to India, mainly Portuguese, were shipped round the Cape of Good Hope on the yearly Lisbon fleet to the Indian west coast. Before 1520 the English government was warned that a Portuguese vessel had just offloaded 200 pieces of 'Indian tin', meaning metallic zinc.[2] This soon became a frequent commodity in the long-distance trade, sought after as a vital component of brass. Two techniques were used for distilling zinc metal.

The Chinese developed techniques for mass-producing cast coins from combinations of copper, tin, zinc and lead (quaternary alloys), which became quite common under Ming emperor Jiajing (1552-1567), but the Chinese did not yet use metallic zinc in making brass, so the zinc content in coins remained at a level consistent with cementation brass.[3] During the late Wanli period (from about 1573) the brass was alloyed twice, with a fresh addition of calamine (zinc carbonate ore) at the second melting. Using this method, the zinc content in coins gradually rose. By 1621, the Chinese were actively distilling zinc by collecting the zinc metal in a pocket inserted at the top of the retort. The distilled zinc was cheaper than the quaternary alloy

---

[1] Hildebrandt 2002: 59-63
[2] Craddock *et al* 1998: 48-50
[3] Bowman *et al* 1989: 26

previously used in minting coins, so high-zinc brass often replaced it.[4] A string of zinc-distillation sites dating back to the Ming period has been excavated along the Yangtze River valley in Fengdu County, Chongqing province, central south-west China.[5]

The Yangtze valley sites included the well-developed Miaobehou zinc distillation site at Yangliusi village, whose ruins show evidence for workshops, ore-crushing platforms, coal-crushing pits and furnaces, and one surviving zinc ingot proved to be 99.2% pure. The retorts were wheel-thrown clay pots with lids. Interestingly, the reducing materials included not only charcoal (vegetable carbon) but coal dust (mineral carbon), placed with powdered roasted zinc ore in the lower part of the retort. Zinc oxide vapour formed under pressure in the lower chamber, where it reacted with the carbon, then steamed up through holes beside a saucer-like ceramic pocket at the top. Condensed zinc collected in the shallow bowl of the pocket, where it could cool to form pure zinc metal ingots.[6] By the late Ming dynasty, Chinese officials had the ingots carried down the Yangtze River to the imperial mints, to produce brass 'cash' coins, by then the main official purpose for distilling zinc. However, a sixteenth-century brass figure of Zhenwu, a bell and an incense-burner have also survived.[7]

In 1513, a London-based Italian wrote home that the Portuguese had just imported a cargo of tin from India. Indians, however, are not recorded as trading tin to London (which had its own local Cornish tin source), so this was probably metallic zinc. Zinc metal had been distilled on a large scale at Zawar, Rajasthan, northern India, from the 10 to the late 16 centuries (then probably disrupted by the Mughal wars) but seems to have been used mainly in northern India, Persia and neighbouring areas. In 1585 China exported a cargo of metallic zinc ingot slabs, each dated and weighing sixty kilos.[8] In total they weighed an estimated six tons and proved to contain 98 per cent zinc.[9]

From 1501, beyond the Himalayas to the west of China, the Shi'a Safavid dynasty ruled Afghanistan, Persia, Armenia, and parts of today's Pakistan. The craftsmen of Herat, formerly in Khorasan (eastern Iran), embellished metalwork with rich Koranic inscriptions, and continued throughout Safavid times to produce small elegant brass ewers or vases, with squat, rounded bodies and long cylindrical necks.[10] At that time, in the Caucasus and Egypt, fine inlaid brass-work was being produced under Mamluk sultans, who were in conflict with the Safavid dynasty. However, by 1516, the Ottomans

---

[4] Dai Zhiqiang and Zhou Weirong, 1992: 52-53;
[5] Liu Haiwang *et al* 2007: 175, 177
[6] Zhou Wenli 2016: 31ff
[7] Cowell *et al* 2003: table 1
[8] Craddock 1995: 318
[9] Browne 1916: 576
[10] Melikian-Chirvani 1982: 260

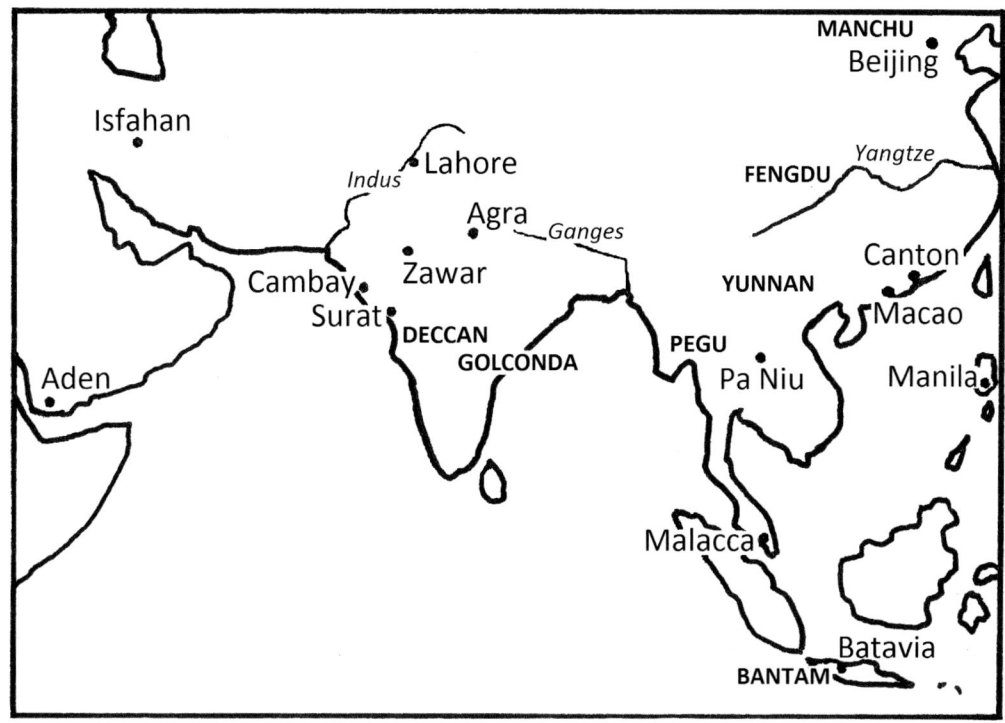

Figure 36. Asia, map showing some sites mentioned in the text

had seized Armenia, Syria, Arabia and parts of the Caucasus[11], so fine inlaid Mamluk brass wares, already reduced in number following invasion by the Mongol army of Temur (Tamerlane, 1336-1405), were no longer produced.[12] During the 16th century, the engraved astrolabes originally invented in Persia were being used throughout the Islamic world, applying science to the practice of religion. From 1539, substantial Islamic pillar-like brass candlesticks, inkwells and wine-bowls were produced,[13] continuing the tradition of decorative surfaces.

The Mughals from Central Asia take up the story. Descendants of Temur and Genghis Khan, they raided into India through parts of today's Afghanistan and Pakistan, eventually reaching Agra and Delhi in 1526, and laying the foundations of a huge empire. Mughals conquered the greater part of India under the leadership of Akhbar, emperor from 1556 to 1605. Smaller sixteenth-century cast brass guns occur in several regions, including an analysed 'Isa Khan' gun from Bengal (Isa Khan, a Bengali leader, used musket-type firearms in naval battles against the Mughals).[14] Emperor Akhbar's vizier, Abu al Faizl, in his '*Ayn-i-Akhbari* (Source of Information), written in 1596,

---

[11] Rayfield 2012: 165
[12] Allan 1998: 29
[13] Melikian-Chirvani 1982: 263, 265
[14] Ray 1956: 209

described three qualities of brass, each containing a slightly higher weight of zinc relative to copper. The first (described as 26% zinc to 74% copper) was malleable in the cold state; the second (30% zinc) malleable when heated, but the third (35% zinc) not malleable, but suitable for casting.[15] The description would be accurate if each of the three qualities of brass contained *less* zinc, so the vizier's scribe probably reversed the explanation by mistake[16].

From 1556 onwards, under the balanced and constructive rule of Akhbar and of his son Jahangir (1605-27), India saw a great flowering of non-religious brass ewers, often taking the form of birds or having spouts ending in an animal or bird. One surviving ewer has fluted spiralling up its body and slender neck, and its handle and spout terminate in stylised dragon heads.[17] Cylindrical candlesticks or oil lamps were sometimes encircled with raised linear spiral decoration or with narrow ridges, interspersed with bands of engraved lotus petal. Because the Mughals were deeply influenced by Persian culture, Islamic brass items like incense-burners, braziers and globular boxes were introduced, and astrolabes were consulted. However, the Mughals tolerated Hindu ritual and invited some Hindus to join the government. A very fine brass horse of the late 16th or earlier 17th century, though Mughal in style, bears Hindu swastika symbols on its flanks.[18] From Gujarat, surviving brass images include the benign Jain goddess Ambika (1521),[19] and the fierce deity, Kala Bhairava (1554).[20] Sixteenth- to seventeenth-century Hindu brass *puja* utensils, used for ritual associated with prayer, were excavated from. Ghantaśâlâ, Andhra Pradesh, where brass artefacts include an image of Śiva and a *dîpa* used for burning camphor before the deity.[21] Hindus used metal vessels in preference to ceramics, since they believed that metal did not pass on ritual contamination if passed from the hands of one caste to another[22]. Mughals in India adopted this usage without being aware of its origin.

Metal was the supreme medium for expressing the traditional Indian solid, balanced and sculptural vision of the world, seen in a late sixteenth-century cast and engraved water vessel from today's Pakistan.[23] Islamic court life now added the refined, decorative elegance of its ewers, flasks and vases, showing abstract decoration with calligraphy, curly arabesques or flowers. Craftsmen used a profusion of flower decoration but now largely avoided portraying humans and animals for fear of being

---

[15] Franklin 1977: 41
[16] The author's thanks to Shirley Northover for noticing this point
[17] Ashmolean Museum, University of Oxford: EA 1976.43
[18] Zebrowski 1997: 99, figures 98-99 and 101
[19] Ashmolean Museum, University of Oxford, EA 2005.17
[20] Biwas 1993: 325
[21] Rea 1989: 42, figures 33-34
[22] Zebrowski 1997: 106, figure119
[23] Victoria and Albert Museum: 1521-1829

called to account on judgement day for attempting to produce images which it was the prerogative of their Maker to create.

Craftsman working for Hindu patrons, on the other hand, eagerly portrayed lush plants, ripe fruits and energetic animals, birds and humans in fleshy exuberance but always with a solid, solemn form. The peacock was a favourite creature to portray in brass, and the types of useful brass objects made in India in the 16th century included oil-holders for lamps, oil lamps on stands, ewers, incense-burners, flasks and pillar candlesticks. By this time there were several types of Indian lighting equipment, including short candlesticks, drum- or bell-shaped pillar candlesticks of medium height and, for Sufi shrines, tall slender stands with an oil reservoir at the top, furnished with several wicks.[24]

During the 16th century, the Portuguese were harrying the western coastal ports of India, especially northern Gujarat, and they gained a foothold in Goa. On their way to India, the Portuguese had discovered West Africa and the copper-rich kingdom of Monomotapa (today's Mozambique/Zimbabwe) on the East African coast, both sources of the gold coveted in Europe. The greatest Portuguese discovery was the spice islands of Indonesia, but on the way, in 1511, they captured the Malay Peninsula. Half way up the eastern Malay coast, facing the South China Sea, lay Terengganu, which had a long tradition of brass casting.[25] Malay craftsmen were celebrated both for Malay brass, and for brass with certain other elements added, including 'white brass',[26] the ternary nickel brass, also known as paktong (a corruption of the Chinese term meaning white copper), which contained about 40-65 per cent copper, 20-50 per cent zinc and 5-20 per cent nickel.[27] China exported paktong tableware and candlesticks through Canton (Guangzhou) from the late sixteenth century. [28]After the capture of Malacca by the Portuguese in 1511, lightweight portable bronze or brass swivel guns (*Lela*), with barrels sometimes two metres long, were cast in Acheh (Aceh) across the Straits of Malacca at the northern end of Sumatra, and in Brunei, then under the rule of the Sultans of Sarawak. The barrels were often ornamented, sometimes representing fearsome decorative dragons.[29]

Before about 1550, the Portuguese obtained ready-made brass articles in Venice and Bruges to exchange for slaves and gold from West Africa.[30] A shipwreck off Gnalić islet, near Venice contained over seventy rolls of brass sheet, one millimetre thick or less.[31]

---

[24] Zebrowski 1997: many pictorial references
[25] Ray 1956: 209
[26] Sheppard 1971: 157
[27] Gilmour and Worrall 1995: 259;
[28] Zhou Wenli 2016: 12
[29] Sheppard 1971: 137-139, 157
[30] Kellenbenz 1977: 337
[31] Fabijanec 2018: 60

BRASS FROM THE PAST

Items for West Africa included brass wire, barbers' basins and candlesticks, and were termed '*merci tedesche*' or German wares, either from Aachen, shipped via Bruges, or from Nuremberg, many of which also went to the Near East.[32] From 1505-7, the Portuguese, supplied 287,813 manillas (tradable bracelets), 1,583 shaving basins, 530 urinals and 3,192 chamber pots to their West African trading post and castle named *São Jorge da Mina* (later Elmina, in today's Ghana) and another 67,095 manillas to their trading-station at nearby Axim. A further 302,920 manillas (about 94.5 tons) were listed on an Elmina inventory in 1512.

The Portuguese sold the brass goods by weight, mainly because skilled West African brass founders and smiths melted down imported brass manillas, shaving-bowls, urinals, cauldrons, water jugs and bells, to re-cast them into articles of African design and of greater meaning to themselves. The Portuguese quickly discovered that what appealed to one group was rejected by another. At first, one slave cost the Portuguese two barber's basins or twelve manillas, but by about 1517 the price of a slave had shot up to 50 manillas or more.[33] For ten years after the 1530s, the preferred metal was brass, but West African buyers were cautious, testing the purity of the brass by ear, by striking two brass items together.[34] Purer, forgeable brass has a sweeter ring. The Portuguese sold manillas of different weights to different parts of the coast, and exercised quality control by returning sub-standard goods.

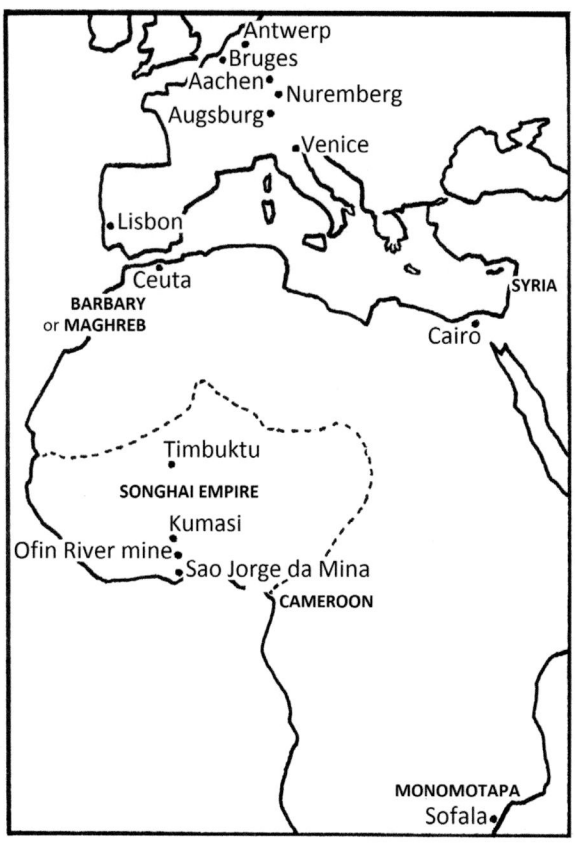

Figure 37. Europe and northern Africa, map showing some sites mentioned in the text

After about 1550, the Portuguese acquired Aachen and Nuremberg brass wares for their West African trade at the ports of Antwerp and Lisbon, trading through the

---

[32] Tucci 1977: 98
[33] Herbert 1984: 126, 128-30, 132
[34] Vogt 1969: 69

Fugger merchant family of Augsburg, southern Germany. By this time, the Fugger merchants, who handled brass, copper and calamine, were exporting brass not only to the West African coast, but to Britain, France and Spain. Although the Portuguese took brass manillas to Sofala in Monomotapa on the East African coast, only a few thin manillas found favour there. On the Guinea Coast of West Africa, however, they remained popular for over 50 years.[35]

Two thirds of the famous Benin 'bronzes' dating to the 16th and 17th centuries are in fact cast from lightly leaded brass. The zinc content of the beautiful Benin brass heads and plaques is comparable to contemporary Arab and European brass, and about half contain enough lead to facilitate casting.[36] By this date, Benin craftsmen probably melted down imported cementation brass pieces from Europe or the Mediterranean region and re-cast them, using the lost-wax process. Lost-wax casting was also practised along the northern rain-forest belt from the Ivory Coast to Cameroon.[37] Gradually, as the Portuguese brought more brass, the Benin heads became heavier and more robust and but with little change in zinc content.[38]

In the late 16th century, the shifting imperial boundaries of north-west Africa and increasing Portuguese trade brought sweeping changes to the region. Moroccan pressure and civil war weakened the western African Songhai Empire until, by the mid-17th century, anarchy had driven Mande and Fulani peoples, including Muslim horseback traders, southwards and eastwards from the old Saharan trading centres. The Portuguese were shipping brass and brass wares to the Guinea Coast, but the efficient Dutch were rapidly learning from them and bringing along a similar assortment of brass goods. The European presence along the still sparsely-inhabited coastal regions gradually shifted brass-trading southwards from the Saharan market cities to the coast.[39]

Brass was among commodities that could be exchanged for slaves, gold, ivory or tropical timbers, but what did the inhabitants of the Guinea Coast use all this brass for? Flemish traveller, Pieter de Marees, who visited the Guinea Coast in 1602, wondered how people used all the European brass artefacts. 'Great heaps of Cauldrons are brought there, which they use a lot for fetching water from Wells and Valleys' and for cooking mealie over a fire. He saw barbers' basins used for washing and shaving. Small rimless beakers were used to cook in, and basins were used as lids to exclude dirt. Ornaments and trinkets were placed in decorated brass basins, and small brass neptunes (flat dishes with broad flat rims) were used to store oil for anointing the skin.[40] European

---

[35] Herbert 1984: 126, 128-130, 132
[36] Willett and Sayre 2000: 168-169, table 2
[37] Shaw 1970: 1, 272
[38] Willett and Sayre 2000: 185-186
[39] Ade Ajayi and Crowder 1985: section 35
[40] van Danzig and Jones 1987, translation of de Marees 1602: 51

sheet brass was turned into shovels, spoons and scale-pans.[41]

These uses were mainly utilitarian and comparatively close to those envisaged by the European makers, but De Marees heard of other uses, less easily explained in seventeenth-century European terms. Large, flat neptunes, for example, were used 'to immure in tombs', or, more mysteriously, to 'carry something or other in'. De Marees saw that very large, open

Figure 38. Seventeenth-century brass pan with lizard decoration

'Scottish' pans, often over 3.5 metres in diameter (and probably intended as salt pans) were used to collect the blood of a slaughtered goat or pig. So many brass basins arrived that they sold at cost price, but he was curious to observe that 'Although these Basins are brought there in such quantities and are not as perishable a Commodity as Linen, one does not see much old brass-ware there; so there must be a huge population in the Interior which uses and employs such quantities of imperishable goods'.[42]

De Marees proves to have been correct – many people were indeed using brass in the interior, but people in different areas used brass in their own distinctive ways. Craftsmen in Akan areas of what is now Ghana cast and hammered brass, using incised, stamped, stippled and repoussé techniques on ceremonial brass containers, to produce elaborate geometric patterns and animal forms.[43] Two surviving large brass pans or basins were found during late nineteenth-century gold mining at Ofin River mine (80 kilometres south of Kumasi). One of them, thought to date to the 16 or 17 century, is embossed with symbolic sun, human and lizard figures.[44]

By the mid-16th century, far fewer of the brass goods exported to Africa were traded from the Maghreb or made in Saharan Africa, but were mainly brought from Europe by the Dutch and Portuguese. This brass export, combined with the steep rise in European demand for brass cauldrons, chafing-dishes, lavers (small hanging cauldrons with spouts and bucket-type handles), wire, candlesticks, instruments and ecclesiastical objects,[45] raises the question as to where all the brass was being produced. Since there is no evidence of large-scale production in the 16 century, it is worth looking at some of the smaller centres that were springing up around Europe

---

[41] De Corse 2001: 131
[42] van Danzig and Jones 1987, translation of de Marees 1602: 51
[43] de Corse 2001: 133
[44] British Museum 1955: AF 225; and 1955 AF 5.226
[45] Brownsword and Pitt 1983: 44-48

at the time. Many brass mills and foundries were closely involved in the European political scene, some preparing weapons, or becoming pawns of the imperial powers. The main interlocking powers were the Habsburgs and their Holy Roman Empire, the Hohenzollerns and the Vasa dynasty. The story of European brass-making among rival empires can be traced from south to north, starting in Milan, in today's northern Italy. By 1510, in Milan, Biringuccio from Siena witnessed glowing crucibles in which calcined calamine was mixed with crushed copper and the city became famed for its output of small brass objects, particularly for bowls and wire.[46] Biringuccio also mentions the use of reverberatory furnaces (discussed in chapter 5). In 1534, in Milan, the second Duke of Sforza saw the advantage of brass production, particularly to make fine yellow brass wire for wool cards to improve his own woollen industry. One Michele Pasquale and his son-in-law entreated the duke's permission to make brass wire 'as good and beautiful as that made in Germany', and in 1548, the duke granted them a privilege to manufacture brass wire.[47] From Italy the story moves to Austria, soon to become a Habsburg stronghold.

A small Tyrolean brass works had been set up in 1481 at Fritzens (11 kilometres east of Mühlau), by Antoni vom Ross,[48] a Venetian-born merchant and financier from Bozen (now Bolzano), who held imperial rights to produce brass in the Tyrol and to attend the emperor's fancy-dress balls.[49] He may also have been the owner a brass works near Prettau (now Predoi, in alpine Italy), about 100 kilometres north-east of Bozen.

In 1485, joint Habsburg emperor, Maximilian I, appointed a Nuremberg master brass-founder Leonhard Offenhauser to prepare a brass-works at another Tyrolean village, Mühlau, now a northern mountainside suburb of Innsbrück. Maximilian I was granted the Austrian Tyrol in return for intervening in a feud between the Tyrol and Bavaria, and he required arms to capture three strategic towns which still resisted him.[50] In a first Habsburg attempt to break Nuremberg's brass-making monopoly in southern Europe, the emperor ordered that his Mühlau brass works should use only Tyrolean copper, but it remained dependent on deliveries of fire-resistant crucible clay from Nuremburg (200 km north of Mühlau).[51] In antiquity, crucibles had been porous, but with better crucible clay the pressure inside a very tightly sealed container increased when it was heated. Atmospheric pressure (e.g. altitude) outside the crucible affects the process, forcing zinc vapour to diffuse into the copper, and can cause zinc to start producing a little vapour from as low as 800°C, well before copper melts.[52]

---

[46] Tucci 1977: 99
[47] Archivio di Stato di Milano: MS. Atti di Governo, Commercio, parte antica, pezzo 198
[48] Ucik 2002: 171
[49] Freydal 1882: page XCV
[50] Rattenberg, Kufstein and Kitzbühel
[51] Palme 2000: 14, 33
[52] Rehren and Martinón-Torres 2008: 168

The Mühlau brass-works overlooked north-east Innsbrück, with a spectacular view across the valley to the great courtyard of the imperial armoury (*Zeughaus*), and in 1503 Leonhard Offenhauser, its technical manager, was paying out for crucibles, timber, calamine and copper.[53] A 1534 map shows the extent of the sloping brass-works site along the tumbling Würmbach mountain stream[54]. From the outset, the Mühlau brass mill, powered by the rapid water flow down to the river Inn below, made only gun-parts. The emperor needed brass gun-barrels for arquebus (*Büchsen*) – either hand-guns (bronze versions had been known for a century)[55] or long, slender-barrelled guns standing on a light support. He also required brass barrels for thicker *Kanonen*, assembled at the Innsbrück imperial armoury. Brass, whose crystals can part slightly under pressure, emits only cold sparks, with a low heat level, so it is termed 'spark-free' and could be safely used in contact with ammunition. Friction, impact and stretching can cause high-tensile metals like iron and steel to generate incendiary sparks hot enough to ignite an explosive material like gunpowder.[56] Brass, being more malleable under stress, was safe for gunpowder storage containers and was used for very early gun barrels, which, over time, succumbed to friction, heat and wear. There is archaeological evidence for brass being used between 1480 and 1518 for heavy products in the royal cannon foundry in Buda (Hungary), managed by Jacobus Marijwerder of Prussia, who operated reverberatory furnaces.[57]

Calamine for the Mühlau brass works came from Nassereith in the Tyrol, 90 kilometres to the west,[58] and high-grade copper came from Prettau in the Tyrolean Ahrn valley, 120 kilometres to the south east.[59] Mühlau obtained timber for charcoal and for fuel from nearby Hall (against stiff competition from the Hall salt-works). In 1505, Mühlau recruited additional Nuremberg brass-founders and smiths, and three more went to Aschaffenburg, about thirty kilometres south-east of Frankfurt, though little more is heard of the Aschaffenburg brass works. In 1508, a visiting Nuremberg brass founder, Stefan Godl, made 1,200 brass-barrelled guns at Mühlau, including 200 arquebus (*Büchsen*) for use in Maximilian I's current power dispute with Venice (settled following pressure from Nuremberg citizens, for whom Venice was a vital trading port).[60]

In 1509, Maximilian I had granted the Höchstetter merchant brothers of Augsburg rights to set up a brass-works at Pflach bei Reutte, about 90 kilometres north-west of Mühlau. Pflach obtained its copper from Prettau and zinc ore from the Fernpass area

---

[53] Tiroler Landesarchiv: MS, Kopialbuch Missiven, folio 13r (1503)
[54] Tiroler Landesarchiv: Karten und Pläne 2873, 1534, map by Jörg Kölderer
[55] www.bkz-online.de, *550 Jahre alte Büchse lag einfach im Gras*
[56] Canadian Centre for Occupational Health and Safety fact sheet. ccohs.ca/ohsanswers/safety_haz/hand_tools/nonsparking.html
[57] Belényesy 2018, 155, 58
[58] Palme 2000: 38
[59] Tiroler Landesarchiv: MS Kopialbuch Missiven, folio 13r (1503)
[60] Palme 2000: 19, 27-28

AGE OF DISCOVERY

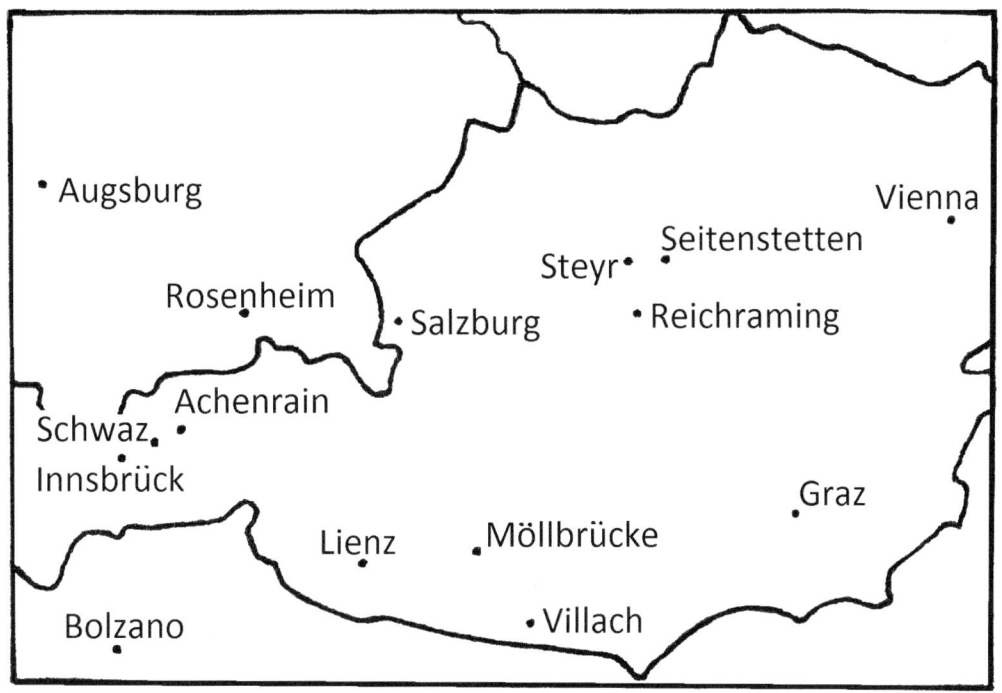

Figure 39. Austria, map showing some sites mentioned in the text

of the Tyrol, and sold its brass through the Höchstetters' southern German markets, particularly Augsburg.[61] Augsburg craftsmen produced brass sundials, terrestrial and celestial globes, quadrants and surveying and drawing instruments, A brass sundial dated c.1521,[62] excavated from a 56-metre-deep well at Logne castle (c.25 kilometres east of Paris), was of cast brass subsequently hammered, divided into seven sections, then skilfully reassembled. It was made for use at latitudes ranging from Paris to Augsburg – where interest in instruments of measurement was particularly strong in the 16th century, notably from the workshop of Christophe Schissler (1531-1608).[63]

In 1519 the death of Emperor Maximilian I provoked an imperial power-struggle. Maximilian's son, Arch-Duke Ferdinand I, gained the Tyrol at the Treaty of Brussels (1521) and much of Austria at the Treaty of Worms (1522), which strengthened his Habsburg Empire. The Spanish Charles V was meanwhile elected Holy Roman Emperor over the Spanish Netherlands and German territories, into which Pflach bei Reutte now happened to fall. This meant it lost its rights to Tyrolean copper from Prettau, so by 1522, its owners, the formerly celebrated Höchstetter merchants, were struggling financially, with one Höchstetter brother locked up in the tower of Augsburg's debtor's

---

[61] Ucik, 2002: 171, Palme 2000, 19-20
[62] Musée archéologique du château-fort de Logne: PU 030327-02
[63] Sausson and Wéry 2014: 79

prison. The Pflach brass-mill foundered and lay unused.[64] A former manager, Georg Hagen, took it on and from 1565 his son brought it into profit, but by 1621 timber shortages and continuing problems with Augsburg merchants brought Pflach brass works to an end.[65]

In 1510, brass-founder Ulrich Kussinger, took over the Mühlau brass works which remained in the Austrian Tyrol, so its brass mill still had rights to the good copper from Prettau, also in the Tyrol. Dr Ulrich Jung, manager of Mühlau in 1528, lived in a substantial house beside the mill, emblazoned with his own heraldic device, which survives. Some members of his staff were well-respected citizens – one skilled worker, a brass-burnisher (*palierer*) was described in a 1533 court case as 'well-born, noble, educated and trustworthy'. By 1539, Prettau copper sources were running out, so brass production declined until supply failed to meet demand. Maximilian's dream of building Europe's best, biggest and most modern brass works using Tyrolean raw materials, remained unfulfilled, so he and his son Ferdinand I lost interest and the Mühlau works declined and closed.[66]

In 1564, in the mountainous southern Tyrol, the extravagantly named Baron Christoff von Wolkenstein-Rodenegg received Habsburg permission to set up a brass works at Lienz, where the River Isel meets the River Drau. He intended to make exclusive use of the plentiful local copper, which had long been mined.[67] The brass works, at the west end of today's Mühlgasse (Mill Lane), was powered by the River Drau (since diverted) and had a factory with four copper and brass furnaces, a foundry, four battery hammers, annealing furnaces and a calamine crushing mill. The mill was set up to produce sheet brass and wire.[68]

In 1550 a mill was set up further north, in Steyr, Kärnten (Carinthia) to produce its own brass, but in 1569-70, 16 kilometres upstream from Steyr, four men took over two former iron battery hammers and established a brass works at Reichraming, near the confluence of the Reichraming River and the broad River Enns. One partner, the experienced Bernard Manstein from Jägerndorf, Bohemia (today's Czech Republic), was the driving force and became sole owner from 1575, taking on a partner, Wolfgang Köberer from Nuremberg.

The Steyr brass works transferred upstream that year, to join the new brass works at Reichraming where timber was more plentiful.[69] The Reichraming brass mill lay within Steyr estate forests, and paid dues to the Habsburg imperial treasury.

---

[64] Palme 2000: 43-45
[65] Ucik 2002: 171-172
[66] Palme 2002: 54, 70-71, 146
[67] Tiroler Landesarchiv: MS Montanwerke, box 1018, Bergwerks Ordnung, Lienz, 1553
[68] Ucik 2002: 172
[69] Landesarchiv Linz: Schloss Steyr: MS 194, 468

The Steyr burghers were ordered to buy their brass from Reichraming, and were threatened with a 50-ducat penalty if they imported it from elsewhere. In 1563 the River Enns was made navigable and in 1567 copper mines were opened at Radmer, eighteen kilometres upstream from Reichraming, providing a ready copper supply.[70] In 1592, after both brass-works partners had died, Manstein's son, Leonard, took over. On the lower mountain slopes above the brass-workshops, he built a great three-winged building around a courtyard, as lodgings for himself and his staff. The surviving entrance arch to this building, the *Meierhof*, bears the date 1597 and Manstein's initials. Backed by family money, Leonard drove the business forward, and in 1598 the Reichraming brass output was worth 8,000 florins.[71]

Several rulers were becoming aware of the potential profits to be made from brass. In 1585 two battery brass-works were established near Salzburg, then an independent prince-bishopric. Wealthy Salzburg merchants, the Stainhausers, set up one mill at Oberalm (near Hallein), and another other at Ebenau, 15 kilometres south of Salzburg city. By 1597, the Stainhauser merchants' two sites were well-placed for water power and timber and for transport routes to the rich south German and north Italian markets. Salzburg craftsmen hammered out semi-prepared brass wares for sale to Nuremberg craftsmen, who worked them into finished objects. After 1600, however, the prosperity of the Stainhauser brothers declined due to their high risk-taking and expensive lifestyles and to getting too closely involved with prince-bishop Wolf Dietrich (1587-1611), who was thrown out and arrested. The Stainhauser fortunes crashed with him, and the brothers were jailed for debt. Creditors took over the two brass-works until they could be sold in 1622.[72]

The independent state of Nuremberg some 230 kilometres north-west of Salzburg, was already becoming well-known for its brass products, including astrolabes. In 1510, a Nuremberg craftsman, Peter Heinlein (*c*.1480-1542) produced a beautiful cylindrical pocket-watch housed in brass, and subsequently supplied presentation watches to Nuremberg City Council as gifts to prominent visitors.[73] In the 1540s Nuremberg craftsmen, including the highly skilled Georg Hartmann (1489-1564), produced numbers of astrolabes.[74] By 1511, Hammer brass mill, Laufamholz, Nuremberg, had three working water-wheels and six crushing-mills. Four years later Hans Mayer from Nuremberg re-built and extended the brass battery-works and constructed himself a small house by the mill.[75] Nuremberg brass-makers bought Hungarian and Tyrolean copper through the Fugger merchants of Augsburg,[76] and at the Frankfurt

---

[70] Aschauer 1953: 314
[71] Brunnthaler 2000: 111-112,
[72] Priesner 1997: 58-59
[73] Germanisches Nationalmuseum, Nürnberg: Wl 1265
[74] Germanisches Nationalmuseum, Nürnberg: 047
[75] Stadtarchiv Nürnberg: MS, Laufamholz 455/1
[76] Palme 2000: 19-20

Figure 40. Astrolabe made and engraved in Nuremberg by Georg Hartmann, c.1540s

fairs they sold Tyrolean copper on to Aachen for brass-making. Aachen merchants in turn brought prepared calamine from the Vieille Montagne near Moresnet (then under Spanish Habsburg rule) to the Frankfurt fairs, to sell to Nuremberg brass makers.

In the 16th century, Nuremberg continued to thrive as a brass-working centre. The illustrated chronicles of the Twelve Brothers charitable trust in Nuremberg show craftsmen of this retirement community preparing gun barrels for light arquebus (*Buchsen*) and rather heavier *Kanonen,* as well as wire candlesticks, candelabra, and chandeliers. Although the illustrated artefacts cannot be analysed, the workers mentioned here are described in the chronicles as working in brass (*Messing*), which has a golden appearance in the illustrations. However, lead and perhaps tin are likely to have been added to the alloy for cheaper, everyday objects, to reduce cost. One brass worker is sitting surrounded by a workbench full of candlesticks, filing off rough parts left after casting, while a long, slender arquebus barrel awaits attention, propped against the wall.[77] Another brass-turner is shown filing unwanted casting-flaws from a cannon-barrel.[78] Nuremberg brass craftsmen had a trading advantage because they were prepared to make just about anything that was in demand – including hour-glass stands, oil-lamps, buckles, key-rings, taps, stop-cocks, plugs, pulleys, thimbles and pins,[79] all portrayed on the workbenches of the Twelve Brothers.[80] Frankfurt took and traded these kinds of goods from any source.[81]

Nuremberg ore sources for calamine were mainly the Altenberg (Vieille Montagne) and to a lesser extent the rich supply from the Silesian uplands (today's Czech-Polish

---

[77] Nürnberg Stadtarchiv: MSS, Landauer 1, Amb. 279, folios 8-9 (1528), 17-18 (1518), 79-80 (1528)
[78] Nürnberg Stadtarchiv: MSS, Landauer I, Amb. 279, folio14-15 (1525)
[79] Dietz 1921: 199
[80] Nürnberg Stadtarchiv: MSS, Mendel I, Amb. 279, folios 4-5 (1425); Landauer I, Amb. 279, ff 8-9 (1516); 59-60 (1586); Mendel II, Amb 317b, 52-53 (1592)
[81] Dietz 1921: 199

Figure 41. Everyday products of the Nuremberg Twelve Brothers: *hanging up, left:* oil lamps, *left to right.* 1525, 1518, *hanging up, right*: key rings *left to right*: 1586 and 1528; foreground, *left to right*, 2 candlesticks 1526; ewer 1544; tankard 1518; two arquebus barrels 1528; tap 1586; candlesticks *left* to *right*: 1525 and 1518; a brass-worker rasping a cannon barrel, 1518

border), an area very tempting to the Hohenzollerns of Brandenburg (today's northeast Germany). In 1495, as mentioned earlier, the mighty Fugger family of Augsburg, in partnership with Hungarian merchant János Thurzó, had formed a copper-trading company, which by 1517 exported roughly 810 tons of copper from Silesia and Upper Hungary (now Slovakia). The Fugger partnership was supporting a Habsburg campaign to seize the crowns of Bohemia and Upper Hungary. This angered the noble Polish landowners of the mining areas, who in 1525 banished the too-powerful Fugger clan and seized back the mines and financial assets. However, the Habsburg Charles V intervened and imperiously allotted one copper mine each to selected Fugger family members.[82]

The Upper Silesian mines lay in the zinc-rich estates around Bytom (Beuthen) and the nearby city of Tarnowitz, both under Polish rule. In 1526, Habsburg Archduke

---

[82] Backowski 2002: 23

Ferdinand I (1521-1564) had gained the kingdom of Bohemia through marriage, and had coveted the zinc and brass-making potential of this area ever since, but the proud exiled Polish noblemen refused to pay the archduke for rights to mine calamine on their estates, or give up their own right to sell a proportion of all the calamine. Ferdinand feared they might try to form a brass-making cartel with Tyrolean copper miners. By 1562, a brass-works had been started at nearby Jägerndorf (now Krnov), then in Habsburg-controlled Bohemia, and in 1569 the King of Poland granted permission for a brass-works at Tarnowitz. Jägerndorf brass-works alloyed Upper Silesian zinc ore with Tyrolean copper to produce 125 tons of battery brass a year. Both Jägerndorf and Tarnowitz were conveniently placed for zinc ore supplies from the Silesian zinc mines at Bytom (Beuthen).

A Habsburg-Hohenzollern calamine struggle ensued, with brass-manufacturers from Hohenzollern territories also clamouring to get their hands on Upper Silesian calamine. The Polish king had already granted sole calamine rights to two brass-maker partners from Küstrin (Kostrzyn) further north on the River Oder, and in 1581 the Rosenberg brothers, brass-producers near Danzig (Gdansk), tried to gain access to the calamine by getting the Küstrin partners' monopoly revoked.[83]

In 1584, Archduke Ferdinand's Habsburg son and heir, Count Georg Friedrich, changed sides through judicious marriage, and became a Hohenzollern. Georg Friedrich at once overrode the Polish nobles and granted sole rights to Silesian calamine to a local Bohemian merchant, who dug numerous new mines but still refused to pay the Archduke for the right to exploit them. This Bohemian merchant supplied calamine to the Krnov brass works, but traded most of it through Breslau to brass makers in Hohenzollern lands, who paid him handsomely.[84] In 1595 the childless Georg Friedrich bequeathed Bohemian Upper Silesia to the emerging Hohenzollern Archduke of Brandenburg, thus ensuring a rich calamine supply to a future Prussia.

In 1555, east of Bohemia, in the independent state of Thüringen (now in eastern Germany), a large brass-mill was built on a site with two mill-ponds beside the River Ilm at Ilmenau.[85] Wolf Weihrach, a merchant who lost his fortune in 1554 in the Markgraf Wars when Schweinfurt was devastated, set up the Ilmenau mill on a former iron working site, to exploit mines of silver-bearing copper ore at Schweina just north-east of Ilmenau. Weihrach's brass mill exploited two former iron battery hammers and a corn mill. Before his son Hans took over in 1569, Weihrach added a brass furnace-house. Between 1579 and 1584, the mine's copper yield halved, but Hans Weihrach bought the mines, and his brass-mill is shown belching forth smoke in a small illustration on a map of 1592. After Hans' death in 1595, his wife Marie took over, marrying her brother-in-law, Bartel Drachstedt, a partner in the business. Drachstedt

---

[83] D'Elvert 1866: 166-167
[84] Karsten 1827: 348
[85] Dietz, 1921: 199

managed the business well and increased copper production from the mines, so in 1609 Marie Drachstedt managed to sell the concern to two Leipzig merchants, who carried on the brass works until 1619, the onset of the Thirty Years War (1618-48),[86] during which one Johann von Bodeck lost a large sum of money on it.[87]

In the mid-16th century, not only war, but the hazards of land transport, were still a major problem for delivering raw materials from source to brass-mill. Zinc ore, being bulkier (though lighter) than copper ore, the other component of brass, is harder to transport, so brass was usually manufactured closer to a calamine source. The fact that calamine ($ZnCO_3$ ore) contained the metallic element zinc was not fully understood in Europe, and the ore was usually described as an 'earth' or 'stone', mysteriously capable of colouring copper and adding to its weight.[88] By 1556, Bohemia (Czech Republic), and Carinthia (*Kärnten*, eastern Austria) had known zinc ore deposits, and Upper Silesian zinc ore from the Krakow region (in today's Poland) had been exploited since the Middle Ages.[89] However, the Vieille Montagne (Altenberg) mine near Moresnet (now in Belgium) remained the major European calamine source. In 1565, the Assay-Master at the English Mint wrote that his principle source of calamine was from 'the mynes of Akon (Aachen) where it is to be had in great aboundance'.[90] The Vieille Montagne was not connected with Aachen, but its ore was traded to England from that city. Nuremberg also supplied calamine to England, probably from Tyrolean or Silesian sources, but border taxes and carriage charges made ore traded through Nuremberg more expensive.

In Aachen itself, the number of brass works steadily increased. In 1528, a local entrepreneur, Aegidius von der Chamen, produced 30 tons of brass wire and 10 tons of brass cauldrons. By 1559, the sixty-eight Aachen brass-mill owners each employed about 17 men collectively producing about 1,500 tons of brass a year, to be exceeded only by the largest brass works over the following two centuries. The unit of measurement, the centner, zentner or hundredweight, approximated to 50 kilos, but fluctuated significantly with time and place of production. By 1562, Aachen brass-makers had customers in the Netherlands, France, Spain and Portugal, and by 1579 they also sold to Austria, Upper Hungary, Poland, Russia and Scandinavia, and ordered Swedish copper from the Falun mine. A contemporary confirmed that they could command exchange credit as far afield as Constantinople.[91]

In 1578, the Aachen artisans, or '*Kessler*', meaning makers of cauldrons (their main product), formed themselves into a guild, once they realised that the entrepreneurs

---

[86] Fiala *et al* 1998: 154-161, map p.60, Thüringen HStA, Weimar, B 15800, folio 268
[87] Dietz 1921: 199
[88] Pettus 1686: 285
[89] Polish Geological Institute, 2004: www.pgi.gov.pl/mineral_resources/zinc_lead.htm
[90] *Calendar of State Papers Domestic* 1547-1580 (1565): 73
[91] Roderburg 1927: 12; Van Eyll 1998: 86

who provided their capital were exploiting them. The Council of Aachen had initially agreed with the craftsmen to prohibit labour-saving mechanical 'battery' hammers, which were trip-hammers enabling one man to beat out sheet brass into semi-prepared wares far more rapidly. The entrepreneurs evaded the prohibition by simply moving their business beyond the city walls to where battery hammers were allowed,[92] particularly to Stolberg, where trip-hammers had been used for beating brass since before 1550, and where brass plate was cast in moulds of Brittany granite, rather than in sand.[93]

From the start of the 16th century, brass was made at Stolberg and its copper came partly overland from Mansfeld, 450 kilometres further east. The nearby Ardennes forests provided timber for fuel and charcoal (the reducing agent). The Stolberg area had seams of calamine (zinc carbonate ore) though of less pure quality than calamine from the Altenberg (Vieille Montagne) mines. For their finest brass, Stolberg brass-makers mixed their local zinc ore with the excellent Altenberg calamine.[94] In 1552, the calamine mines of the small town of Iserlohn, tucked into the forested hills about 20 kilometres south of Dortmund, were re-opened after lying idle for 50 years. The mines produced calamine declared comparable in quality to Altenberg calamine. Iserlohn zinc carbonate ore lay at a geological boundary, the oxidation zone where Middle Devonian limestone met impermeable slate[95]. By the late 16th century, Iserlohn calamine was sold mainly to merchants from Cologne and Soest (a Hanseatic town about 40 kilometres away), who traded it on to Hamburg, Bremen and the Netherlands.[96]

Calamine from the Altenberg or Iserlohn was used for sixteenth-century monumental slabs, from brass probably produced in the Rhine-Meuse area but cast in the Netherlands, London, or more rural foundries. Brass memorial slabs to English families and individuals quite often portrayed rows of children of descending age, and sometimes several wives, all accompanying the father. Written brass memorial inscriptions became both longer and more common.[97] By the late 16th century, English memorial brasses contained markedly increased amounts of zinc, rising from an average of about 21 per cent in 1500, to about 32 per cent by the final quarter of the century and steadying at a similar level during the following century.[98] As already mentioned, metallic zinc was regularly distilled in China by the later 16 century, though not yet used there for making brass. It had long been used on the Indian subcontinent for brass making, so it seems likely that by 1599, when a weight of 35.5 per cent zinc

---

[92] Haedecke 1970: 18-19
[93] Van Eyll 1998: 90
[94] Schleicher 1974: 13, 17, 36-37
[95] Klostermann 1996: 47-48
[96] Klostermann, Rolf, 2013: 'Historische Einordnung', unpublished presentation at Iserlohn Museum, 3, 7
[97] Norris 1978: 58-59
[98] Author's calculation based on metallurgical analyses in Calver 1990: folios 68 and following folios

was analysed from an English monumental inscription,[99] imported metallic zinc was being added to some European brass. Since sixteenth-century interest in technology was increasing, it has also been argued that European higher-zinc brass might perhaps have been produced through better technology alone.[100]

In 1482, mining was revived at the former Roman source of pure zinc carbonate ore (calamine) at Gorno, in the Alps north-east of Milan.[101] Leonardo da Vinci, chief engineer to Cesare Borgia, was invited to Milan in 1506 by Charles d'Amboise, French governor of the city, and made a detailed sketch map indicating Gorno and all the other mines in the Serio and Parina valleys. Although the main target was silver, in 1590 a Venetian named da Lezze mentioned the transport of zinc oxide from those high valleys to Milan.[102]

One writer on sixteenth-century Milanese brass technology was Vanocchio Biringuccio (1480-1539), mines supervisor for Tuscany and the alpine valleys. In his *Pirotechnia* published in 1540, he noted that Flanders, Cologne, Paris and Milan were important centres of brass-making. By 1540, Milan was processing calamine (ZnO_3) to make the brass articles for which Milan became famous.[103] Biringuccio saw brass there, variously hammered and cast into buckles, cups, bells, thimbles, spoons, basins and candlesticks. He observed a Milan brass works which either used Valencia refractory clay for making crucibles (involving a thousand-kilometre sea voyage, plus 150 kilometres overland from Genoa), or bought crucibles ready-made from Vienna. In the Milan area, 10 kg of copper fragments were placed in several crucibles, each capable of holding about 25 kg, which were then filled with calamine (and no doubt charcoal). The whole was covered with a layer of powdered glass, which melted at 700ºC (below the melting-point of zinc), forming an airtight seal, and the crucibles were further lidded with clay.[104] The efficient double-seal procedure described by Biringuccio would have improved the uptake of zinc within the copper and could have raised the zinc-to-copper weight ratio to slightly above 30 per cent.[105]

The term zinc was used by the Swiss-German physician and scholar, Paracelsus (1493-1561), writing around 1520-41, who mentioned *zinck*, from Carinthia (eastern Austria).[106] Georgius Agricola (Georg Bauer, 1490-1555), explained in 1546, in his *De Natura Fossilium*, that furnace-flue zinc oxide could, like mercury, be placed in a container and heated until it sublimated to a black, brown or grey metal with corrosive

---

[99] Calver, 1990: 68 and following folios
[100] Welter 2003: 29, 34
[101] ecomuseominieredigorno.it/cms/index.php/percorsi/miniere-e-minatori
[102] Furia 2012: 43-45
[103] Haedecke 1970: 28
[104] Smith and Gnudi, 1959: 72-73, translation of Biriguccio, 1540
[105] Welter 2003: 28-29
[106] Webster 1671: 339

properties, called *cadmia sublimata*. This *cadmia*, he added, was the same metal which the people of Rhaetia and Noricum (today's Switzerland and Tyrol) called *zincum*. In 1556, at the Goslar lead and silver furnaces, the zinc deposited on furnace-flue walls was called *contrefey* or *counterfeht*, because, when introduced into copper it produced a metal (brass) counterfeiting gold. Agricola mentioned that it was also abundant in Silesia.[107] By 1556, therefore, both zinc oxide and some sublimed metallic zinc were being systematically retrieved from furnace flues in the Harz, Upper Silesia, and Carinthia. Agricola himself was based at St Joachimstal, Bohemia (now Jáchymov, Czech Republic), a silver-rich valley in the Erzgebirge (Krušné Hory) mountain chain between today's Germany and Czech Republic. Furnace-deposited zinc metal, however, though useful, was not as pure as the distilled zinc metal produced in Asia.

In 1542, in the Harz Mountains, Earl Wolfgang had started a brass works on his land near Wernigerode to exploit copper from Mansfeld lead ores, from which he already extracted silver. Under the management of Peter Kogelmann, an experienced Hesse brass maker (probably from Kaufungen, near Kassel), the Wernigerode mill produced brass wire and cauldrons. After two years the mill was transferred to Ilsenburg, nearby. In about 1550, at the Rammelsberg in the Harz, Erasmus Ebener (1511-77), a scholarly Nuremberg-born diplomat, recognised hardened drops of sublimed metallic zinc among the zinc oxide collected on the lead and silver furnace-flues.[108] In 1552, Earl Wolfgang allowed him to convert a Goslar copper mill to exploit the zinc furnace residues. This venture led to a trade in furnace zinc metal and zinc-oxide, which by 1555 was delivered south to brass-works at Kaufungen (Kassel, Hesse) and Ilmenau (Thuringen).[109] The Harz brass mills at Goslar, Bundheim, Ilsenburg and Wernigerode were roughly ten kilometres apart, along a line south-east from Goslar over the mountains.

By 1556, zinc-oxide powder was being systematically retrieved from furnace flues in the Harz, Upper Silesia, and Carinthia. From 1562, Erasmus Ebener, by this time described as a silver-producer at Bundheim in the Harz,[110] managed a *Saigerhutte* or liquation plant, to separate silver from copper. Liquation was an important innovative process from the 16th to 17th centuries, described in detail in 1556 by Georgius Agricola, who devoted the eleventh book of his *De Re Metallica* to the subject. The liquation method used heat to co-melt raw copper with lead. As they solidified, the copper, having a higher melting point, became spongey and porous, allowing the more liquid lead and silver to gather in its pores. On a second heating of the cooled ingot, liquid lead flowed out of the copper structure, and silver, having an affinity for lead, poured off with it. The lead was then heated to a high temperature till it oxidised, and the silver, (which

---

[107] Hoover and Hoover 1950: translation of Agricola (1546-1548) book 11, 368 and footnotes
[108] Aitchison 1960, volume 2: 481
[109] Priesner 1997: 23
[110] Dennert 1979: 1a.6.90, 22.09.1571

does not form an oxide) separated from it.[111] The copper, meanwhile, had to be refined to remove impurities before it could be used for brass, and refined even more if it was to be hammered into sheet-brass or drawn out to wire.[112]

In 1571, Bundheim in the Harz had a brass mill with a furnace, battery hammer and annealing oven for shaping cauldrons, a calamine mill and five undershot waterwheels driven by two falls of water. A wire-drawing workshop was added, and the manager and workers were provided with housing. In 1572-3 the Bundheim brass mill made a 2,400-florin profit. In 1572, Ebener listed some brass objects being made in the surrounding Brunswick region, including bird-cages, spinning-wheel parts, money tills, camp-beds, hunting horns and book-clasps. A further list mentioned brass used to embellish harness, swords, rapiers and even Hungarian sabres.[113] In the Harz, production fluctuated, but mineralogist Lazarus Ercker wrote in 1594 that, at Goslar, zinc-oxide accretions scraped from the lead furnace flues grew to a hand's-width thickness over ten to twelve months. The zinc oxide was roasted and ground up fine – one part of zinc-oxide powder being combined with two parts ground charcoal. After being dried and then soaked with a little alum, it was heated to reduce it to zinc metal, and finally raked through. Goslar zinc-oxide accretions were still supplied to Ilsenburg and Kaufungen brass works.[114] Ilsenburg brass works, which exported through Lübeck, continued with fluctuating output, producing about 49 tons of coarse brass wire and cauldrons a year, but it foundered during the wars of the following century.[115]

Kaufungen brass-house and battery works had been set up in 1527 by Count Philipp the Magnanimous in a former grain mill in the Kaufinger Forest south-east of Kassel, Hesse. The brass works and wire-drawing mill lay on the constantly flowing River Losse, with a few local names still commemorating it.[116] Kaufungen mill received a delivery of Harz zinc metal in 1594,[117] but it was to founder early in the 17th century, under threat of war. In the 16th century, rivalry between the two powerful ruling families of Habsburg and Hohenzollern at first led to an increase in the number of European brass works, because both dynasties coveted the profits currently to be gained from brass-working. They also desired brass for gun-parts, family memorials, personal adornment, household embellishment and scientific instruments. In their quest for brass, the two dynasties each strove to gain control of supplies of calamine – the zinc carbonate ore so essential for cementation brass-making.

---

[111] Hoover and Hoover 1950: 491 and footnotes, translation of Agricola (1546-1548) book 11, 491
[112] Jean-Marie Welter, personal communication
[113] Beddies 1996: 178-179 and footnote
[114] Ercker 1594: 111v
[115] Priesner 1997: 22-24
[116] *Lohmühlweg, Kupferhammer*
[117] Ercker 1594: 111 v.

Across the North Sea, in England, no viable sources of calamine were discovered until the end of the 16th century, so little brass was alloyed there from native raw materials. England, rather than prospecting for calamine within its own shores, had relied on supplies from the Altenberg (Vieille Montagne). However, the European Eighty Years War (1568-1648) left this Moresnet calamine source under the Habsburg rule of Philip of Spain, whose policies threatened England, so the English monarchy, as enemies of Spain, could no longer rely on direct calamine deliveries from the Altenberg.

Faced with Spanish hostilities, England urgently needed to produce its own brass for the brass wire needed for wool-cards for the very lucrative English wool trade, and for gun parts. The English queen, Elizabeth I, requested Daniel Höchstetter (agent to the Augsburg merchants who succeeded the Fugger family) to send over Tyrolean prospectors from Innsbrück, Schwatz and Gastein in the Inn Valley.[118] In August 1566, the prospectors found a copper mine at Newlands near Keswick (Cumbria),[119] unworked since the reign of Henry III (1207-72),[120] and in September that year Johann Fugger sent over 20 Tyrolean miners. Remains of Tyrolean wooden railway tracks and miners' trucks (*Pergtruhen*) excavated in old hillside mine shafts near Keswick exactly resemble those used by contemporary miners in the Inn Valley.[121] A Company of Mines Royal was founded to grant and control royal mining rights. The 'privilege' (sole rights) to mine copper ore at Newlands was granted to Höchstetter and the queen's clerk, Thomas Thurland, but the landowner, the Duke of Northumberland, hotly disputed the Crown's right to extract copper from his private land. Crown lawyers, however, defeated the duke's claim, by declaring that the ore also contained silver and gold, to which the Crown did have rights.[122] In October 1566, furnaces and smelting houses were erected, and mining proceeded.[123]

In 1568, Elizabeth I authorised Höchstetter to mine, wash, process and smelt copper ore; and drain or divert water for copper mills in many named counties of England.[124] A foreign mining expert detected copper in Cornwall, and there was a local protest at the higher wages paid to 'Dutch' and 'German' workers. By 1640, however, the Cornish ores at St Just, Perran Sands and St Ives were largely abandoned as unworkable.[125]

During October 1566, when England's first copper smelting houses were erected at Brigham, Keswick (Cumbria), the unfortunate Tyrolean miners suffered 'assaults, murders and outrages' from local inhabitants. Despite all, in September 1567, Queen

---

[118] Collingwood 1912: iv and 2, spelt Hechstetter in English documents
[119] *Calendar of State Papers Domestic* 1547-1580: 276
[120] Camden 1637: 822
[121] Allison *et al* 2010, 61-9: rails and truck displayed in Brixlegg Museum, Tyrol
[122] Hulme 1896: 147
[123] *Calendar of State Papers Domestic* 1547-1580: 279, 287;
[124] British Library: MS Loan 16/1 folios 7, 8, 39
[125] *Calendar of State Papers Domestic* 1582-1583: 134; 1584-1586: 163, 164, 170, 244; Rees 1968: 439-440

Elizabeth received a report that they had finally produced 'fine and perfect copper'.¹²⁶ In 1568, some copper-workers and equipment were sent to Keswick from Schwaz and other Tyrolean villages, and some from Styria (Steiermach, eastern Austria). A further copper smelting house that burned wood and peat was constructed, together with a crushing-works, lime shed, water-wheels with cogs, water-courses and a weir. In 1569, four mine pumps were imported, and a copper refinery was built.¹²⁷ The relatively poor local copper ore must have daunted the Tyrolean smelters, who were

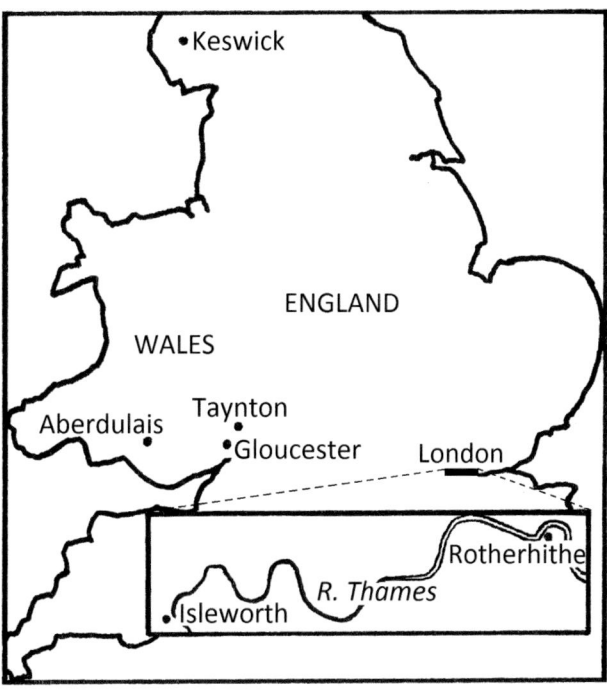

Figure 42. Map of England and Wales, showing some sites mentioned in the text

accustomed to enriched alpine ores. Camden in 1586 mentioned that the Keswick miners' workshops beside the 'forcible' River Derwent had 'ingenious inventions' (possibly trip-hammers) to power their bellows-driven foundries and forges. The unfortunate workers sent requests home for luxuries like ginger, dried fruits, spices and paper.¹²⁸ Meanwhile, in 1584, German copper smelter, Ulrich Frosse, established a works for smelting Cornish ores, at Aberdulais near Neath, South Wales, which ceased production in 1638.¹²⁹ Keswick copper-smelting works continued until the English Civil War (1642-51), when it was sacked by Scottish troops.

The search for English zinc carbonate ore was even more challenging than copper, but in 1566 a local tip-off led the prospectors to a source of almost pure calamine in the Mendip Hills, Somerset.¹³⁰ This discovery would, a century later, enable England to become a leading brass producer. Tintern,¹³¹ on the picturesque Welsh border, was nominally the site of the first English brass wire works, but its brass production proved experimental rather than commercial. When Cristofer Schutz (or Schultz)

---

[126] *Calendar of State Papers Domestic* 1547-1580: 279, 280, 287-380
[127] Collingwood 1912: 26-27, 31-32, 44, 53-54, 6
[128] Camden 1637, 1971 edition: 822, footnote 3
[129] Rees 1968: 432
[130] Day 1973: 17
[131] National Grid Reference SK 526-9 008

from Saxony and William Humfrey of the English Royal Mint, searched for a site to set up a brass-mill near the Mendip calamine source, most were already overcrowded with mills devoted to the thriving local wool trade.[132] Daniel Höchstetter found a suitable site on a fast-flowing hill stream, the Angidy, which flows into the River Wye at Tintern Abbey.[133] In 1565, Elizabeth I chartered a Royal Society of Mineral and Battery Works, which granted the Tintern partners a 'privilege', meaning a permit, patent or sole rights, to exploit zinc or copper ore, make brass and draw brass or iron wire in several named English counties.[134] The 'privilege' was not free but required an annual payment. The queen promoted this society to encourage brass making but in the next century, under weaker monarchs, it would prove more of an obstacle to enterprise than an encouragement.

In November 1566, Humfrey and Schutz, set up water wheels, forges and annealing furnaces at Tintern, but only after a long delay could Humfrey finally manage (in February 1568) to send Queen Elizabeth a first sample of Tintern brass wire made using Mendip calamine.[135] In July, Humfrey asked for more money, because he still had no battery-works, foundry, rollers or casting-stones,[136] indeed, the only outgoings shown in the Tintern accounts for 1565-70 are for discovering, mining and roasting calamine and making crucibles for trial brass-making.[137] Over 20 tons of Mendip zinc carbonate ore were delivered to Tintern for the use of a foreign craftsman named Hinkins, who imported crucibles, built furnaces and produced an alloy, but it was not malleable enough for wire-making. Brass for fine wire (as opposed to casting-brass) needs very pure copper. Hinkins blamed the Mendip calamine for his failure, but the Keswick copper ores were probably the problem. In subsequent years, other foreigners tried brass wire-making at Tintern, but 'gave it up as not feasible'.[138] On 7 July, 1569, 'charges for the Calamyne' are even mentioned as a burdensome debt and the Tintern works turned to making iron wire instead.[139] In March 1573 Sir Richard Martin, London goldsmith and Assay-Master at the Mint, leased Tintern wireworks, and employed a foreign expert, Barnes Keysar, to change all the machines to making iron wire.[140]

In December 1580, Martin and his partner Hanbery separated the Society of Mineral and Battery Works' Tintern 'privilege' into two parts, one for making iron wire at Tintern, and the other for rights to mine calamine and make and forge brass and

---

[132] *Calendar of State Papers Domestic* 1547-1580: 278
[133] National Library of Wales: MS, Badminton Manorial 2/2445 folios 24-25
[134] British Library: MS, Loan 16/1, folio12 (*c*.1568-1603)
[135] *Calendar of State Papers Domestic* 1547-1580: 281
[136] *Calendar of State Papers Domestic* 1547-1580: 311
[137] British Library: MS Lansdowne 76/34 (1577)
[138] British Library: MS Lansdowne 81/1, column 1 (1596)
[139] British Library: MS Loan 16/1 folio 92 (*c*.1568-1603)
[140] British Library: MS Lansdowne 76/34 (1577)

brass wire. This latter privilege was passed 'all in a canvas bagge' to Martin and to Humfrey Michell, both members of the Society,[141] who in 1582 invested the brass-making privilege in a brass mill set up that year by chemist, John Brode, at Isleworth on the River Thames south-west of London. Brode undertook to pay the Society £50 a year for the 'privilege'. Tintern brass-making equipment was sent to Isleworth where, in 1576, Michell and others had improved the water-flow to Queen's Mill by diverting various upstream watercourses.[142] Brode later rode over to Tintern Abbey, hoping to scavenge some of the reported thirty tons of unused zinc carbonate ore, but he found it had all been thrown into the River Wye to repair a dilapidated fish weir.[143] In 1593, a passing traveller noticed the Isleworth 'copper and brasse mill', where ore was 'melted and forged' using many 'artificial devices', and reported that its calamine came mainly from Worley Hill in the Mendip hills.[144] Although Brode's 'privilege' allowed him to mine calamine from the Mendips, it prohibited him from using the same type of mining equipment, engines or tools as those introduced to Britain by Humfrey and Schutz. In 1594 his 'water-driven battery works' forged and drew brass wire for making pins and 'other necessaries'.[145]

Why were pins made from brass wire in demand at that time? Wool cards (described in chapter two) consumed the most, but fashion now dictated that men and women at higher levels of society wore elaborate lace ruffs and cuffs. A glance at any portrait of Elizabeth I indicates that lace-makers, who employed vast quantities of pins for their craft, were major consumers. Brass wire was preferable to iron wire, both for lace-making and wool-carding, because it was rigid and did not rust. Iron corroded and left rust-marks, so brass pins were also the only available clean way to hold papers together.

John Brode's brass works at Isleworth, started up in 1582 close to the River Thames, initially employed skilled foreign workers who brought in 'their fellow-countrymen'.[146] He later pointed out that he was not alone in employing foreign workers, instancing the Tyrolean copper-miners at Keswick, and the London wine-glass makers. By 1596 he claimed that his workers were 'Englishe borne' and that he intended to 'keep the arte in this his native country'. After over eight years' practice, he claimed to have devoted £3400, and fourteen years study and effort, to perfecting brass-making at Isleworth.

---

[141] British Library: MS Loan 16/1, folios 101, 122 (c.1568-1603)
[142] The National Archive: MS E 178/1399
[143] British Library: MS Lansdowne 81/1, column 3 (1596)
[144] Norden 1593, 1723 edition: 41
[145] British Library: MS Add 12497 (1613), folio 438
[146] British Library: MS Lansdowne 81/3, folio 2 and folio 6, note 3

In 1598, Brode opened new calamine mines at Wrington on the Mendip Hills,[147] and at least some of his copper seems to have come by sea from 'Barbary' or 'Maroc' (Morocco),[148] shipped up the Thames to a wharf at Isleworth, the furthest place upstream to which sea-going vessels then plied.[149] At that period, vessels arrived at Isleworth from Germany, the Netherlands and Belgium.[150] Moroccan copper ores were still imported into Britain in the next century.[151] Brode managed to make malleable brass from the ore,[152] probably later using the newly discovered copper from Devon or Cornwall. Brode claimed to produce one and a quarter tons of brass a week (about 65 tons annually), charging his crucibles with 18 hundredweight of copper and 36 hundredweight of calamine (evidently only seven hundredweight reacted) to produce 25 hundredweight of brass, possibly fit for wire drawing as well as casting.[153] As elsewhere in late sixteenth-century mainland Europe, brass craftsmen used brass for casting cauldrons, basins, taps, warming-pans, smoothing-irons, cooking pots and nozzles for fire-fighting equipment.[154] The surfaces of brass cooking vessels were sometimes given a thin coating of tin, to prevent the copper element in the alloy from reacting, over time, with alkaline foodstuffs, or the zinc with salt.

John Brode, a determined entrepreneur living in a spacious house,[155] met his downfall in 1596 when, faced with official constraints and financial demands, he took to fighting law-suits. The Society of Mineral and Battery Works forced him to close his fully operational Isleworth brass works because he owed them 'privilege' money – a debt which Brode blamed on his partners' failure to invest capital.[156] The Society ordered Brode to stop employing his experienced workers. His influential sleeping partners (both members of the Society) also turned against him, and contrived to be granted a 21-year lease of the Isleworth brass-works at £100 a year. Brode refused them access to his copper stocks, or to fresh Mendip calamine. He withheld the secrets of his process, and refused to let them employ his skilled brass-workers. The partners, rendered helpless because they were ignorant of the technology and the trade, turned to the powerful Society of Mineral and Battery Works. On March 21, 1596, the Society threatened Brode with imprisonment, ordering him to release all the 'callomyne, copper, mater, stuffe, artificers, workemen, instruments and necessaries' for making brass. They also ordered his suppliers to sell and deliver to the partners.[157]

---

[147] The National Archive: MS, E 134/41 Eliz/Trin 8
[148] British Library: MS Lansdowne 81/1 and 84/7
[149] *Calendar of State Papers Domestic* 1547-1580: 491
[150] Isleworth Community Council 1991: booklet, *Isleworth: a guide and some of its history*, 9
[151] Riksarkivet Stockholm: MS Bergskollegium, E III, 4-5 (1691-1703), Odelstierna, folio 4
[152] British Library: MS Lansdowne 81/3 and 81/6 (1596)
[153] British Library: MS Lansdowne 81/3 and 81/7
[154] Haedecke 1970: 105
[155] Glover, Moses, 1635 map, Stanford facsimile 1880: Survey of Istelworth Hundred,
[156] British Library: MS Lansdowne 81/2 and 81/4
[157] *Acts of the Privy Council* 1595-6: 322

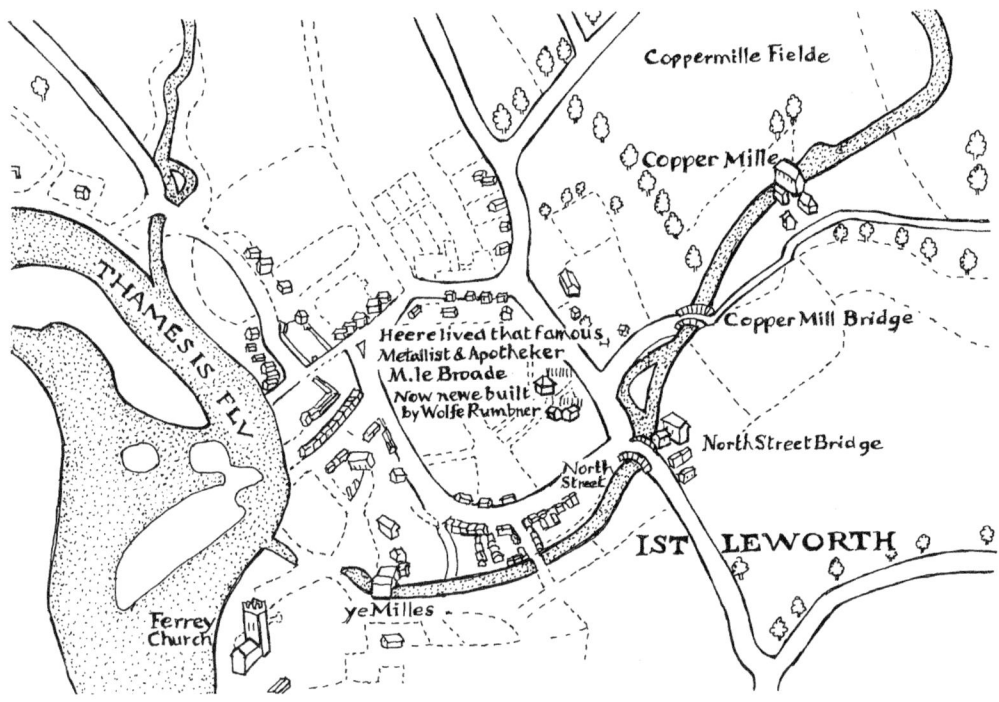

Figure 43. Isleworth mills and Brode's (Monsieur le Broade's) house, 1635

In June 1597 Brode was again warned to pay up or make no brass.[158] The partners accused him of contributing little to the brass-making process; but he countered that he had personally refined 43 tons of coarse Barbary copper into malleable, workable metal.[159] Brode had almost won his legal case against the Society, when, in June 1598, the Privy Council, ignoring all else, overruled every issue, insisting that Brode had no legal right to calamine, tools or Tintern machines, and that he would be imprisoned.[160] This was the end of a pioneer industrial achievement. By 1607 the mill had become a paper mill.[161] On a 1635 map, John Brode's house is captioned, 'Here lived that Famous Metallist & Apotheker M. le Broade' – a fitting epitaph.[162]

In 1598, shortly after Brode was forced from Isleworth, Dutch merchant Abraham van Herwick and partners gained a 'privilege' to start a brass-mill at Rotherhithe,[163] a Thames peninsula then about six kilometres east of London.[164] Former Isleworth workers who transferred to Rotherhithe knew the calamine source at Wrington in the

[158] *Acts of the Privy Council* 1597-8: 235
[159] British Library, MS, Lansdowne 81/ 3 and 81/6 (1596)
[160] British Library, MS, Add 12497, folio 438 (1613)
[161] Reynolds, Victoria County History, Middlesex, volume 3: 113
[162] Glover, Moses, 1635: map, Survey of Istelworth Hundred, Stanford facsimile, 1880
[163] British Library: MS Lansdowne 81/1 (1596)
[164] National Grid reference TQ 3561 8010

Mendip Hills,[165] which now began to supply Rotherhithe Mill. Copper was delivered there by sea from Devon and Cornwall.[166] The brass mill lay beside a small Thames tributary, a channel on the west side of the peninsula, into which converged streams and canals that drained Rotherhithe marshes into the Thames. Later maps still show a mill or mills at this site,[167] on the north bank of the channel, just west of where the small Surrey Water pool now stands.[168] Abraham van Herwick tried to persuade local farmers to let him deepen and enlarge the watercourses to improve water flow to his mill, but by 1598 he was involved in a legal dispute. In court, van Herwick promised that he would not inconvenience local farmers, that he needed the water and would provide employment, but the farmers dug their heels in, refusing to yield. The Commissioners of Sewers heard the case, but passed the problem on to the Society of Mineral and Battery Works, who told Van Herwick and the farmers to sort it out between themselves.[169] It is not certain how long the mill lasted, but the next English brass mill would start at Taynton, west of Gloucester, in the Forest of Dean.

Meanwhile, back across the North Sea, several late sixteenth-century brass mills operated in the region that stretches northwards from Hamburg to Schleswig and across to Lübeck. South of Lübeck, near Ratzeburg, in an extensive lake district surrounded by woodland, lay the Mannhagen brass works. It was one of three started up by Duke Christopher of Mecklenburg (1537-92). On returning from religious wars in 1570, he turned to the peaceful pursuit of mineralogy and chemistry, built himself a house near Mölln (then part of the state of Mecklenburg) and set up a laboratory to examine local ores. In 1573, he employed an expert mineral prospector from Schneeberg near Zwickau, to search for gold and silver, encouraging him to dig timber-supported shafts into the Steinberg, a prehistoric fortified hill below Mannhagen. On finding no gold or silver, the duke established a brass mill at Mannhagen, on the site of former iron- and copper-works.[170] The fast-running Steiner Brook fed the Mannhagen mill-ponds (today in dense woodland) and then flowed down past the Steinberg into the River Stekenitz (now canalised), a tributary of the River Trave.

Another mill among the idyllic lakes south of Lübeck was at Bäk near Ratzeburg. In January 1590, a Lübeck citizen bought a mill on a stream called the Bäk (small river), which flowed fast down steep wooded slopes from Lake Mechow into Lake Ratzeburg below. By 1591, a brass works was established on the Bäk, with battery and crushing mills, managed by brass smith Rottgard Münter.[171] In 1592, Duke Christopher of Mecklenburg added a brass battery works to his existing copper works at Neustadt-in-

---

[165] The National Archive: MS, E 134/41Eliz/Trin8, item 1
[166] Acts of the Privy Council 1597-8: 491
[167] Stow, John, 1720, 1756 edition: Survey of London parishes, Redriffe
[168] Roque 1741: Carte géographique des Villes de Londres; 1747: Citie of London and Westminster
[169] *Acts of the Privy Council*, 1597-1598: 491
[170] Lisch 1842: 61-62 and footnote
[171] Kock 1981: 78-79

Figure 44. Ratzeburg and its lake in 1586, looking north (Bäk is at the top left corner)

Mecklenburg, south of Schwerin. Run by a tenant, it had a furnace-house, foundry and cauldron workshop.[172]

In 1546, a Lübeck brass foundry was set up on a spit of land between the river Trave and the moat encircling the Lübeck city walls; and by 1550 a brass-founders' guild existed.[173] Matthius Benningk, official city brass-founder from 1561, created a memorial brass for Lübeck cathedral, commemorating Prince-Bishop Johannes Tiedemann, and measuring about three metres by two. In 1575, Lübeck burghers set up their own copper furnaces and refineries, but the city was forced to import copper from Sweden and Upper Hungary (now Slovakia).[174]

From 1586, Frau Barbara Rantzau, an early female brass-mill proprietor, operated a copper hammer on the Höltenklacken estate in Holstein, north of Lübeck. In 1591 she added a brass mill and two further battery hammers, employing furnace men and

---

[172] Lisch 1842: 61
[173] Gunther Meyer, personal communication
[174] Jürgen 1914: 30

apprentices, brass scrapers, wire-drawers, cutters and a copper refiner and beater. She recruited her works manager, Isaac Soldener, from Stolberg, and her battery-workers operated trip-hammers for shaping cauldrons and smaller vessels. Her son Heinrich Rantzau had meanwhile started a brass mill at Nütschau on a tributary of the river Trave, four kilometres north-west of Oldesloe in Holstein.[175]

Copper for both Rantzau brass-mills was imported from Sweden by Lübeck merchant Marcus Mewes (or Moos). In July 1593, Barbara Rantzau signed an agreement for an annual supply of three and a half tons of calamine from Brilon (High Taunus hills, central Germany), to be imported through the port of Hamburg. The brass-works produced sheet brass and brass wire to be exported mainly through Hamburg.[176] In 1595, 111 rings (tied-up coils) of brass wire, plus two tons of brass metal, were sent to various merchants but, on at least one occasion, shortage of water to power the mill caused Barbara Rantzau to fail to deliver an order. Copper works and refineries for processing copper ores from Sweden and Upper Hungary, were numerous in the Holstein area at that time, thanks to vigorous copper trading through Hamburg and Lübeck.[177]

In 1590, Hermann Oldenhoff built several brass mills at Trems about four kilometres north of central Lübeck, on a mill-stream leading directly off the River Trave and flowing into the Trems Lake. The works included a battery-hammer, a furnace house, annealing workshop, grinding mill and polishing workshop. In dry summers, the water ran very short, but in rainy seasons an extra water-wheel operated a slitting-mill to cut brass sheet into thin strips ready for wire-drawing.[178] A French mines official wrote 150 years later that he was not sure if the brass-works in the Lübeck and Hamburg areas were still in operation, but he knew that they used copper and scrap brass to make cementation brass. And that they imported calamine from Aachen through Bremen, and from Poland through Lübeck, and imported granite casting-moulds through Bremen.[179]

The Swedish brass mills, like those on the Holstein peninsula, emerged only late in the 16th century. Sweden's problem was that, despite possessing vast copper and timber resources, no viable calamine was available for brass-making. Copper production at the Swedish Falun copper mines increased sharply during the 15th to 16th centuries, and in the later 16th century a new level of richer ores was discovered.[180] However, for lack of calamine, Swedish craftsmen worked with brass imported from Germany or the Netherlands. King Gustav I Vasa (reigned 1523-1560) faced an extra dilemma

---

[175] Jürgen 1914: 32-33
[176] Jürgen 1914: 33
[177] Kreis Lauenberg 2013: www.muhlen.lauenburg.de/18/html, 1
[178] Warncke 1938: 215
[179] Galon 1764: 55
[180] Forss, T. 2006: Falu Mine. www.geonord.org/shows/falueng.html, 2

because he had banished all Hanseatic merchants from Stockholm in 1535, following a trade war with the Hanseatic port of Lübeck. This left him unable to import zinc ore from the rich Altenberg calamine mine near Moresnet. The Altenberg mine was leased from the Spanish Habsburg king by the Schutz merchant family of Antwerp. Their biggest calamine customers were the Aachen brass-works owners, who dreaded brass-making competition from copper-rich Sweden. Gustav I's better relations with the Fugger copper magnates (who took his side against the Hanseatic League) were of no help because the Fugger firm needed calamine for its own brass mills. The only option left was low-grade Polish calamine, no use for brass-making without the addition of better ore.

In 1524, a Swedish bishop, Hans Brask, felt that Swedish workers could learn to make brass, so he sent an envoy to Germany to find out more about brass making.[181] In 1550, however, German merchants put renewed pressure upon King Gustav I to import small ready-made Nuremberg brass wares. In the 1550s it was even decreed that all Stockholm citizens should possess a brass fire extinguisher,[182] a plan that benefited Nuremberg brass craftsmen but put Sweden in a dependent position. However, a Bohemian mines-delegate, Caspar Richter, from Schlackenwald (now Horni Slavkov, Czech Republic), wrote to Gustav I in 1559 to inform him that Sweden's brass-making potential was being discussed in southern European metal-working circles.[183]

King Gustav's plans to make Swedish brass met with animosity from both Nuremberg and Aachen. In 1550, Aachen's city council even banned imports of Swedish copper and refused to stamp it with the eagle that showed that its quality was adequate for brass making.[184] Antagonism from competitors forced King Gustav of Sweden to enlist the help of Caspar Richter, the Bohemian brass expert, to assess Sweden's capacity to manufacture brass using its own calamine.[185] Richter, who was deeply involved in trading metal to Nuremberg, dared not deal overtly with the Swedish government for fear of alienating Nuremberg, and on entering Swedish service he did indeed become banished from Nuremberg.[186]

In 1559, Caspar Richter wrote to King Gustav, advising him to choose a brass mill site with good water flow, plentiful timber for charcoal, and a budget for copper, iron (for tools), transport for bulky calamine and 100 barrels of charcoal for alloying and annealing. Richter himself calculated costs for equipment, raw materials and a staff of 12 skilled brass battery-workers with five apprentices. He proposed four battery-hammers, and workshops for drawing wire, etching and polishing; and suggested

---

[181] Ekström 1985: 178-179
[182] Forsgren 2010: 9
[183] Ekström 1985: 179
[184] Forsgren 2010: 9
[185] Jaroslav Zapletal, personal communication
[186] Malmsten 1939: 152

that Swedish brass should be marketed in Nuremberg, Liège (Lüttich) and Lübeck.[187] However, King Gustav and his advisers procrastinated, so Caspar Richter returned to Bohemia to await the outcome. In 1560, hearing of King Gustav's death, he gave up and left Bohemia to work for an Augsburg trader, at his Tyrolean brass-works north-east of Lake Garda.[188] Whilst there, Richter received a royal letter recalling him to Sweden, so he gathered together a team of skilled brass-workers, arriving with them in Sweden in 1561-2. No brass-works, however, would start there for another ten years, so Richter went instead to an iron works near Arboga, central Sweden.[189]

The trigger that finally started the Swedish brass industry was the Netherlands Freedom War (1555-72), in which Dutch Protestants under William of Orange tried to break free from Spanish Catholics. Both sides touted for support from Sweden's two wealthy crown princes, which led to the surprising origin of Sweden's first brass mill. Crown Prince Johan of Sweden promised William of Orange a major battle ship, the *Troilus*, armed with 20 cannons, 728 cannonballs and plenty of gunpowder – in return for everything needed to establish a Swedish brass mill. Envoys were sent to Aachen, the Netherlands and the Harz for the promised trained brass craftsmen, anvils, battery hammers with alternate strike, tongs for wire-drawing, casting stones, crucible clay and a large consignment of the very good zinc ore ($ZnCO_3$) that Sweden lacked. Through this unusual warship bargain, Sweden gained valuable links with the brass-trading centres at Aachen, Antwerp and Amsterdam. An emissary, dispatched to hunt for suitable mill sites, chose Vattholma, on Crown lands north of Uppsala, where Sweden's first brass mill was set up in 1571.[190] The start of the Swedish brass industry, at Vattholma, was an event that would later have significant impact on world markets. The highway from Uppsala to Dannemora, which crossed the Fyrisån, the small swift river at Vattholma, provided reasonable communication,[191] and the brass-works stood by a weir to the west side of a small island (*holm*) that gave Vattholma its name.[192]

In 1571, the task of managing Vattholma brass works and procuring the promised skilled Aachen workers fell to two Netherlanders, Caspar Johansson and Virgil (Gilius) Pacquet. The factory was built at record speed, with four battery hammers, a furnace house, wire-drawing workshop and charcoal house. It was provided with several barrels-full of crucibles, ten wire-drawing machines, two pairs of metal shears and almost seven tons of unrefined copper from the Swedish Falun mines.[193] Pacquet had promised to acquire Swedish calamine from zinc-containing rock further north in Dannemora. His promise, however, failed because the deposits proved too low in

---

[187] Malmsten 1939: 144-146, transcription of Kunglige Arkiv, Stockholm, Bergsbruk 124, 1559
[188] Near the Tyrol border with Walska, then an independent church state
[189] Malmsten 1939: 152, transcription of Riksarkivet MS 2 I-O (*c*.1561-1562)
[190] Forsgren 2010: 10-11
[191] Forsgren 2010: 7
[192] Malmsten 1939: 162, sketch map of Sahlsta district;
[193] Malmsten 1939: 145-146, translation of Kungliga Arkiv, Stockholm, Bergsbruk No 124

zinc to be profitable (rock is classed as ore only when the mineral within it proves profitable to mine, transport and extract).[194] Pacquet refused to accept this setback, and dragged his feet for months before importing foreign calamine.[195]

By August 1572, the trip-hammers of the Vattholma battery-works were in operation, beating out sheet-brass and shaping cauldrons and smaller hollow objects like jugs, buckets and bowls.[196] In 1569, Virgil Pacquet had started to build a copper refinery at Uppsala, for re-heating the copper several times to remove impurities.[197] With refined copper, Vattholma produced much better brass. After it had been alloyed, liquid brass was poured from crucibles into the gap between two large stone casting-slabs, to set as sheet brass, ready to cut into strips for drawing to wire.[198] By using Swedish copper to make brass, King Gustav I's heir, Karl IX (1550-1611), at last reduced Sweden's brass imports and increased its export income (for a map of Swedish sites, see the next chapter). Vattholma brass-works produced innovative items of cast and hammered work, including moveable braziers and fire-pans for the Swedish court, and warming pans and wine-coolers for the royal household. Early Swedish braziers and foot-warmers included an iron inner bowl filled with hot stones or sand, inserted through a door in the side of a deep round brass container, punched with holes, to let out heat.[199]

Vattholma deliveries to the royal Swedish island castle of Stegeborg included small artillery guns (*falkoner*), two cannons, two military drums for Duke (later king) Sigismund; cooling vats for the royal cider (with a ring and two lion's heads on either side) and a cooling vat for 'Mistress Elizabeth'. Although these are not analysed artefacts, their high-status use suggests that they were of true brass. Ingot brass was despatched to the royal candlestick-maker to create fashionable chandeliers for the king's hall.[200] The king had first call on the mill's brass products but the mill-managers were allowed to market any surplus items.[201] A tax survey showed that 66,000 Swedish households possessed about a thousand brass objects between them, most of them in the 300 houses of the wealthy. A later map of Sahlstra district, dated 1683, marks the brass mill by the lower dam, west of the Fyrisån (the river), with the 'Crown Mill' indicated on the middle pond.[202] In 1574, foreseeing future charcoal shortages for the brass works, the king ordered local landowners to plant oak trees.[203]

---

[194] Ekstrom 1985: 181
[195] Forsgren 2010: 10
[196] Malmsten 1939: 174-175
[197] Ekström 1985: 181
[198] Forsgren 2010: 8
[199] Forsgren 2010: 166
[200] Ekström 1985: 183
[201] Forsgren 2010: 10
[202] Malmsten 1939: 162, 175
[203] Lundgren, Helga, April 1991: Skultuna bruk och skogen. *Hembygds journalen Västmanlands föreningar* www.hembygd.sew/index.asp?lev=15124

Vattholma mill flourished under its technical manager, Lambrecht Bark, an Aachen master brass-maker and dealer, but not for long. By the late 16th century, the mentally unstable Swedish King Erik XIV, already deeply in debt to the Hanseatic League, removed Virgil Pacquet for failing in his promise to find and use only Swedish calamine.[204] Pacquet's partner, Caspar Johansson, appears to have continued independently, but he gave up in 1580, the year when the state temporarily withdrew free access to Swedish copper from privately funded enterprises such as Vattholma mill,[205] which therefore ceased to operate.[206]

A new brass-mill was planned in 1562 near Höjen ('mound', in fact a local tumulus) just west of Arboga and on a navigable river connecting the town to the port of Stockholm 125 kilometres further east. The site had strong water-power, and extensive nearby forest timber for fuel and charcoal (which the king shared out between competing factories). Valuable casting stones and 140 crucibles from Vattholma were shipped to Höjen from Stockholm.[207] Bulky calamine was delivered by water and, in winter when hard ground made transport easier, copper was brought by sledge over frozen lakes and rivers from the Falun mines to Västerås, and from there by boat to Arboga. The mill lay where an electricity generating works exists today, on a stretch with a powerful water flow, hard to harness in times of spate. In 1588 the mill-manager, Thomas Schademandel, probably from the Harz, was favoured by King Johan III with free shelter, food, beer, workshop equipment, labourers, coal, wood and three dollars spending-money a week.[208] He was joined by Lambrecht Bark, the former master brass-maker from Vattholma. In 1589, the king ordered 170 kilos of Falun copper to be brought to Arboga.[209] By that time, the mill had four furnaces, a calamine calcining house, crushing mill, charcoal house and crucible pottery. Johan III, already coveting the brass that the factory might produce, took impatient personal interest in its preparation.[210]

Although Höjen brass-works mainly produced brass wire, it was probably the only brass works ever to pave a royal palace. In 1589, Johan III ordered the first 384 brass flooring-sheets and brass nails for his Stockholm palace,[211] the accounts for which still exist in the Royal Library. 2040 of the 3060 metal floor-plates were of brass, alternating in a grid pattern with copper plates. In total, the brass tiles weighed about a ton and three-quarters, and brass nails were used to attach them to an underlying wooden floor, so that yellow brass and reddish copper gave the floor a tiled pattern

---

[204] Forsgren 2010: 10
[205] Gadd 2007: 4
[206] Riksarkivet, Stockholm: MS, Bergsbruk, 111. Uppl. Handl. 1587, number 28
[207] Riksarkivet, Stockholm: MS, Bergsbruk, 124, 7 November 1589
[208] Corin 1978: 166
[209] Anders Jägbring, personal communication
[210] Corin 1978: 166-167
[211] Forsgren, 2010: 11

Figure 45. Stockholm Castle interior, 1616, with brass and copper floor tiles. The king is receiving the Dutch ambassador

in contrasting colours. An idealised image of the Royal Palace, painted in 1616 before the palace burnt down, shows an interior scene with the two-tone tiled floor.[212] Two years after Sigismund III had ascended the Swedish throne in 1592 he retreated with his navy to his mother's native Poland, leaving a letter announcing his sale of the royal Arboga brass-works, with its staff, zinc ore, copper and casting-stones to a man last recorded as working in a Stockholm beer-cellar. In 1599, Höjen brass mill closed.[213] Sweden supplied its valuable copper to northern Europe, but kept the secrets of its brass production shrouded in mystery lest they slipped out to the Germanic free traders who dealt in Swedish copper but protected their own German interests. Competition from Sweden had caused the income of other northern European brass works to decline,[214] and reduced Hanseatic League trading through Lübeck.

---

[212] Kungliga Biblioteket, Stockholm: Stockholm Royal Castle interior, copper engraving, (1616); the author's thanks to Lennart Stenfeld for drawing my attention to this work
[213] Riksarkivet, Stockholm: MS Bergsbruk 124, (7 November 1589), 'om mässingsgjutningen wedh Högen'
[214] Forsgren 2010: 8

The demand for brass was meanwhile mounting in mainland Europe, not least for scientific instrument-making. In 1583, the Habsburg Holy Roman Emperor, Rudolf II, a patron of art and science, moved his court back to Prague in Bohemia, after a temporary spell in Vienna. He headed a trend for collecting and using scientific instruments and other brass curiosities, which would soon spread to the aristocracy of Europe. From 1603 his collections were displayed in 37 cabinets in an early purpose-built private museum (*Kunstkammer*). Rudolf II, whose zeal for collecting rarities seems to have exceeded his leadership qualities, would eventually let his empire slide into the Thirty Years War (1618-48), which would profoundly influence European brass-making.

World population growth led to demand for more commodities, which in turn encouraged entrepreneurs to make more brass. Many brass works simply produced only ingots or sheet-metal brass, using water-powered mills to crush calamine and to work bellows to heat furnaces. Some, however, also produced semi-prepared cauldrons and basins, wire or finished articles. The range of goods now increased with the demands of expanding markets, and consumer demand was stimulating improvements in technology.

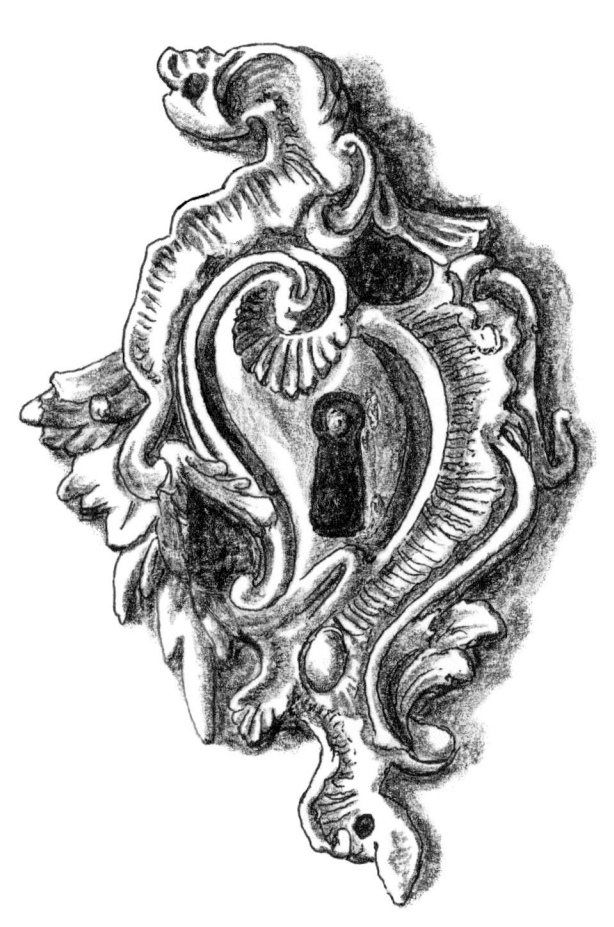

Pestle and mortar cast in Bristol, 18th century. Inv. 3243, photo © author, courtesy of Blaise Castle Museum, Bristol.

Chapter 5

Merchants and migrants
c.1600-1650

During the 17th century the merchant classes rose to wealth, power and importance through expanding long-distance maritime trade. This led them to increasing aspirations, motivating the desire for possessions to demonstrate their new-found status in society. War affected Europe, West Africa and the Far East, but conflict provokes migration, and skilled experienced migrant workers were the lifeblood of emerging brass industries. At this period, while the Manchu were making aggressive forays from the north-east into Ming China, much of India enjoyed a stable period of creative craftsmanship under strong emperors, and was probably the greatest contemporary producer of beautiful and useful brass objects, some of them influenced or made by Persian craftsmen.

The Mughal Empire was still forcibly adding to its Indian territories and attempting to exert control over the societies and finances of its ever-expanding population. As the markets, roads and major cities of this wealthy merchant society grew, the role of taxes, money and trade increased in social importance. Indian craftsmen produced decorative brass incense-burners, oil lamps, architectural finials, ewers, basins and pilgrim flasks. Ornate inlays of brass and silver were applied to bowls and other objects. The warm glow of brass was seen on Hindu, Jain and Buddhist shrines around India,[1] including a portable Rajasthani brass shrine to the Jain saviour, Parshavanatha, dated 1633.[2] A fairly high-zinc brass image of a Chauri-bearer, or female spirit holding a fly-whisk, survives from seventeenth-century Gujarat. The Dutch (and from 1617 the British) East India Companies were permitted to trade Indian goods to Europe, so refurbishment expenses for the British royal apartments in 1705 included 'new gilding [of] the brass work of a large Indian chest'[3].

Persia to India. Between 1570 and 1650, many elaborately engraved Islamic brass astrolabes, composed of fine metal discs placed one on top of the other and signed with the makers' names,[4] were cast and engraved at Isfahan and Lahore. 40 analysed brass astrolabes include six examples containing 40 per cent zinc, made in the Lahore region from 1603 to 1662, clearly using metallic zinc directly mixed with molten copper.[5] Brass containing over 35 per cent zinc was very hard, which made it easier

---
[1] Zebrowski 1997: 95-98
[2] Ashmolean Museum, University of Oxford: EA OS 109
[3] *Calendar of Treasury Books* 20/2, 1705: 463
[4] Museum of the History of Science, Oxford: 47376, 42730
[5] Newbury *et al* 2005: 359 (analysis of remaining elements not available)

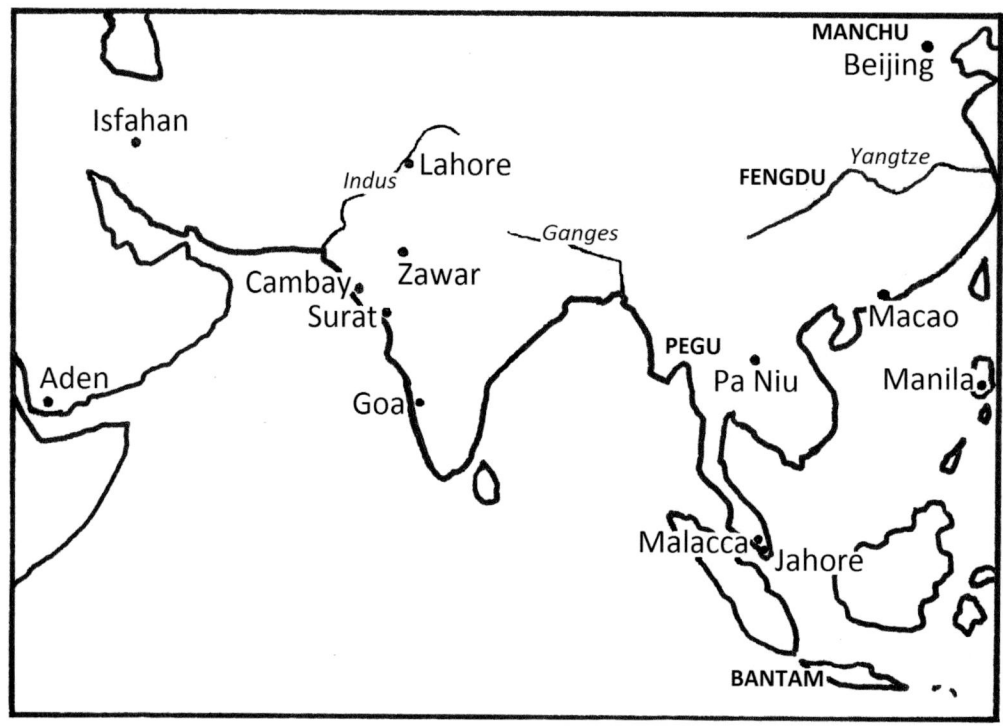

Figure 46. Asia, map showing some sites mentioned in the text

to incise neatly with a sharp tool. The Lahore astrolabe makers used lower-zinc brass for cast parts but harder, high-zinc brass for parts to be engraved. This allowed astronomical measurement details to be finely and precisely engraved onto the plates and rete, which produced a more accurate astrolabe. The high-zinc astrolabes from Lahore and Isfahan suggest that the Zawar zinc-distillation plant supplied metallic zinc to Persia (Iran) and northern India (including today's Pakistan).[6]

Metallic zinc was also used in producing black *bidri* cast wares, composed of a little copper, 76-98 per cent zinc, and varying amounts of lead. The original mix was burnished to a deep black with local soil from the Indian fort of Bidri, mixed with ammonium chloride and water. A warm solution of one part potassium nitrate and four parts ammonium chloride causes a black patina to appear immediately on the surface of the alloy. It is thought that the ammonium chloride dissolves the zinc from the bidri surface, and that the now copper-enriched surface is oxidised black by the potassium nitrate.[7] Pale, whitish-coloured brass (about 35-40% zinc), was another popular Indian alloy, but brass containing over 36% zinc needs to be hot-worked (hammered while hot). Brass and silver were both used for fine decorative inlays on Indian and Persian copper or *bidri* articles.

---

[6] Newbury *et al* 2003: 360, 366-367
[7] La Niece and Martin 1987: 97

In the time of Shah Abbas I (1571-1629) of Persia, the *huqqa* (water pipe for smoking tobacco) became popular, though the Shah himself apparently favoured other crafts than brass-work. Most early Indian *huqqas*, with globular bases and short flaring necks, were of black *bidri*, richly inlaid with brass and silver designs. The wealthy *huqqa*-smoking clientele used brass or *bidri* caskets or boxes, sometimes for spices. Specific to India and southern Asia, the habit of chewing betel, or *pan*, was (and is) another pleasure enjoyed after meals for its alkaline digestive properties. Among the wealthy, this habit was accompanied by brass or *bidri* betel-containers called *pandans*, and by brass spittoons to receive the saliva the betel engendered. The Indian merchant classes who, like the elite, pursued the pleasures of a fleeting world, offered betel to their guests. The *pandans* followed the shape and style of boxes that had been made in Iran from the 13th century onwards.[8] In south India, analysis suggests that copper alloy objects contained very little zinc,[9] but in West Tibet, zinc contents rose in the 16th to 17th centuries, while lead remained more or less constant. By the 18th century, zinc would be low or absent in Tibetan copper alloys, meaning little brass[10].

China and South-East Asia. Further east, China entered troubled times in the first half of the 17th century with the gradual approach of Manchu conquerors from the north, however China's trade in metallic zinc was about to develop fast. Wang Chai's *San Thsai Thu Hu* of 1609 declares that zinc from the south-eastern province of Guangdong was the best (for distillation), with inferior ores coming from Pa Niu, southern Shan mountains. He also mentions that zinc metal (*aen*), was added when making brass vessels. The zinc metal was cast into plates over 30 cm long, 14 cm wide and less than 2.5 cm thick. Medicinal ointments in petal-like zinc oxide flakes (*yakuken*), were a secondary product.[11] Zinc ores were exploited early in the mountainous ruby-bearing Shan states straddling north-east Burma (Myanmar) and western China,[12] and zinc-working residues have been found in Chongqing province, central China.[13] In the period 1576-1621, the distilled zinc contents in Chinese brass coinage rose to 37 per cent, but in general the values settled down to around 30-35 per cent.[14] Lower-zinc brass however, was used in constructing the Xiantong Temple Hall in the Wutai mountains (1572-1620).[15]

China, which had started distilling zinc metal under the Ming dynasty (1368-1644) was now probably the most prosperous place on earth, trading in luxurious porcelain, silk and tea. Zinc metal was used as ship's ballast. A Europe-bound ship, wrecked in

---

[8] Zebrowski 1997: 225, 263-264, 286,
[9] Riederer 2003: 160-161
[10] Riederer 2002: 155
[11] Needham, 1974, volume 5, part 2: 212
[12] Pascoe 1950, volume 1: 336, a pical, one porter's load, then weighed roughly 60 kilos
[13] Liu Haiwang *et al* 2007: 177
[14] Dai Zhiqiang and Zhou Weirong 1992: 53
[15] Zhou Wenli 2012: 48

1585, carried ballast of 1000 Chinese ingot slabs of 98 per-cent-pure metallic zinc, weighing an estimated six tons,[16] and in 1605, a Dutch sea-captain seized a Portuguese carraque off Malacca, loaded with Chinese porcelain and carrying a ballast of 265 tons of metallic zinc.[17] By 1618, the end of the Wanli period (1573-1620), the Portuguese regularly used Chinese zinc metal, instead of sugar, as ballast when trading from Macao (Macau) to Japan, India, Manila, Malacca, Siam, Chongqing, Cochinchina and Cambodia.[18] This created new markets for metallic zinc around India and South-East Asia. It was partly because metallic zinc was used as heavy ballast to counter-balance tall-masted European sailing ships leaving China, that, for Indian brass makers, Chinese metallic zinc could supplant Zawar-distilled zinc.

Zinc metal, then relatively cheap, remained the major ballast item in Chinese maritime trade throughout the 17th century. After 1621, Chinese coin-minters were adding metallic zinc to molten cementation brass, giving a higher but stable zinc content.[19] Various terms were used for zinc metal, a relatively new and unfamiliar import to Europe. The Dutch called it *spelter* or *spiauter*, which also meant 'tin' or 'pewter'. The Portuguese called it *tutenaga*, possibly originating from the Persian word *tūtīyā* (smoke or vapour). Both Indian and Arab merchants, whose forebears had traded to China since the 10th century,[20] used the terms *calaem*, *calaim* or *kalin*, derived from an Arabic word *kalai*, meaning tin.[21] A Chinese text written in 1609, mentions that *tutenag* was a term introduced to them by foreign (presumably Arab or Portuguese) merchants.[22] *Tutia* or *lapis tutia* from the Middle East, made from sublimed zinc oxide, was mainly intended for medicinal purposes, and was listed in cargoes alongside such Arabian items as myrrh and coffee.[23]

Europeans were confused about the source of zinc metal. In 1597, German physician, Andreas Libavius (*c*.1545-1616), received some *calaem* or 'east Indian zinc' via the Netherlands, calling it 'Malabar tin' or Malabar lead.[24] In 1644, however, Portuguese vessels were importing Chinese goods to Europe from Malabar, including 'Callaim and *Tutenaga*',[25] so Malabar was therefore a trading post, not the producer of the zinc metal.

In January, 1609, a Dutch sea captain who seized three Portuguese vessels off Malacca, chose Chinese 'spelter' from their cargoes as a gift for the Rajah of

---

[16] Needham 1974: volume 5, part 2: 212
[17] Valentijn 1724-1726, volume 5, part 1: 329
[18] Souza 1991: 293, 297, 301
[19] Dai Zhiqiang and Zhou Weirong 1992: 53
[20] Chaudhuri 1990: 121
[21] Dawkins 1950: 31
[22] Needham 1974: 212, translating the *San Thsai Thu Hu* of Wang Chai
[23] British Library: MS, IOR B/45, 226
[24] Dawkins 1950: 17, 30-31
[25] Bocarro 1644/1937: 354

Jahore (Singapore), who used zinc metal for bullets to defend his river against the Portuguese.[26] Shooting with zinc bullets is also mentioned in 1648 by Father Antonio Gomes, a Portuguese visitor to Manyika in the copper-rich East-African Monomotapa Empire (now Mozambique/Zimbabwe). He wrote, in Portuguese, that Manyika people alloyed their copper 'with a metal like tin that in India is called Calaim. But it differs from tin in one particular way – its value in carats is deemed to be greater and better than that of tin. The locals in Goa go and use it up for all this shooting'.[27] Tin was evidently inferior in value to *calaim* (metallic zinc). The Dutch used small vessels to fetch '*Tutanague*' (zinc metal), ivory and other goods from Pegu and Jahore (Singapore), where it could be bought cheaply for selling elsewhere.[28] Arabs had for centuries traded goods between Aden, Cambay, Manyika (Monomotapa) and Manila. Despite the Portuguese and Dutch, their vigorous, centuries-old Indian Ocean trade continued. In the 1650s, they traded scrap brass from Surat (western India) to the Monomotapa Empire for making armlets or anklets.[29]

Chinese early seventeenth-century zinc metal production, as already mentioned, was for minting copper-alloy coins. The Ming emperor Wanli (1572-1620) fell into disputes with his officials and withdrew from the public gaze, leaving his eunuchs in control, so the Ming Empire began to decline. A seventeenth-century Chinese text describes metallic zinc being produced from calamine ($ZnCO_3$) in central China by placing five kilograms of prepared ore [and a reducing agent] in a sealed clay crucible, fired by layers of coal and charcoal briquettes, over kindling. Once the jars were red hot, the metal melted and fused into a mass. The cooled zinc, having lost 20 per cent volume, was retrieved by breaking the jar.[30] The resulting zinc metal was easily vaporised by heat and mixed with copper to make brass, and because 'it is similar to lead, yet fiercer in nature, it is called Japanese lead'. Analysis of distillation-retort sherds from Yangliusi zinc-distillation site in Fengdu county, Chongqing province, central China, confirms that coal fuel, not wood or charcoal, was used for this zinc distillation method and that the reducing agent was powdered anthracite.[31]

Under the late Ming Chongzhen Emperor (r.1627-44) the average zinc content for coins was 20-30 per cent, and they were cast in sand moulds (the 'sand' used in the West for sand moulding was a mixture of quartz-sand, loam, horse manure and calves' hair pounded together to bind the substance,[32] and the Chinese recipe may have been similar). Coin minting was no longer the sole Chinese use for metallic zinc. The

---

[26] Valentijn 1724-1726, volume 1: 198
[27] Gomes 1648, 1959 edition: 196
[28] Lockyer 1711: 72
[29] Buckeridge 1654, 1973 edition: 32
[30] Li Ch'iao-p'ing 1948: 44
[31] Zhou Wenli 2012: 221-222, 272
[32] Haedecke 1970: 37

Huangji Hall of the Forbidden City (now the Palace Museum, Beijing), was mentioned in the Chongzhen period as having brass in its construction – recently confirmed by analyses of brass on its roof.[33] This reign, characterised by a terrible earthquake, exceptional cold and floods, saw an excessive flow of silver into China from Spanish South America, which weakened the Chinese currency. The Manchu were meantime spilling into China over its northern borders.

In 1609, the Dutch East-Indiaman *Mauritius* bringing 122 tons of 98.5%-pure Chinese metallic zinc back via Bantam (now Banten, Java) to Europe,[34] was shipwrecked off Gabon (west coast of Africa). The ship was probably bound for Amsterdam, which traded raw materials to brass works. In 1576-85, Antwerp had been crippled by a Spanish siege, so brass trade had been transferred to Amsterdam and Rotterdam, inspiring a revival in the creation of artistic brass works in the northern Netherlands.[35] From the late 16th century, Europe had workshops capable of producing brass solders containing a high percentage of metallic zinc. As early as 1629, thick copper cladding sheets, covering the iron barrel-linings of some Dutch East India Company cannons, were brazed together at the edges with brass solder containing over 32% zinc.[36] Brass cannons were sometimes produced in South-east Asia, perhaps with a puddled wrought iron tube inside them, but brass was not the easiest alloy for casting cannon barrels, and, although spark-free, it proved to corrode rather faster than other alloys in contact with burning gunpowder.[37]

Sweden. In this chapter, European brass works will be described starting from the north because the 17th century proved to be Sweden's great brass manufacturing period. As its economic power rose, Sweden became the world's leading copper exporter (for a map of Swedish sites see the previous chapter). Although it lacked workable calamine, Sweden owed its prosperity to its prolific Falun mine of high-quality copper. King Karl IX (1604-1611), an enthusiast for industry, decided to build Sweden's wealth on national assets, not on export income, so he inspired craftsmen to innovate and to develop ways of using Sweden's copper at home. The king welcomed rich merchant financiers from the Netherlands, and leased out his royal lands and industries to private entrepreneurs.[38] In 1607, Karl IX set up a brass works at Skultuna, a hundred kilometres west of Stockholm.[39] Workers and leftover stock were transferred to Skultuna from the former Höjen factory at Arboga, but, as ever, true brass-making expertise had to be imported, so in March 1612 a Netherlands brass-master, Jacob Johansson, took the

---

[33] Zhou Wenli 2012: 46
[34] Welter 2003: 49
[35] Haedecke 1970: 31
[36] Gilmour 2000, 89-90
[37] Peter Northover personal communication
[38] Forsgren 2010: 7, 13
[39] Gadd 2007: 5

Skultuna lease.[40] He recruited skilled Stolberg workers from among the thousands of refugees fleeing the conflicts between Protestants and Catholics – particularly during the Netherlands Freedom War (1572-1648).[41] By 1630 Skultuna was producing fine 'latten' sheet brass.[42]

Skultuna brass mill had good water-power from its rushing river, the Svartån (dark river),[43] and charcoal from thousands of acres of mainly pine and spruce forests. Heavy demands on timber drastically reduced the trees, raising woodland values and causing a rush to buy up nearby forested estates – land-owning mill-operators spared their own timber by buying it from neighbouring forests. In 1695, the Skultuna brass works hired a forest warden to guard timber, act as gamekeeper and distil quantities of pine tar to seal the wooden walls and roofs of the brass-works buildings.[44] In winter, when the streams were frozen over, copper ore was delivered to Skultuna by sledge from Falun copper mine, and calamine was shipped in through Västerås harbour and then by reindeer sledge across frozen lake country. Calamine came initially from distant Poland or Upper Hungary (Slovakia), and later from the Altenberg (Vieille Montagne) five kilometres west of Aachen. Seasonal drawbacks occurred with failure of water power for the mill when the Svartån froze in winter or dried up in summer.[45]

Under Jacob Johansson and his skilled recruited Stolberg staff, the Skultuna workshops and forges produced sheet brass, wire, candle-sticks, pots, pans, bowls, clocks, wall-plates, mortars and chandeliers. Up to 30 individual moulds might be needed for the parts of one chandelier, and the oldest known Skultuna chandelier, dated 1619, hangs in the church of Our Lady, Enköping. Johansson eventually ran out of capital to pay the Crown for costly copper deliveries, so in 1620 he forfeited his lease. A fine 1625 three-tier Skultuna chandelier and a 1629 wall light in Västerås cathedral date from the time of his successor, Arnold Düppengiesser, a refugee from religious persecution. Düppengiesser, who was already owner of a brass mill in Stolberg, was better capitalised, but he met a premature death through stressful foreign financial speculations. When his expert brass workers left Skultuna for other brass mills, they carried Duppengiesser's innovative industrial practices with them.[46]

In 1632, a Dutchman, Winandt Nacken, started a royal foundry at Skultuna,[47] where he personally hand-crafted the moulds for fine brass chandeliers. Early seventeenth-century Dutch and Flemish craftsmen cast baroque chandeliers in

---

[40] Västerås stadsarkiv: Skultuna Akta, box 55 E.109,E3,
[41] Forsgren 2010: 13, 14
[42] Gadd 2007: 10
[43] Riksarkivet, Stockholm: MS map, Skultuna bruk, 2 Angåenden kungsgården och kungsladagården 3, folio 334
[44] Lundgren, Helga, April 1991: www.hembygd.sew/index.asp\/lev=15124
[45] Gadd 2007: 6
[46] Forsgren 2010: 15
[47] Västerås stadsarkiv: MS Skultuna Akten, Box 55, E.109.E3

Figure 47. Chandelier by Winant Nacken, 1633, in Skultuna church

many styles, including the early type with spherical framework and baluster shaft. The skilled craftsmanship of Winandt Nacken so impressed his royal patrons that he was granted personal use of certain Crown copper sources.[48] In 1633, he presented a chandelier to Skultuna church,[49] and in 1643 a small engraved brass pyx,[50] and the factory also made domestic objects like brass spoons and candle snuffers. Training was a key State requirement, so Winandt Nacken recruited French and German workers to teach their skills to Swedish men and boys under a four-year apprenticeship scheme. After 1643, Skultuna was run by a silk mercer of German stock who supplied brass and brass wire to Stockholm for making pins, clasps, staples, candlesticks, chandeliers and semi-prepared cauldrons, dishes and basins.[51] For a map of Swedish sites, see the previous chapter.

In 1610, the enthusiastic Karl IX of Sweden, established a brass mill at Nyköping, ninety kilomentres south-west of Stockholm (*ny* means new, and *köping*, shopping or market). Willem de Besche (1573-1629), trained in mechanical engineering and architecture in his native Netherlands, was engaged to build it.[52] He introduced Walloon brass-smiths to Sweden, with support from the powerful financier Louis de Geer (1587-1652).[53] Born a Walloon himself in Liège (Lüttich), Louis de Geer became a rich Amsterdam merchant supplying arms to the Swedish king, who invited him to Stockholm to re-vitalise Swedish industry.

Disaster struck Nyköping in 1622 when the brass works was destroyed by fire. Although it was re-built, the mill closed again in 1627 because Willem de Besche was away acting

---

[48] Riksarkivet, Stockholm: Register, fol. 144.2, 17 September 1640
[49] Forsgren 2010: 15
[50] Gadd 2007: 18
[51] Forsgren 2010: 16
[52] Riksarkivet, Stockholm: typescript, register, folio 351:1, 11 Dec 1673
[53] Forsgren 2010: 54

as Louis de Geer's right hand man, simultaneously managing three ironworks,[54] and setting up a cannon foundry at nearby Finspång and a brass mill at Norrköping ('north market'). The neglected Nyköping brass-works remained closed for almost twenty years, until Louis de Geer's son, Laurens, rebuilt it and guaranteed it a loan of four years working capital. It re-opened in 1646, and twenty-year-old Queen Kristina (1626-89) granted it the use of her adjacent castle yard, plus a ten-year tax exemption,[55] and the right to work battery hammers. Nyköping exported brass wire, mainly to France, England and the Netherlands.[56]

In 1627, Louis de Geer took Swedish nationality and came to live in Norrköping, about sixty kilometres west of Nyköping.[57] Louis de Geer, related to the powerful Trip brothers, Amsterdam merchants, had managed to divert Swedish exports away from hostile Hanseatic ports like Lübeck, and towards Amsterdam and the west.[58] The brass works financed by Louis de Geer at Norrköping, started production in 1627 with Willem de Besche from Nyköping brass works as technical partner. In 1629, De Geer became sole owner, and had a magnificent stone château built for himself. He also built a school for the children of the many Walloon- and German-speaking metal craftsmen who had settled there. The brass mill and its battery works were the largest of many local industrial units (producing steel, clothing, ships, rope and paper), which together, by 1629, employed nearly 100 households.

The Norrköping brass works, named Holmens Bruk (island mill), lay south-west of the town, on an island on the vigourous River Motala that flows through Norrköping and out to the Baltic Sea. The mill was powered by three great falls of water, held back by weirs. The mill channel no longer exists, but the roar of the fall of several metres of powerfully flowing river water remains deeply impressive. The nearby Östgötland forests provided charcoal, and Norrköping had a good harbour, though it was far from the Falun copper mine.[59] Norrköping, with 18 furnaces – more than any other Swedish brass mill – produced mainly wire and brass plate, from which cauldrons and basins were produced. In 1634-5 a passing traveller noted with amazement the horrendous racket of the water-powered trip-hammers beating out brass plates. He watched workers put plates on an anvil, cut them into long strips with shears, then carry them to a rotating wheel to be drawn out to wire – a process that created a screeching sound so terrible, intense and ear-splitting as to deafen the workers. Thin brass sheets, looking like gold, were polished and cast into hollow-shaped wares, including spoons and ladles.[60]

---

[54] Österby, Lövsta and Gimo, Uppland
[55] Riksarkivet, Stockholm: MS Cop Boken Lit. N. 23, folio 971, 29 March 1647
[56] Forsgren 2010: 54-56
[57] Nörrköping, 2006: www.kungsgard.norrköping.se/_comenius/European_migration
[58] Kumlien 1977: 257
[59] Forsgren 2010: 84-85
[60] Ogier 1656: 314-315

From 1642-7 Sweden exported almost twice as much brass as copper,[61] and the industry was growing competitive. For a time Norrköping faced fierce competition from a brass works at Nacka near Stockholm, which enjoyed royal permission to sell cauldrons nationwide at reduced prices. Louis de Geer acted quickly to obtain similar benefits for Norrköping.[62] De Geer resided mostly in Stockholm, leaving a broker in Norrköping to order raw materials and ship out the brass,[63] and a works-manager to cope with workers striking over poor housing and low wages. The craftsmen noticed the higher wages paid at nearby Nyköping mill, and many defected there. In retaliation, De Geer ordered factory owners not to hire any worker previously dismissed for insubordination or insolence, or lacking a pass from his former employer. The exodus of master-craftsmen from Norrköping to Nyköping stopped once de Geer equalised the wages of the two mills.[64]

Figure 48. Sweden, map showing some sites mentioned in the text

Vällinge brass works, 34 kilometres south-west of Stockholm, lay in a secluded setting between two lakes, among rolling wooded hills. It was powered by a small river only 750 metres long but with an impressive ten-metre drop in level between Lakes Bornsjön and Mälaran.[65] In 1624, shortly after the onset of Europe's Thirty Years War, King Gustav Adolf (1594-1632) granted Arnold Düppengiesser a 'privilege' to build a brass-works at Vällinge. Arnold shortly died, leaving his son Arent and a manager in charge. In 1626, with mills on two weirs, the works had battery hammers, furnaces, foundries and wire-drawing workshops.[66] Arent inherited an agreement which made tax demands that threatened the mill's survival during the severe national economic downturn of 1629 (due, as usual, to war). Vällinge mill burned to the ground shortly afterwards, so Arent Düppengiesser resolved to obtain a new manufacturing licence

---

[61] Kumlien 1977: 256-257
[62] Forsgren 2010: 85
[63] Helmfrid 1965: 331
[64] Forsgren 2010: 87-88
[65] Swedish Land Survey: map, 1636
[66] Kunglige Biblioteket, Stockholm: typescript, 1988, Mellander, Frederika,'Vällinge 1535-1908'

from the king, freeing Vällinge mill from taxes. Gustav Adolf had parted for the Thirty Years War (1618-48), but Arent pursued him to Germany, to request the necessary changes in person. Arent managed to gain an audience and obtained the king's verbal agreement, but King Gustav Adolf was killed at the Battle of Lützen immediately afterwards, so Arent never had the agreement in writing.

Arent Düppengiesser nevertheless rebuilt Vällinge mill with the financial support of relatives. With four furnaces he re-started production in about 1640, but the Regency, which harboured grievances against Arent's father, refused his requests to increase the number of furnaces from four to six, and to buy copper and calamine at reasonable prices. They even denied him the required documents for receiving a major shipment of supplies, so he imported them on a different ship – which was promptly hijacked by privateers. Local landowners also rose against him, refusing him the right to dam the Bornsjön stream to increase water-power, so long droughts reduced water-flow to Vällinge brass-works.

When, in September 1640, Arent Düppengiesser gave up the mill at Vällinge, the government refused to let him transfer the business to Willem Momma (an Aachen-born Amsterdam merchant), insisting that the brass-works must have a Swedish owner. In 1641 Momma had to sell the mill for a pittance to a Stockholm consortium, who expanded Vällinge brass works to six furnaces, increased the wire output and produced a range of other wares. In 1654, one of the group, with interests elsewhere, became wealthy enough to lend money to King Johan III (r.1658-92), who had grand plans for setting up Europe's greatest brass-works at Vällinge, which, however, came to nothing.[67]

Another brass works even closer to Stockholm was at Nacka, just five kilometres south-west of the city, built in 1625 by two Dutch experts, Abraham Werden and Abraham Meliss. Nacka had plenty of forests, water-power and industrially-experienced local workers and Louis de Geer raised the capital, so they hoped to fulfil the ambitions of king and state for domestic consumption of Swedish copper through home production of brass. The state granted tax exemptions and the right to tear down buildings and replace them with brass furnaces and workshops,[68] which King Gustav Adolf intended to extend.[69] To be profitable, brass for home consumption needed to have value added, for example by converting plate and ingot brass into dishes, pots, basins, chandeliers, carriage-fittings or horse-harness; or by fashioning brass wire into pins and clasps. However, to achieve this, Sweden still relied on know-how from abroad, and capital from foreign financiers.

---

[67] Forsgren 2010: 66-68
[68] Riksarkivet, Stockholm: MS copiae 13: Lit.F: N. 22, , folio 757, 1361, 21 April 1625
[69] Riksarkivet, Stockholm: MS copiae: Boken. Lit.F: N..22, folio 1266, 17 June 1632

Production at Nacka began, but by 1632 Sweden became involved in the Thirty-Years War (1618-48), then raging in most European states, and the brass mill was overspending and running at a loss. One of the original managers died, to be replaced by the unscrupulous Winandt Nacken, who learned all he could at Nacka, before defecting to Skultuna where, as already mentioned, he proved an enterprising leaseholder and very gifted chandelier-maker. For Skultuna, he poached nine expert workers, originally recruited for Nacka from Aachen, luring them with promises of better pay. Nacka mill had become so neglected through bad management that it risked reverting to the Crown, so Sweden's financial saviour, Louis de Geer, was called in to lease it for only four years, giving it up in the early 1640s.[70] In 1647 Queen Kristina granted one Abraham Werden permission to revive Nacka mill, allowing him six per cent of her copper for three years.[71]

Northern mainland Europe. The Thirty Years War, in which Sweden participated, kept mainland seventeenth-century Europe divided into feudal territories ranging in size from independent city states to the Habsburg Empire, and ruled by emperors, kings, prince-bishops, dukes and counts. The disputes and wars of these ruling power-seekers disrupted copper and zinc ore supplies to brass-mills, and hampered deliveries of manufactured brass. Routes to and from Hamburg, Antwerp and Amsterdam were diverted, economic conditions became complex and some brass mills and their workers suffered serious military attack. The brass works north of Hamburg were affected by sporadic local battles between individual rulers as much as by the more widespread Thirty Years War.

Some 40 kilometres south of Lübeck, the beautiful Ratzeburg lake district with thickly wooded hills had two separate brass-working sites. From 1602, Marcus Mewes from Lübeck leased out a brass mill at Mannhagen, east of Mölln, and in 1645, the two Leers brothers (from Lübeck and Hamburg respectively) bought the Mannhagen brass mill and copper mill, with a farm-house and land, near a hill named the Steinberg.[72] The other, Ratzeburg, brass works lay twenty kilometres south of Lübeck, between lakes on two different water-levels. The distance between the upper Lake Mechow and lower Lake Ratzeburg was just two and a half kilometres but the height difference between the two lakes totalled an astonishing 28 metres. A dam and sluice gate stood between the outflow from the Mechow Lake, and a strong stream called the Bäk, which flowed down a steep wooded valley to Lake Ratzeburg, with enough force to power at least five battery hammers. The top sluice-gate lay just within the state of Mecklenburg.[73] In 1642, the Leers brothers, already making brass at Mannhagen, built a brass mill on the powerful Bäk, replacing a former oil-mill.

---

[70] Forsgren 2010: 73-5
[71] Riksarkivet Stockholm: MS Års relation, folio 35.2, 1711
[72] Lisch 1842: 62
[73] Kock 1981: 83

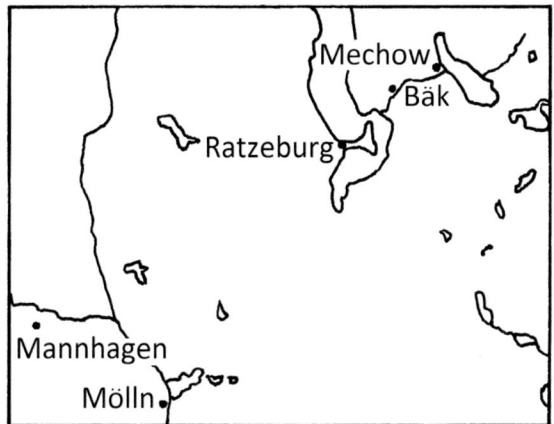

Figure 49. Map of Ratzeburg lake area, showing some sites mentioned in the text

A description of contemporary brass-making, 230 kilometres south of the Lübeck lake district, survives from 1617, the year before the Thirty Years War broke out. Mining expert Georg Löhneyss (1552-1622) described the cementation process at the Goslar brass-mill beside the Harz Mountains,. The circular brass furnaces were in pits in the ground with two holes underneath for drawing in a draught. This type of furnace, commonly used in Europe, held eight empty crucibles, heated until hot enough to receive their contents. Into each crucible were placed 3.89 kilos of powdered calamine in carefully measured ladles-full (31 kilos total for eight crucibles), on top of which was placed 3.63 kilos of broken copper (29 kilos total).

The crucibles, continued Löhneyss, remained in the furnace at a constant red heat for nine hours before being checked by stirring the brass lightly with an iron poker. The brass was left in its flux to refine (an iron flux or purifying agent prevented the surface of molten copper from oxidising), then lifted from the furnace and poured into a shaped hollow in the ground to form ingots. If a piece was broken off, a beautiful golden brass colour appeared at the point of fracture. For making wire or hollow vessels like cauldrons, the crucible contents were poured between large flat casting stones to set as brass plates or sheets that could be shaped under battery hammers or slit and drawn out to wire.[74] The casting-stones used at Goslar, called Brittany stones, were cut from grey granite at Les Champs du Boule (12 kilometres from Villedieu les Poêles, Normandy, where brass was recycled) and shipped from the port of St Malo, Brittany.[75] In 1617, Löhneyss wrote that Goslar obtained calamine from the River Ach in Schwabia (north-west of Lake Constance) and from the Tyrol, but mainly from the Vieille Montagne.[76]

It is clear that brass workers in the Harz, Sweden, Lübeck and its lake district were all dependent on close connections with the Aachen and Stolberg area for expertise and calamine supplies. By the early 1600s the brass workers of Aachen had earned themselves a greatly improved quality of life, but the high standing of Aachen would soon be transferred to Stolberg. Prosperity had provoked jealousy, leading to disputes

---

[74] Löhneiss 1617, 1690 edition: 83 r, v, 112
[75] Galon 1764: 102
[76] Löhneiss 1617, 1690 edition: 112

and religious bigotry, followed by major conflict. The brass-makers temporarily in control of Aachen city were Protestants. Emperor Rudolf II (1576-1612), descendant of Emperor Maximilian I and, paradoxically, a well-known collector of brass objects, had decided in 1593 to outlaw the obdurate Aachen Protestant brass-makers, to force them to adopt catholicism. To this end, he blocked Aachen's copper and charcoal supply routes and enlisted the combined help of the King of Spain, Elector of Cologne and Bishop of Liège to help him cripple the Protestant brass-workers into religious compliance.

The Aachen brass-mill proprietors implored their business contacts, the Elector of Saxony and the Mansfeld copper suppliers, to influence the Emperor in their favour. However, even their Protestant allies in Brandenburg and the Netherlands were powerless before Spinola's Spanish Catholic two-thousand-strong army, which besieged their city early in 1614. Expert brass-makers scattered as far afield as Sweden and England, but most settled in the Netherlands or German cities, and several in the nearby Stolberg valley only ten kilometres east of Aachen. Stolberg lay in the Duchy of Julich, a territory awarded in 1614 to the Protestant Hohenzollern rulers of Brandenburg. Soon after the expulsions and flight away from the city, the Aachen brass-making industry collapsed.[77]

At Stolberg the narrow, rapid River Vicht provided strong water power for brass-mills, and the nearby Ardennes forests provided timber for charcoal, firing and annealing. Until the middle of the Thirty Years War (1618-48), Stolberg's copper supplies came overland from Mansfeld, 300 kilometres away, across difficult terrain. An English merchant named Gerard Malynes (1565-1641) wrote that in 1629, when this copper source was cut off, Stolberg brass-makers turned to the Falun mines, copper being 'very aboundant in Sweaden',[78] with peak production around 1650. More surprising alternative copper suppliers to Stolberg during the Thirty Years War (1618-48) were Chile and Peru (imported through Spain).[79] Local Stolberg seams of zinc carbonate ore (still visible today) were less pure than calamine from the nearby Vieille Montagne (Altenberg) at Moresnet. So, for their finest brass, Stolberg brass-makers mixed local and Vieille Montagne ores.[80]

The Duchy of Julich, paradoxically (since the Stolberg brass-makers were Protestant), had Catholic landowners, the Duke of Julich and the Abbot of Cornelimünster, who found it hard to agree over brass-making laws but, scenting profit, found accord, and encouraged the Protestant refugee brass industry. Protestant Anabaptists in the Duchy of Julich-Cleves were granted freedom of propaganda,[81] which acted in the

---

[77] Roderburg 1927: 12-13
[78] Malynes 1629: 265
[79] Schleicher 1974: 36
[80] Schleicher 1974: 17
[81] Cohn 1970: 258

Protestant brass-workers' favour. As shrewd business men, the Stolberg master brass-makers retained their former Aachen trading advantages by each keeping on a house in Aachen. Whilst the brass-makers moved to Stolberg, most of the hand craftsmen remained in Aachen, relying on Stolberg for brass supplies.[82] By 1648, Stolberg's 65 family-run brass furnaces produced 1000 tons a year, three operated by the Momma family. Each substantial brass-works was grouped around a defensive courtyard. Stolberg brass products included sheets or plates, wire, and 'semi-prepared' large hollow vessels, especially cauldrons. These essential coarser brass wares were always to be Stolberg's main products.

Throughout the seventeenth century, cauldrons, pots and pans were shaped using water-powered mechanical battery hammers (trip-hammers). They were then partially finished under different hammer-heads and the 'semi-prepared' hollow shapes were dispatched to hand-craftsmen, often in Flanders or the Netherlands, to be finished as cauldrons, bowls, buckets, jugs, flagons or barbers' basins, with added handles, lugs or legs.[83] In the 17th century, cauldrons or kettles (German *Kesseln*, Swedish *kettlar*, English *cauldron*, French *chaudron*) were produced in a range of sizes. In the days before piped water, they were essential for liquid storage and some could heat up to 90 litres of water at a time. They normally had three added stumpy legs, two handles for suspending them over a fire to heat water or stews, and a wide, flat pouring-collar at the lip. English cauldrons bulged wider below the mid-line, to expose a greater area to the fire.[84] For cauldrons, brass had some advantages over copper, being lighter and harder, so easier to lift and less readily damaged. During the 1600s, skillets, flat-based saucepans and frying-pans with handles and three legs, developed from the cauldron shape. Their widespread use in Europe is glimpsed in Flemish and Dutch Old Master oil paintings.

West Africa. The type of items being prepared in Stolberg and Flanders at the time, including cauldrons, beakers and barbers' basins are often mentioned in receipts in the 1645-7 log-books of the Dutch fort of Elmina (in today's Ghana).[85] Throughout the 17th century, the Dutch shipped European brass to the Guinea Coast of West Africa, in exchange for gold and slaves, and Stolberg was the nearest manufacturering centre capable of fulfilling the high demand for brass in West Africa. Although the Portuguese kept their trade as secret as possible, the Dutch had, by 1612, established a fort just sixteen kilometres east of the larger Portuguese fort of São Jorge da Mina, which the Dutch siezed five years later, re-naming it *Elmina*.[86] From 1635, Dutch merchants were urgently bartering for African slaves for their Brazil sugar plantations and, with their more intimate access to Stolberg brass and Flemish craftsmen, they could acquire cheaper, better-quality, brass items than the Portuguese.

---

[82] Haedecke 1970: 30
[83] Schleicher 1974: 34-35
[84] Haedecke 1970: 66
[85] Ratelband 1647: 1953 edition: 400, appendix
[86] de Corse 2001: 23

Figure 50. Figure of a Portuguese soldier, 15 to 16 century

Once brass entered another culture, it might be viewed and employed quite differently. Brass items had a value, and could be exchanged for something of corresponding value, in this case slaves and gold. A commodity is produced as a culturally recognisable and serviceable item – a barber's basin, for example, was known and used by Dutch or Flemish barber-surgeons for shaving and blood-letting. When introduced into West African cultures in exchange for slaves, such commodities often lost the meaning they had possessed in the original culture, and gained a new, perhaps deeper and more spiritual aura.[87] So the mercenary motives of the slave trade were energetically fanned by very genuine but varied special relationships with brass in different areas along the Guinea Coast. In some areas brass was used mainly as currency, though in Ghana, where brass was sacred, this use was rejected. Others saw it simply as suitable for arm and leg ornaments, whereas in Benin brass was a symbol for displaying wealth and for confirming power over others, and in the Akan region of Ghana, brass connected people to a spiritual world.

In the ancient kingdom of Benin (now in Nigeria), the king had rights to the possession of brass to enhance his royal might. Imported Portuguese or Dutch brass was re-cast in the royal Benin brass foundry next to the king's palace.[88] This was a rich civilisation – in 1686 a Dutch visitor noted that the king's walled palace beside Benin city covered as much space as the Dutch city of Haarlem. Fine palace galleries were supported by wooden columns encased in brass, carrying depictions of the king's victories. Most royal buildings were roofed with wooden tiles or palm branches, with a brass bird, its wings outstretched, decoratively perched at the tip of each pyramidal corner tower.[89] Many symbolic designs in brass adorned the altars, walls and roof of the palace at Benin, including statuettes and plaques of early seventeenth-century Portuguese soldiers in military uniform, carrying muskets[90] or playing horn-like instruments[91] –

[87] Kopytoff 1986: 64, 68
[88] Strieder 1933: 258
[89] Dapper 1686: 308
[90] National Commission for Museums and Monuments, Nigeria: 54.19.1
[91] National Museum of Scotland A 1985.630

symbolising the power of the ruler of Benin over the European intruder. Some brass images are thought to signify the legendary defeat over Olubun. The Portuguese were initially believed to be messengers of Olobun, the legendary white-faced god of the sea, defeated in battle and stripped of his great wealth.

In Akan areas of what is now Ghana, brass was imbued with a more mysterious aura. As in Benin, Akan craftsmen cast and hammered brass, using sophisticated repoussé, stamping, stippling and engraving techniques on brass containers to produce elaborate geometric and animal patterns.[92] The 'Akan' were a series of peoples, including the Ga people, related by language group and culture. The Ga kingdom lay in Ayawaro district, within today's greater Accra, and Akan relatives lived in the coastal and south-eastern hinterlands of Ghana. From the 15th century, Ghana's lowland Akan people held sacred the river god, Densu, who appeared at dawn, so Ga libation through the medium of brass, an act of communion, brought harmony between a natural and a supernatural being, through a living intermediary – water.

The earliest Ga ancestors of the expert brass workers of Akan, Ghana, had met with the legendary Nii Gua – god of thunder in smithing – who came to the Ga people in a brass bowl – *ayawa* – suspended on a long chain, and taught them the craft. The fourth major god, still celebrated in urban Accra in the 1960s, was Gua, the smith, creator of certain stars.[93] Excavations in the Ayawaro suburb of Accra indicate that some Ga craftsmen produced pottery with complex artistic motifs and stylised figures and others melted down brassware in clay crucibles, to create new articles. The Akan, or Ashanti, peoples of Ghana used lost-wax brass casting techniques, and religion still inspires the use of brass, so some Ga families remain brass-workers.[94]

The central religious role still accorded to brass by the Ashanti at a much later date is illustrated by Captain Robert Rattray a traveller through the Akan region in the early 1920s. The Ashanti word, *obosom*, meant a god or a shrine consisting of a brass pan, capable of temporarily harbouring a supernatural spirit or spirits. Once consecrated, the shrine contained certain ingredients to make it acceptable to a spirit but it only became an *obosom* once a spirit from water, embodying the supreme life-giving force, had entered it. The sacred lake Bosumtwi, about 30km south-east of Kumasi, harboured an energetic spirit, so no brass pans, fish-hooks or canoes were allowed near it. Along the river banks of south-west Ghana, Rattray saw altars of sticks, each surmounted by a brass pan. Ceremonies relating to the river spirit Tano, held at a ford or a river source, involved placing in a brass pan 'something from the river' and the blood of a sacrificed white chicken.[95]

---

[92] de Corse 2001: 133
[93] Kilson 1969: 171
[94] Josiah-Aryeh, N. A. 1997: website: http//members.tripod.com/tettey/Gapart2.htm
[95] Rattray 1923: 74, 90, 145, 199, 200, 210

In the Ghana and Gold Coast regions, brass pans and lidded brass pots, known as *kuduo* were vehicles of divine power. Several ancient examples of *kuduo* survive from the area, although their dating is not secure.[96] At one sacred grove ceremony, Rattray witnessed a procession where an old woman bore on her head a brass *kuduo* containing sacrificial sheep's blood and meat, which were later transferred, using a brass spoon, into eight cups. Anyone possessed by a spirit prepared a brass pan and placed in it water with medicinal and other items, while incantations were made and sacrificial animal blood added.[97] Several fragments of brass vessels (by then called *forawa*) have

Figure 51. Examples of kuduos, *top right:* kuduo, royal grave, Kumasi, Ghana; *lower left and right:* Ashanti kuduos from the Gold Coast

been excavated from mid-eighteenth-century contexts at Elmina fort, though *kuduo* originated much earlier. There, the brass vessels contained gold dust to place beneath a wealthy person's head at burial, or for mixing with other substances as a powerful magical cure.[98]

Further Guinea Coast evidence included a Ghanaian procession of priests with uplifted hands carrying brass pans representing shrines to a local god, Ta Kese. Each brass pan was later cleansed, marked with three fingers on the outside, and sacrificial blood was dripped into it. Rattray reported that the chief had rights to all same-sex twins, from their birth, and had them carried to his palace in a brass basin. Almost everyone possessed a personal *kuduo*, used for rites, for example at the birth of a child. The father would put mashed plantain in his *kuduo* and pour sacrificial chicken's blood over it, with an incantation for the baby to emerge peacefully. Relatives ate soup around the *kuduo*, while husband and wife performed rites to form a spiritual connection with it. In Rattray's day, old people placed offerings in brass *kuduo* during domestic religious rites; and if the owner died, gold dust and aggrey (blue, glass-like) beads were placed in their *kuduo*, to be buried with them.[99]

---

[96] British Museum 1956; AF 1.1; 1947 AF 12.36; 1948 AF 21.25; 1952 AF 7.15
[97] Rattray 1923: 125-128, 146
[98] De Corse 2001: 133
[99] Rattray 1923: 51, 99, 155

In the earlier 17th century, the Portuguese and Dutch brought the brass worked and used all along the west African coast. By the mid-17th century, however, the Dutch faced competition on the Guinea coast from the merchant fleets of France, England, Denmark and Brandenburg, who copied the Netherlands but, lacking experience, traded less efficiently. The other countries had to obtain most of their brass through Antwerp and Amsterdam, from the same Stolberg manufacturers and Flemish craftsmen as the Dutch, but without their local and family contacts. In 1631, the English king, Charles I, granted the London-based 'Guinea Company' a 31-year monopoly over other English traders, from Cape Blanco to the Cape of Good Hope (a coastal stretch reduced by Cromwell in 1651). In December 1657 the English East India Company bought the remaining eight years of Guinea Company monopoly, to let their India-bound ships exchange cloth and brass for the African gold and ivory demanded in India,[100] but in 1664 the Company withdrew from the Guinea trade.

Those highly competitive Dutch or Germanic merchants who lived and traded in London guarded their own connections with the lucrative Guinea Coast trade. They also raised strong objections to selling English-made brass in London – because England was a profitable market for Dutch and German brass. After the English Civil War (1642-51) the nascent English brass and copper industry was left very weak. Seventeenth-century English kings related to their brass industry far less favourably than the Swedish royal family with their Falun copper source,. Most Swedish monarchs, whilst keen to milk their industries in times of stress, strongly supported and encouraged the Swedish copper and brass industries.[101] Mid-seventeenth-century English Stuart kings, on the other hand, gave minimal support to England's brass industry, viewing it mainly as a source of income, and 'suppressing' any brass mill that failed to pay heavy 'privilege' dues to the Crown. In this atmosphere, early British brass-makers often failed, but they grew less dependent on the Crown than did their Swedish counterparts.

Britain. In the 17th century, the English Society of Mineral and Battery Works was a pawn of the Crown, leasing out annual royal 'privileges' to brass-makers, which gave them a monopoly on production in a particular area. Income from 'privileges' passed directly to the king, rather than being a tax imposed by an elected parliament. In 1627, therefore, when King Charles I urgently demanded money to fund his brief involvement (defending Huguenots) in wars in France, the Society's mainly Royalist committee hastened to detect and fine any brass-works operating without a fully paid-up privilege. These works were forced to pay all arrears in cash, or were fined for defaulting and summarily closed down.

A brass-mill near Maidstone in Kent was an early victim of the Society of Mineral and Battery Works (for English and Welsh sites, see the map in the next chapter). By the

---

[100] British Library: MS, IOR E/3/85, folios 105 and 128 (1658-9)
[101] Pohl 1977: 236

late 16th century, Maidstone had a substantial population of Walloon and Netherlands refugees. The town had petitioned to take 60 families competent at making various goods. Some of them established fulling-mills on the River Len,[102] and others evidently made brass, because before 1621 the Court of the Society of Mineral and Battery Works advanced more than £200 towards setting up a 'latin' (latten, meaning sheet brass) battery works 'near Maydstone in Kent', and in 1621 demanded the cash back. The mill's partners had already shared out part of the original money, but they found £200 for the Society.[103] Neither archaeological excavation in Maidstone[104] nor documentary research into the River Len mills,[105] have so far revealed further brass-mill evidence. In 1627, the Society of Mineral and Battery Works also targeted Tewkesbury brass wire mill on an earlier corn mill site near the confluence of the Rivers Avon and Severn in Gloucestershire.[106] They fined its owners – who had evidently been keeping their brass-making activities quiet to avoid payment.

Once the English Tintern works had abandoned brass wire-making in favour of iron, its 'privilege' to make brass and to extract calamine from the Mendip hills was available to be transferred elsewhere. In January 1630, two London members of the Society of Mineral and Battery Works, Bayldon and Webb, were granted the 'privilege', to mine calamine, produce and cast brass, operate battery hammers and draw brass wire. Bayldon chose to operate a brass battery works at Taynton in the Forest of Dean, and in 1631 he passed on his 'lease of the latten Battry' to Sir Charles Powell.[107] Taynton brass-works lay within a northern Forest of Dean district,[108] near the Welsh-English border east of Gloucester, where two fields are still named Brass Mill[109].

In Brass Mill Field, Taynton,[110] finds of raw copper and zinc-permeated purplish-blue clay crucibles, some streaked with copper-green, indicate that fresh brass was produced as well as worked at this site. The mill was located at the south-east corner of today's larger Brass Mill field.[111] Near the former mill-stream, a late Tudor two-ounce lead disc trade weight was discovered, associated with pure copper waste, brass scrap and small pieces of coal.[112] The use of coal in English brass-making would have been unusual at this date. Managing coal fuel to avoid contamination from coal fumes suggests the use of reverberatory furnaces, known in the medieval period and experimented with (unsuccessfully) in the 17th

---

[102] Clark and Murfin 1995: 43, 46
[103] British Library: MS, Loan 16/2, folio5, Court Books, Society of Mineral and Battery Works
[104] Philp, Brian 1999: 9, Kent Archaeological Rescue Unit, evaluation excavation report, Romney Place, Maidstone,
[105] Goodsall 1957: 106-129
[106] National Grid reference SO 893 311
[107] British Library: MS, Loan 16/2, folios 50v and 59
[108] Rudder 1779: 28-29
[109] Gloucestershire Archive: MS P 1812/177 and TRS 224/177 (1840), Taynton parish, tithe map
[110] National Grid Reference SK 7343 2047
[111] National Grid Reference SK 734 206
[112] Gloucestershire Sites and Monuments Record: 4008/3/19-20

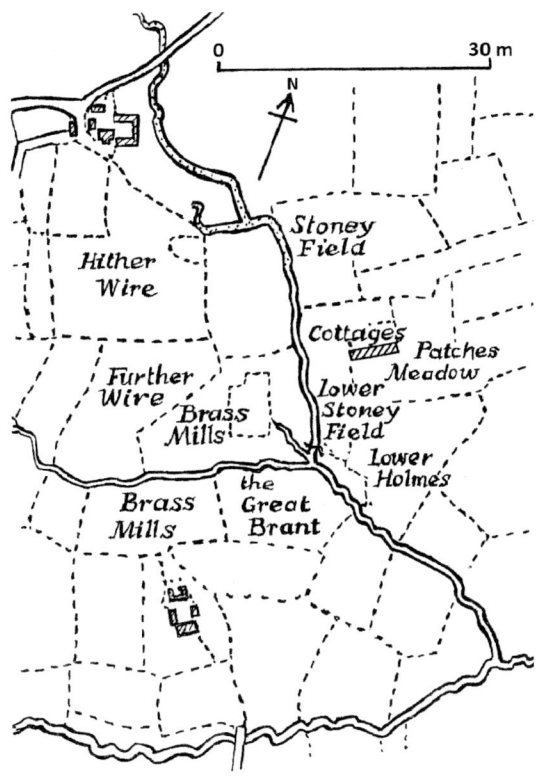

Figure 52. Taynton field names, including Brass Mill Field

century for iron.[113] Coal fuel was used in the glass industry from at least 1612,[114] and its use in reverberatory furnaces was mandatory for British glass making from 1615, to conserve timber.[115] A glass-maker known to be at Newent glass works (adjacent to Taynton) in 1607, claimed to be using coal fuel, but archaeological analysis has so far only found evidence for wood fuel on Newent crucibles.[116] Coal emits sulphur fumes that contaminate copper, but domed reverberatory furnaces can prevent this by being built in two sections, fuel in the outer part and ore in the inner. The heat from coal burning in a largely enclosed lower hearth is drawn across the upper part of the furnace by a draught, and reverberated (deflected) down from a low vaulted roof or cupola, onto the ore and charcoal in a separate compartment immediately beneath.

Calamine for Taynton was shipped to Gloucester port from the Mendips (mining rights to Mendip calamine were included in the 'privilege'). At this date, good copper ore could have come from Ecton in Staffordshire or from Sweden, but accessible Cornish supplies had now diminished. Other mainland continental sources were ruled out by the Thirty Years War (1618-48), which cut off even Stolberg from its Mansfeld copper supply, forcing it to look to Sweden. Four Swedish *öre* copper coins dating to about 1629 (one partly melted),[117] were found near former workers' cottages beyond Taynton brass mill.[118] The coins probably represent Swedish copper scrap to add to crucibles.[119] In 1620, the European copper market had slumped, due to a glut of silver from Spanish America, causing the European monetary standard to transfer from copper to silver 'pieces of eight', so financial value was now based on silver.

---

[113] Barker 1966: 34
[114] Neri 1611, 1662 edition: 239
[115] Barker 1966, 35
[116] Dungworth 2010, 10
[117] Gloucestershire Sites and Monuments Record: 4008/3/1, 10; 17, 18, 21, 22; coins in private hands
[118] Taynton tithe map, 1840; National Grid Reference 733 209
[119] Sherratt, Don 2000: local history booklet, History of Taynton, 26-27

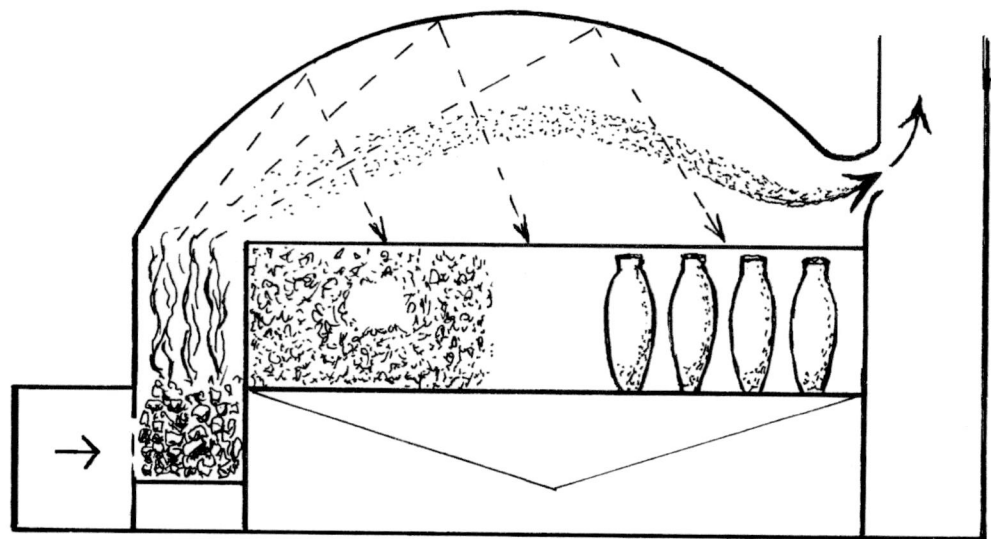

Figure 53. Diagram of a reverberatory furnace

A Swedish copper monetary standard was therefore introduced to ensure that Sweden consumed more copper at home, to keep prices high and avoid swamping the European market with now-cheap copper. The size of Swedish copper coins was increased to keep their value in parity with silver coins. As the larger, heavier, Swedish copper coins grew too unwieldy to handle or even transport by sledge, Queen Kristina (1644-54) decided to abandon copper coinage (once her taxes had been collected) and, some decades later, paper currency was issued.[120] The 1629 Taynton Swedish coins, are therefore likely to represent surplus copper, but it is worth remembering that many skilled Stolberg refugee brass craftsmen fled to Sweden during the Thirty Years War (1618-48), and some may have have accompanied bulk shipments of Swedish brass scrap back south again. Taynton clearly had to recruit expert foreign brass-makers to start up successfully, and their presence is confirmed by records, dated 1634-43, of several 'dutchmen' living at Taynton,[121] (at this date 'dutch' or 'deutsch' included Netherlanders or inhabitants of neighbouring German states). These 'Dutch' workers could have come direct from Aachen or Stolberg.

At the Taynton site, finds of triangular offcuts of flat brass sheet, a partly sliced strip, and snippets of wire indicate that ingot brass was cast into sheets before being sliced for wire-drawing. A heavy iron bar, iron wedge and tongs found there were perhaps used for hoisting up and wedging the lids of casting-slabs (a fragment of a brass-casting slab was found by the author on site). An English coin of 1625-49 helps to date the brass works.

[120] Forsgren 2010: 7
[121] Juriča 2010: 334

One Taynton brass customer, John Tylsley, a Gloucester pin-maker, originated from Bristol. In this Cotswold Hills area of England's great sheep-pastures, brass was used mainly for rust-free pins for wool cards used by the burgeoning English wool trade. Tylsley came to Gloucester in 1625, and borrowed an initial £20 from Gloucester Corporation, who provided him with a workshop to employ 30 poor boys at 'Wyremakinge'.[122] In 1631 Gloucester City granted Tylsley £100, plus £100 annual loan, to establish a factory providing pin-making work for more than 80 boys and girls, where there had been no work before. Tylsley's brother in Bristol was also prospering at pin-making, and they both bought their wire from the Taynton brass works. By 1631 the Gloucester business was worth about £2,000,[123] a huge sum for those days, showing that Taynton was producing very significant quantities of good brass and brass wire. Fields at Taynton still carry the names Further Wire and Nether Wire. Brass wire did not rust (as iron did) so it was much better for making 'pins' for wool-cards for the flourishing Cotswold wool trade, as well as pins for haberdashery and lace-making and for securing documents. Other brass finds from Taynton include the bowl of a brass spoon (c.1580-1650),[124] a dome-headed pin, and a seventeenth-century button.[125]

A wide variety of brass alloys was produced and used at Taynton, some showing deliberate additions of lead and tin to facilitate casting. Zinc contents were generally consistent with cementation brass, though the average was certainly higher than most contemporary brass, with two samples containing 30%[126] and 36% zinc respectively,[127] which is consistent with English late sixteenth- and early seventeenth-century monumental brasses, which contained, on average, over 32 per cent zinc (eight of the analysed examples containing from 33 to 36 per cent zinc).[128] The high-quality brasses produced at Taynton, suitable for casting or for wire-drawing, must have been prepared from well-refined copper, probably Swedish. A few English jettons (trading tokens) also exceeded 33 per cent zinc by the 1660s,[129] so zinc metal, which was entering Europe from China, may have been added. Taynton mill may have foundered during the English Civil War (1642-51), though in 1658 a pin-maker was living there and in that year a Taynton field is described as lying 'near the brass mills'.[130] However, in 1661 the former brass mill was let out as a corn mill, and was pulled down in 1706.[131]

England had hitherto imported most of its arms and armour, but in 1629 a member of the Ordnance Office set up a sword factory at Hounslow, west of London, using brass

---

[122] Gloucestershire Archive: MS, GBR/G.3/SO/1, folio125 (1626)
[123] The National Archive: MS, SP 16 205/11 (undated fragment, 1631), Tilsley of Gloucester, pinmaker
[124] Morton 2009: 274, analyses by Chris Salter
[125] Gloucestershire Sites and Monuments record, 4008/3/6 and 8
[126] Morton 2009: folio 274, analyses by Chris Salter
[127] Dungworth *et al* 2010: 6
[128] Calver 1990: folios 68 ff
[129] Pollard and Heron 1996: 216
[130] Gloucestershire Archive: MS, D 5554/10 (1658), (fragile, read in transcription)
[131] Sherratt, Don, 2000: local history booklet, History of Taynton, 24

for sword-hilt pommels cast in shapes like bird or dog heads. The initial skilled brass-founders were recruited from Solingen in the Rhineland, where swords had been made since the Middle Ages.[132] The Hounslow sword-makers escaped fines by casting imported brass, rather than making their own. However, a London armourer, who made and forged his own brass near London, was not so lucky. In 1634, the Society of Mineral and Battery Works heard that 'the Brazier of the Helmatt (helmet) in Cornhill', London, had set up his own unauthorised brass battery works near Erith in Kent. Many seventeenth-century braziers produced armour or military horse trappings, including various combinations of curb and snaffle bits for bridles, often embellished with decorative escutcheons or cheek-pieces. The Society fined the Brazier of the Helmatt, whilst deciding whether or not to suppress his brass mill.[133] The site was probably three kilometres south-east of Erith on the River Cray, a tributary of the Darent, just out of reach of Thames tides.[134] The site is described on a 1777 map as iron-mills (which frequently succeeded brass-mills).[135]

London was the major seventeenth-century English port for export, so most brass-mills were located nearby, particularly in Surrey, which was served by an early network of transport canals called 'navigations'. Wandsworth brass-works, eight kilometres south-west of London, lay near the confluence with the River Wandle and the Thames.[136] It was run by Cornelius van Hallen (1581-1617), a Flemish immigrant who fled to England in 1609 from Malines near Liège (Lüttich). This Flemish city, sacked by the Spanish in 1572, was renowned for beating out brass pans, cauldrons and other hollow veassels. In 1611 an account book of the Flemish community church at Wandsworth recorded 'Received of Mr Holine (Hallen) for Christmas, a quarters rent, £10'. The baptismal register for April 1653 listed, 'Cornelis, the son of William Hallen' (of whom more later).[137]

In 1604, one Thomas Steere started a Surrey brass wire works at Chilworth,[138] about fifty kilometres south of London, but only after he had 'intised or perswaded' expert Tintern workers to work there.[139] At Chilworth, Tintern-designed water-wheels and machinery were installed, and Steere proposed to produce brass chains, clasps, fish-hooks, knitting needles, mouse-traps, bird cages, window lattices, buckles and curtain rings. The Society of Mineral and Battery Works, concerned that Steere's factory

---

[132] North 2005: 24-25
[133] British Library: MS, Loan 16/2: folio 69,
[134] National Grid Reference TQ 5275 7545
[135] Andrews, John and John Dury, 1777: map of 'London Environs'
[136] National Grid references SU 168 760 and TQ 255 749
[137] Hallen 1885: 15-20
[138] The National Archive: MS, E 134/2Jas1/Hil 12 (1603-4)
[139] Brandon 1984: 84

had easier access to London markets than their own Tintern factory, suppressed it in 1606.[140]

Two other Surrey mills, Abinger and Wotton, lay on the private estate of seventeenth-century landowner, Richard Evelyn of Wotton House, ten kilometres east of Chilworth.[141] In 1622, Richard Evelyn had partially converted Abinger powder mill, also on the River Tillingbourne,[142] to lease out as a battery mill producing cauldrons, pans and scales, especially those needed for gunpowder manufacture, where brass, being spark-free, was widely used.[143] In 1627, Richard Evelyn leased out a former gunpowder mill in the grounds of his Jacobean mansion to Peter Brocklesby, a pewterer from Holborn, London.[144] In 1628, Peter Brocklesby also took the lease of Abinger mill, where he could refine copper sufficiently to produce brass suitable for wire-making.[145] The brass was made at Wotton, where it was cast into plates before being returned to Abinger to be beaten into finer sheets to cut into strips for drawing to wire.

The private grounds of the country mansion at Wotton kept Brocklesby's brass mill out of sight, so he managed for several years to evade paying for the Society of Mineral and Battery Works annual brass- and wire-making 'privilege'. However, late in 1633 he was detected, and had to pay for a 10-year 'privilege' to use one or two furnaces to make and cast plate brass, but he was banned from making wire without written permission from other wire-makers. Brocklesby went ahead anyway, and by 1639 he had 12 years' experience in brass wire-making, selling it successfully to pin-makers in London and elsewhere.[146] The devious James Ledges, one of Brocklesby's London financial partners, quietly noted that Brocklesby had never gained the required permission from other Surrey brass wire-makers before starting his mill.

Brocklesby had initially employed foreign brass-makers at Wotton, being a pewterer himself, but by 1644 his foreign experts had trained several English workers.[147] John Evelyn, the diarist, later recalled that Brocklesby's brass-mills had battery-works, a foundry and wire-drawing workshops with a water-powered *ingenio* from Sweden. This dynamic device for drawing wire, was operated by men sitting 'harnessed in certain swings, taking hold of the brass thongs fitted to the [draw-plate] holes, with pincers fastened to the girdle which went about them; and then with stretching forth their feet against a stump, they shot their bodies from it, closing with the plate again'.[148] John Evelyn's 1653 sketch of the Tudor-chimneyed Wotton House, depicts

---

[140] Stringer 1713: 192-198
[141] National Grid reference TQ 1206 4695
[142] National Grid reference TQ 1107 4710
[143] Brandon 1984: 21
[144] British Library: MS, Loan 16/2, folio 84, Court Books, Society of Mineral and Battery Works
[145] Aubrey 1710: I, xlviii-1
[146] *Calendar of State Papers Domestic* 1639: 216
[147] British Library: MS, Loan 16/2, folios 83-84v,
[148] Aubrey 1710: 1, xlviii-1

the Pigeon House pond north-west of the house. From the south-west corner of the pond, a stream flows between the two simple brass-mill buildings, one with a smoking chimney, before dropping to flow beneath a track.[149] A later undated drawing of the redesigned house shows a tall 'Brewhouse' standing on the same site.[150] By 1667 the Abinger copper refinery had stopped functioning.[151]

In 1638, James Ledges bartered with the Society of Mineral and Battery Works to gain himself a 21-year privilege for making brass battery and wire.[152] He leased a mill at Ember, Surrey,[153] near Thames Ditton, consisting of a four-bay house and mill, three-bay barn, stable, two corn mills, orchard, garden and pasture.[154] It lay astride the Ember River between Thames Ditton and East Molesey parishes, at a bifurcation of the River Mole, about 250 metres downstream from the landowner's mansion, Imber Court. In 1639, Ledges carried his first annual £100 'privilege' payment to the Court of the Society of Mineral and Battery Works in London, while the king was away trying to suppress a rebellion on the Scottish Borders. The 18 Society members who happened to be present that day simply shared the cash out between themselves as they sat around the table.[155]

In 1639, James Ledges claimed to the Society that the wire he made at Ember was better than imported wire, though Brocklesby retorted that Ledges had no idea how to make brass wire. Ledges indignantly reminded the Society that Brocklesby was forbidden from making brass wire at Wotton and Abinger without written permission from other makers, and was contracted to sell it only to him – Ledges. He asserted that Brocklesby not only *made* wire but 'sold it in hugger mugger unto others'. Brocklesby countered that he had spent huge sums on equipping his wireworks, and that Ledges had neither wire-making experience nor any right to stop others from making or selling it.[156] During the English Civil War (1642-51), Ledges cunningly managed to dodge all payments for his Ember brass-making privilege.

Around 1630-37, on the English country estate of the Byron family, at Newstead just east of Nottingham (200 km north of London), brass battery and brass wire were produced at sites along the River Leen – at Wier (wire) Mill near the river's source,[157]

---

[149] British Library: MS, Evelyn 78610 G and H (*c*.1653-4)
[150] British Library: MS, Evelyn 78610 J (*c*.1653-4)
[151] Brandon 1984: 18, 21
[152] British Library: MS, loan 16/2. folio 79r/v, Court Books, Society of Mineral and Battery Works
[153] National Grid reference TQ 1459 6724
[154] Surrey History Centre: MS, 2629/16 (1607), survey of Imber Court estate
[155] British Library: YD 2005 b.234: 2, Greenwood, G.B., 1980, unpublished typescript, Elmbridge Water Mills
[156] British Library: MS, loan 16/2, folio 83; *Journal of the House of Commons*, 1639, 216-217
[157] National Grid reference SK 4358 3511

and Forge Mill,[158] a former iron mill for the canons of Newstead Abbey.[159] In 1615 young (later Sir) John Byron (1592-1652) had leased Forge Mill (Bulwell Forge iron works) for 21 years.[160] In 1631, John Reeves of Fleet Street, London, an 'ingenious Fownder', who knew Sir John Byron and was 'well known to his Majesty', was summoned before the court of the Society of Mineral and Battery Works. They had heard that Reeves used water power to manufacture brass at Papplewick on the River Leen seven kilometres north of Nottingham, without paying the Society for a 'privilege'. Reeves argued that he was not usurping the Society's 'privilege', because his equipment, which was capable of drawing 1.32m of wire at a time, was all his own invention.[161] When Reeves did apply for a 'privilege' to use scrap brass and calamine to make battery brass in Nottinghamshire, the Society declared the sums of money he offered 'so meane that they would produce little proffytt to the Company'. Nevertheless, they found Reeves 'an Ingenious man' and from December, 1631, Reeves was permitted to produce brass battery and wire within a 14-kilometre radius of Nottingham.[162]

A month later, a joint 21-year 'privilege' was granted to Sir John Byron, John Reeve and a third partner, to make wire from brass, steel or any other metal, using 'Engines devised or to be devised' by them.[163] By 1634, Byron had invested £800 in his brass and wire works, having located calamine only 40 kilometres away in Derbyshire. In 1634 he petitioned the Society to give British wire a chance to compete, by controlling imports of cheap foreign wire. Byron promised that if the Society applied this control, he could, within two years, set up enough works to supply the entire kingdom, and would only raise his wire prices if raw materials prices rose.

Sir John Byron was a new style of brass-works owner, who claimed to have brought 'great industrie and changes' to brass-wire manufacture. In 1635, he declared that his brass wire works were up to standard and could melt and cast metal and draw it into wire 'as good as any foreign wire'. The investors he had tried to attract had been discouraged by three things – the scale of investment needed; the Society's high charges for 'privileges', and competition from unchecked import of cheap foreign wire.[164] The Society granted Sir John a cheaper 'privilege' for battery, foundry work, calamine mining and wire-drawing. Through his royal connections, he found a market in France, where his brass wire sold well and was 'approved to be good wyer'.[165]

---

[158] National Grid reference SK 4547 3472
[159] Nottinghamshire Archive: Newstead Abbey estate map, 1817
[160] Derby Local Studies Library: MS, Kerry 4681, folio159-160
[161] British Library: MS, Loan 16/2 folio56, Court Books, Society of Mineral and Battery Works
[162] British Library: MS, Loan 16/2, folio59v, 61v
[163] British Library: MS, Loan 16/2 folio63v, 64v, 66
[164] British Library: MS, Loan 16/2, folio 68
[165] British Library: MS, Loan 16/2, folio76v

When Sir John Byron's venture was threatened by cheaper foreign imports, two court ladies, Dame Mary Hamilton and Dame Elizabeth Savage, sprang to his support. They petitioned King Charles I on behalf of shareholders who had invested large sums in Byron's brass works, arguing that the plentiful newly-discovered Derbyshire calamine should be fully exploited. They protested that Dutch and German merchants had for years tried to 'beat the English out of this trade' by buying up all the Swedish copper while selling quantities of Continental brass goods in England at reduced prices. The court ladies demanded higher taxes on imported brass and brass wares. Lady Savage persuaded the king to impose an import tax on brass for 21 years at £100 per annum, rather than paying her the £1,000-worth of pension arrears he owed her for 'long service done to his Majesty and the Queen'.[166] The king agreed to the bargain of this spirited woman so, in November 1635, the Society of Mineral and Battery Works petitioned for restraint on the import of foreign brass wire.

Late in 1637, when Sir John Byron gave up his brass-making 'privilege', it became available to others, and was granted to James Ledges at Ember Mill. The following month, with the Civil War looming, Sir John Byron became Deputy Governor of the Board of the Society of Mineral and Battery Works,[167] at the time probably more lucrative then than brass-making. Byron's short-lived but significant brass industry was remembered in 1668 by a fellow-member of the Society of Mineral and Battery Works, as an 'excellent Brass Manufacture in Nottinghamshire', which 'employed no less than 8,000 men daily' (including miners, carters, traders, loggers and so on), an employment figure which might have been true of many European brass works. By 1670, however, the brass works had decayed.[168] The Nottingham brass industry was ambitious in scale, when so many economic and political odds were stacked against it.

From before 1629, a William Berry ran a brass wire mill probably at Baptist Mill near Stapleton, three kilometres north-east of Bristol.[169] In 1610 'Baptist mylle' straddled the River Frome north-east of Bristol.[170] William Berry's name is absent from Bristol parish registers, so he is likely to have been a Protestant Dissenter, probably an Anabaptist. In 1573, a William Berry, alias Kyes or Keyes was a Bristol wheelwright,[171] so William Berry, a Bristol brass-maker who in 1629 employed one or more former Tintern brass-workers, may have been born Keyes, or related to Keysar, a former immigrant Tintern brass-maker.

In July 1629, the court of the Society of Mineral and Battery Works caught wind of Berry's brass works making wire 'and other things', using water wheels and 'other

---

[166] *Calendar of State Papers Domestic* 1634-5: 471-472
[167] British Library: MS, Loan 16/2, folios 70-74 and 77v
[168] Pettus 1670: 33
[169] National Grid Reference ST 6029 7427
[170] Bristol Archive: microfiche 04480, Kingswood Forest map,
[171] Bristol Archive: MS, FC/BB/1(a), 1-2, index to Bristol Burgess book

Engines and Tooles', and employing a Tintern wire-drawer. In November 1629, William Berry appeared before the Society's court, and was fined for making brass wire without a 'privilege'. He paid up and was permitted to make 'yellow [brass] wier' for ten years.[172] By the early seventeenth century, Anabaptists were particularly strong in Bristol, and in 1639 an Amsterdam pastor visited to preach to them and hold the first of several mass river baptisms and thousand-strong Dissenter meetings by the small River Frome beside Baptist Mills.[173] By 1675, some Anabaptists were convicted and imprisoned, causing rioting among their followers, but two-thirds of the Bristol population were thought to be sympathisers.[174] Stolberg brass workers also favoured the Anabaptist religion, suggesting a possible connection.

Brass, being a strong, hard, rust-free alloy, was especially suitable for making wire for use in wool-carding, lace-making and paper-fastening. As mentioned earlier, brass wire was in heavy demand for pins for wool-cards in the Cotswold sheep-farming area, where wool for cloth was produced, spun and woven for export from the port of Bristol. Wool cards consisted of a pair of flat rectangular wooden implements, each with a wooden handle, with tightly spaced, stiff pins (wires) protruding through tough leather covers on one side of each implement. Wool fleece was combed between the two cards in each direction to straighten, sort out and gather up the wool staple preparatory to spinning it into woollen yarn. Because iron wire could rust and spoil the wool, brass-wire wool cards, though costlier, were preferable. During the 17th and 18th centuries, different thicknesses of brass wire were made and used throughout Europe to manufacture pins, wire for wool cards, needles, chain-links and rods.[175]

Central Europe. Rust-free brass wire was also in high demand for lace-makers' pins, particularly along the mountain range known as the Erzgebirge or Krušné Hory. These mountains formed the border between Saxony and Bohemia (today's German-Czech border). Below their northern slopes, in Saxony, ran the Auer river valley (*Auertal*). The first Auer Valley brass works was at Rodewisch where two smaller rivers, the Goltzsch and Wermesbach, met. This region had good transport connections by the historic Charcoal Way or '*Kohlerstrasse*' to the extensive forests of Bohemia, passing over the Erzgebirge, where the thriving lace industry required huge quantities of brass pins.

In 1593, Peter Ficker, a raft-skipper from Zwickau, bought a long-derelict iron battery mill at Rodewisch, with its collapsed weir, rotten water wheels, unworkable hammers, and shredded bellows. Ficker, already an ironworks proprietor, swept in like a new broom, and by 1599 a new weir, battery hammer and wire-drawing workshop were in operation. The transfer from iron-making to brass manufacture was profitable in times of prosperity, but could prove ruinous when economic conditions turned

---

[172] British Library: MS, Loan 16/3 (1629-30), folios 37, 38, 42
[173] Ivimey 1811: 2, 523-524
[174] *Calendar of State Papers Domestic*, 1675-6, (1975 edn): 10; 1677-8 (1977 edn): 426
[175] Schleicher 1974: 35

BRASS FROM THE PAST

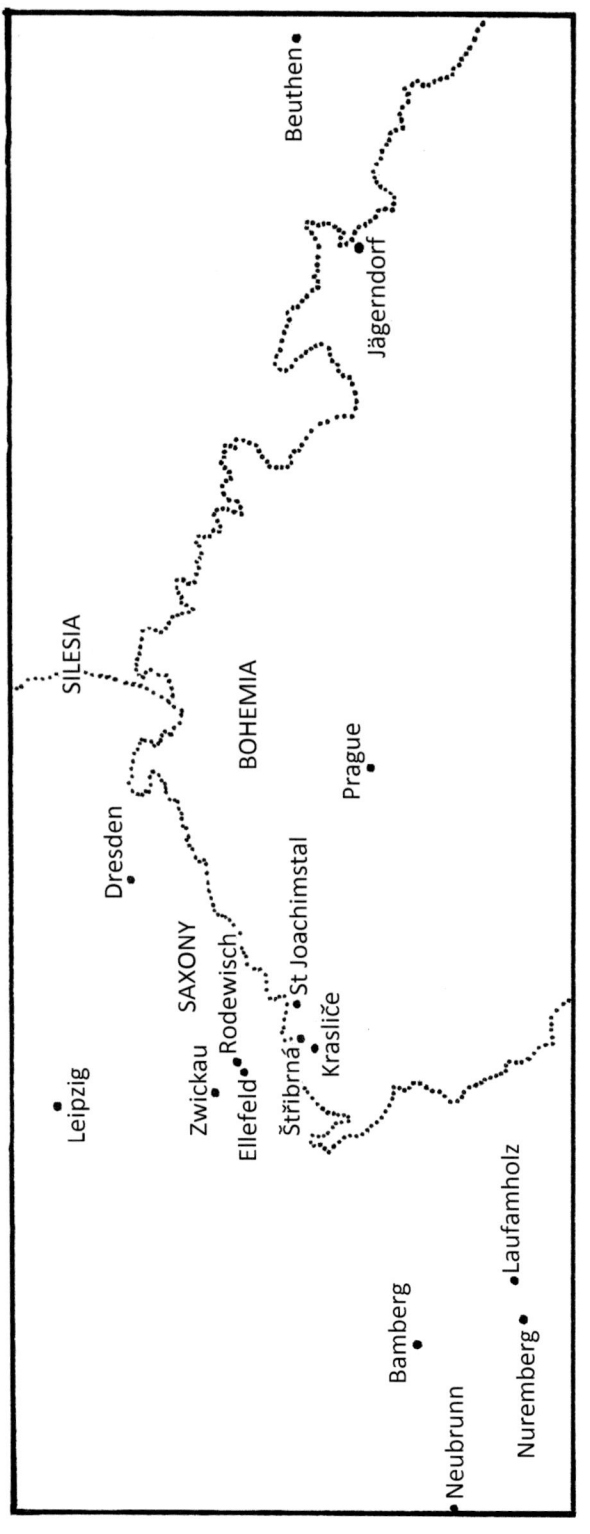

Figure 54. Map of the Bohemia area showing some sites mentioned in the text

harsh. From 1603, Peter Ficker gained a 15-year monopoly on brass production in the region, to keep competition at bay and guarantee some regional protection from taxation. Such 'privileges' encouraged chosen businesses to grow unhampered, but ultimately discouraged free competition.[176]

Initially, Rodewisch brass works struggled to make a profit. Three and a quarter tons of brass were produced in 1610-11, but less than two tons were sold, so Peter Ficker took on three partners to share the financial risk. One of them was Matthius Gnaspe from Ilsenburg, a Harz brass-works that was declining due to poor leadership and copper shortages, and was forcing its skilled workers to seek employment elsewhere. Gnaspe recruited experienced Harz brass-makers, battery-workers, wire-drawers; scrapers, etchers and polishers. and contracted for copper deliveries from the Harz and from Krasliče copper-works in nearby Bohemia. A few workers from Neubrunn (between Frankfurt and Nuremberg), Lübeck and Hamburg were hired by the hour. The Rodewisch works doubled its sales within two

---

[176] Gericke 2008: 41-42, 53

years but still failed to reach the 25-ton yearly sales needed to survive. However, between 1611 and 1618, just before the onset of the Thirty Years War (1618-48), the annual brass output rose to just over 54 tons.[177] The 25-ton annual brass output estimated for survival at Rodewisch must have been valid for many small brass works, and those which produced less clearly struggled. Ficker finally dismantled the old plant at Rodewich, moving all the equipment to Ellefeld six kilometres up the valley.

As the Thirty Years War approached, copper deliveries became uncertain so, between 1618 and 1622, a backlog of smelted copper supplies grew, because the Nuremberg and Leipzig markets would not guarantee advance payments to their producers. In the early 17th century, only Mansfeld (southern central Germany) produced copper in large quantities, but the Thirty Years War disrupted its production, and led to a general copper shortage. In 1615, Ficker obtained zinc furnace accretions from Goslar, and calamine from Brilon in the Sauerland hills, Westphalia. However, by 1642 Brilon supplies were declining,[178] and Iserlohn's calamine mines had caved in, caught between opposing troop movements in the Thirty Years War, so its calamine was not available either.[179]

Performance at Ficker's Ellefeld mill was therefore poor, so, around 1619, Matthius Gnaspe deceitfully started up his own unauthorised battery-works at Ellefeld, drawing off many skilled brass-workers, but it soon ceased production. By 1632, when Peter Ficker's son set up a new, authorised, brass-works in Ellefeld, the Thirty Years War was ravaging southern Saxony. Hostilities and plague struck the Auer valley hard. Lying close to the border between Saxony and Bohemia, the brass-works was plundered, the workers were half-starved and five units of Croatian troops were billeted on local families. In 1640, an attempt was made to re-start Rodewisch brass mill, but only a nucleus of experienced workers remained.[180] In the late 1670s, Christian Körber, a former Rodewisch partner, lured Ellefeld workers to his new brass-works much further north, at Grünau in Thuringen, on the Loquitz stream five kilometres south of Leutenberg. It was placed under the same management as brass-works at Hockerode, and Arnstadt, respectively five kilometres north and 40 kilometres north-west of Leutenberg. Not until 1688, well after the Thirty Years War, was permission gained to re-start the Ellefeld works.

About 500 kilometres further east, in Silesia (now Czech-Polish border), lay Jägerndorf (Krnov) brassworks, supplied with calamine from nearby Tarnowitz and Beuthen (Bytom). In 1595 Joachim Friedrich, later Archduke of Brandenburg, had inherited Jägerndorf and its brassworks. But around 1631, when Habsburg Emperor Ferdinand II, a militant Catholic, drove out all Protestant calamine miners and removed their

---

[177] Gericke 2008: 44-45, 52, 55
[178] Gericke 2008: 41, 44, 56-57, 60
[179] Hildebrand 1983: 33
[180] Gericke 2008: 56, 58-59

churches, calamine supplies dried up and the Tarnowitz and Beuthen calamine mines fell into disuse. Not until 1660, long after the Thirty Years War, was calamine once again extracted there.[181]

The Vienna-based Catholic Habsburg campaign (c.1590-1600) to eradicate Protestantism from Austria forced Protestant brassworkers to emigrate, some to Bohemia (Czech Republic), which possessed rich ore deposits in the Kružné Hory (Erzgebirge, or ore mountain range). From about 1550 to 1792, the Bohemian brass mill at Stříbrná (Silberbach) stood at the western end of the Kružné Hory, three kilometres north-east of the copper-mining town of Krasliče (Graslitz). The original mill on the Stříbrná (*Silberbach* or silver stream, named after the silver that was once panned along its bed) had one mill-pond fed from the fast-flowing mountain stream, later dammed at a narrow rock gully to produce an upper mill-pond to ensure a constant flow. The north side of the steep, forested Stříbrná valley held igneous rock, but the mountain slopes on the south side held metamorphic rock, including seams of zinc carbonate ore ($ZnCO_3$). The nearby Krasliče (Graslitz) copper-smelting works supplied copper, transported three kilometres uphill in horse-drawn carts. Krasliče, a free mining town since 1571,[182] was second only to St Joachimstal, further east along the mountains, which produced the silver for Taler (dollar) coins.

By this time, numerous European brass works, like those in Saxony, Silesia and Bohemia, were producing brass wire, plate and sheet brass or semi-prepared brass vessels to be turned into cauldrons, bowls or basins. A smaller number of brass works also produced finished brass artefacts, which were more often crafted in the numerous individual foundries and forges. Nuremberg craftsmen were famed for producing outstanding brass scientific instruments including hour-glasses, sundials, quadrants, equinoctial rings and nocturnals. Around 1600, ivory diptych sun dials were made with brass fittings, and later in the century came horizontal dials and drafting instruments like dividers. Thomas Tuher and Paul Reinmann were two important instrument makers. Mechanical brass clocks displaying the planetary motions were made for Habsburg Emperor Maximilian II (1564-76), and a bass trombone was produced in Nuremberg in 1612,[183] probably, it is thought, for the City Council to show to potential buyers visiting this new free-trade city.

The aristocratic vogue for Cabinets of Curiosities undoubtedly encouraged the production of finely crafted brass and other objects for private collections. The Habsburgs, particularly Emperor Rudolf II (1552-1612), who had a passion for beautifully made scientific equipment, inspired other European princes to collect unusual clocks and rare scientific and astronomical instruments, sundials and astrolabes.[184] In the

---

[181] D'Elvert 1866: 166-167
[182] Jaroslav Zaplatal, personal communication
[183] Germanisches Museum, Nuremberg: Ml.168
[184] Haedecke 1970: 85

17th century, scientific research was carried out under the patronage of university and court circles. Most early seventeenth-century brass scientific instruments and clocks were created in Nuremburg and Cologne,[185] which naturally created a demand for yet more high-quality brass.

Wire, weaponry, including gunners' sights, and a great range of everyday brass objects were produced in Nuremberg. The lively manuscripts of the Twelve Brothers community of retired Nuremberg craftsmen, show a wire-drawer using two hands to turn a handle to wind two spools (fixed to a bench) to draw wire through holes of decreasing size.[186] Brass weights for markets, first cast in Nuremberg, consisted of a bucket, about 20 cm high, containing successively smaller nesting buckets, each of a specific weight and found on a number of seventeenth- to eighteenth-century archaeological sites.[187] London traders petitioned the Corporation of London for all weights to be made of brass (because soft lead was too easily tampered with) and their request was granted in September 1614.[188] The weights ranged from ½ oz (14 grams) to 56 lbs, (25 kilos) and were universally used on weighing-scales for commerce and trade, gradually giving way to flat disc weights.[189] Privately issued metal tokens, or jetons, often made of brass, sometimes replaced coinage for limited or local transactions.

Nuremberg is famed for its high-quality finished brass objects, but its mills also produced brass from ores. In 1614, 96 tons of calamine were delivered to Nuremberg from two ships berthed at Bamberg, north of the city (the furthest point up the River Main reached by shipping).[190] The calamine was probably for Hammer brass mill at Laufamholz in outer Nuremberg, run by four generations of the Kanler family, who had a subsidiary battery hammer at Simmelsdorf.[191] Timber supplies for fuel and charcoal were constantly being negotiated with the forest authorities.[192] In 1619, a merchant named Georg Loos (probably from Aachen or Stolberg) introduced to this brass mill a 'Netherlands' method (*Niederlandische Art*) which economised on wood fuel.[193] It involved embedding the brass crucibles up to the rim in ground-up charcoal within the chamber of a reverberatory furnace. After four to seven hours, depending on the total cementation time, a fireproof-brick bung was released and another tub of charcoal was poured in to replace any that had smouldered away.[194] From 1597 to 1636, Nuremberg was supplying sheet brass and finished goods to English merchants.[195]

---

[185] Schleicher, 1974: 35
[186] Nurnberg Stadtarchiv: MS Mendel, Landauer I, Amb, 279,.2°, ffolio119-120 (1649)
[187] Colonial Williamsburg Foundation, Virginia: CWF 2 J, set of nesting weights
[188] Connor 1987: 330
[189] Colonial Williamsburg Foundation, Virginia, weight, CWF 0119 1344 1634-19A
[190] Dietz 1921: 199
[191] Priesner,1997: 26
[192] Stadtarchiv Nurnberg: MS 455/14, 1640.
[193] Wittek 1984: 121
[194] Hachenberg and Ullwer 2013: 181
[195] Unger 1966: 65

At the outset of the Thirty Years War (1618-48), Nuremberg brass manufacture was flourishing. Between 1621 and 1626 Thomas Kanler at Hammer Mill, Laufamholz, added two new calamine crushing-mills to his existing three. Threat of war, however, meant that Hammer Mill, seven kilometres outside the Nuremberg city walls, had to be protected with its own wall. A 1628 map shows the River Pegnitz, recently canalised and equipped with a weir close to the industrial settlement, which is enclosed by a square wall. Enclosed by the wall are the factory, battery works, administrative buildings and powerful water wheels. In front of the west gate is the great barn. An extensive garden fits between the works and road, and across the river lies the mill-yard.[196] The small industrial settlement of Hammer is still discernible today. Export opportunities (outside war zones) were excellent for this brass works, because major trade routes passed through Nuremberg, leading from Hamburg to Venice, from Breslau to Geneva and from Vienna to Brussels.

After the death in 1630 of Thomas, the last Kanler at Hammer, circumstances at the brassworks deteriorated rapidly. Between 1631 and 1636, famine and wartime ravages reached the industrial area on the Pegnitz, and production declined. The battery works, mill and dwelling houses were destroyed by the ravages of both the Thirty Years War and the local Markgraf Wars of 1640 (and more recently and seriously in the Second World War).

Following the deeply traumatic setbacks of the Thirty Years War, the Hammer battery works at Laufamholz were reconstructed as walled, fortified factory premises. The brass-works business grew into a flourishing industry which, over the following three centuries, would send its products well beyond Europe. Most of the surviving buildings were built at this time, with the far-sighted intention of supporting the social structure of the settlement,[197] with workers' apartments, an inn, stabling, lodging-houses, a corn mill, furnace-house and foundry. In 1647 the brass workers even petitioned for a school for their numerous children. The workers' accommodation was free – a concession extended to widows and dependents or to sick and aged men. The innkeeper at Hammer was forbidden to hold wedding or baptismal parties, but was obliged to provide beer for the brass workers.[198] Charcoal and fuel for the brass furnaces demanded so much timber from the nearby forests that it had to be used more cautiously. Nuremberg was a free city under Habsburg influence, but to its south lay the decidedly Habsburg region of imperial Austria (for map of Austrian sites, see previous chapter).

Austria. Timber at Reichraming brass mill in its steep eastern Austrian alpine valley about 350 kilometres south-east of Nuremberg, was also becoming exhausted. In 1604, the forestry authorities complained that Reichraming brass works was always

---

[196] Staatsarchiv Nurnberg: Rep 58, Nr. 410, map, Johan Bien, 1628
[197] Hussennether 2010: 5
[198] Wittek 1984: 118-122

demanding more wood, so they introduced new forestry regulations to share out timber supplies between industry, on the one hand, and house building and repair on the other.[199] Mountain passes were constructed to allow the import of timber from more distant valleys.[200] By 1607 the brass-workers' great thirst, caused by heat from the furnaces, caused the mill-owner, Leonard Manstein, to start brewing beer for them.[201] However, in 1614, when Hanns Köberer (son of a former partner) ran Reichraming brass works, forestry officials objected to his tax-free handouts of beer and wine. A pencil sketch dated 1613 shows the brass-works buildings and water-wheels.[202]

The Thirty Years War (1618-48) brought untold misery to millions, but benefited the makers of military brass buttons, horse accoutrements and gun parts.[203] Reichraming brass works, however, failed to exploit the situation because its main customers – the knife-makers of Steyr – were too specialised. As knife production slackened, less decorated brass knife-hafts were needed. Köberer's income dropped,[204] workers were laid off, and in 1623 the brass works was temporarily closed. Hanns Köberer was fined 100 taler for financial fiddling, and was locked up in the castle.[205] In 1624, brass started fetching a slightly better price, so brass sheet and wire were transported downriver, mostly to Vienna, where, however, sales profits were heavily taxed.[206]

In 1628, Hans Egger bought Reichraming brass-works for 6,200 florins, acquiring extra forest leases for timber and adding wire-drawing workshops.[207] Emperor Ferdinand II granted him a 'privilege' as sole brass supplier to the province of Steyr,[208] but in 1639, the craftsmen of Steyr complained that they could import better casting-brass for only 40 florins but were forced to buy Reichraming brass at 44 florins.[209] By 1651, however, Egger had bought more forest (for charcoal) and could compete on price. By clearing and burning forest and using horse power to move timber, he earned a considerable landlord's profit. In one day, one of his men with four strong horses could shift timber from forty thousand square meters of forest.[210]

Lienz brass mill, about 200 kilometres further south, in the mountainous southern Tyrol, used Tyrolean calamine. Around 1600, Lienz brass production gradually increased

---

[199] Aschauer 1953: 320
[200] Aschauer 1953: 315
[201] Brunnthaler 2000; 123
[202] Brunnthaler 2000: 18, reproduction of pencil drawing dated 1613
[203] Westermann 2002: 86
[204] Aschauer 1953: 315; one Vienna centner weighed 56 kilos
[205] Brunnthaler 2000: 113
[206] Stiftsarchiv Seitenstetten: MS, 53/V (1624)
[207] Landesarchiv Linz: MSS, Herrschaft Steyr, 195/5-11 (1628)
[208] Aschauer 1953: 315
[209] Landesarchiv Linz: MS, Herrschaft Steyr, box 999 (1607-61)
[210] Brunnthaler 2000: 113

Figure 55. Lienz brass battery works in ruins after the fire of 8 April 1609

from about 11 tons to 93 tons annually.[211] A plan view of the town, dated 1606-8 shows the brass works just before it was destroyed by a great fire in 1609,[212] which ruined the manager's house, furnace-house, battery-works, scraping workshops, calamine mill, trading offices and timber warehouse. A contemporary pen and ink drawing records the surviving remains of the brass-alloying, battery and scraping workshops.[213] After this blaze – a wall of flame driven by strong winds – four fire-fighting points were installed along the main street – Messinggasse or 'Brass Lane'. In 1643 the brass works returned to Tyrolean state ownership.[214]

About fifty kilometres due east of Lienz, at Möllbrücke in Kärnten (Carinthia), another mountainous part of today's south-eastern Austria, Ludwig von Dietrichstein received a 'privilege' in 1599 to set up a brass-works. The area was at risk from snow, storms and forest fires, but in 1600 he produced just over four and a half tons of brass, doubling it to nine tons by 1617. Transport must have been problematic, but calamine came mainly from Raibl and from the Jauken mountain range. Copper came initially from Gail-valley mines and the Finkenstein area, but later from Steiermark and Upper Hungary (now Slovakia). Around 1628, the Pachmann family from Villach bought the Möllbrücke brass mill and still made brass there in 1636. Further south-

---

[211] Ucik 2002: 172
[212] Vienna Haus-, Hof- und Staatsarchiv: Kartensammlung Ke3-6/5, codex W 231/9, Lienz
[213] Tiroler Landesmuseum Ferdinandeum: pen and ink sketch of Lienz fire, dated 1613
[214] Heinricher 2006: 10

west, in the fertile olive and vine-growing Lecce area of northern Italy, a metal-works with furnaces and water-powered battery hammers was forging iron, copper and brass.[215] It is clear that individual factories were producing brass throughout much of Europe in the mid 17th century, especially where calamine, copper, timber and water-power were readily available. The other workers involved with brass were of course the craftsmen who wrought and cast the alloy into individual items with appeal to specific customers.

Most brass craftsmen worked in quite small workshops, foundries and forges scattered throughout the cities, towns and countryside, some making specific objects, but most producing a wide variety. A valued piece of kitchen equipment was the spit-jack with decorative brass front, for turning meat to roast over a fire. Northern European households counted warming pans among the most valuable of their possessions. These flat pans with a single-hinged lid and long handle were filled with hot embers and a servant slid the pan back and forth across the bed for up to half an hour, to dry and warm it. English brass warming-pans had longer handles than Dutch examples, while American-made warming pans had a crook-like hook at the end of a slender handle.[216]

America. From the 1640s, some American settlers from the Netherlands made music from simple mouth-harps with brass flanges that twanged, and others traded brass cauldrons and cooking pots to Native Americans in return for furs.[217] As discussed above, Netherlands and Aachen brass workers were under duress at that time, but Stolberg was increasing its brass output. In 1647, an early brass founder in America, Joseph Jenks, an English immigrant from London, opened an iron works at Lynn, 15 kilometres north of Boston, Massachusetts, where he subsequently added water-mills and made brass wire for wool carding.[218] His move to America coincided with the English Civil War, when almost all British brass works were forced to close. To make brass wire at Lynn, he must have imported good Stolberg brass.

The brass artefacts of Europe and America were similar, because the people who possessed the most brass artefacts in colonial America and the Caribbean originated from Europe, as did most of their artefacts. Many brass consumers were now motivated by an urge to achieve a recognisable self-identity and to belong to a successful social group, with interests in science and related subjects. Seventeenth-century European artefacts include many household, maritime, scientific and industrial items for which brass was chosen for its special qualities. The spiritual appeal and power of brass was still visible in church, temple and mosque furnishings, decoration, tools and images. The mysterious or magic dimension is also evident from the disappearance

---

[215] Frumento 1958: 101
[216] Higham 2007: 40-41
[217] Huey 2009: 18
[218] Calcutt, January 2000: 1

into the interior of a large quantity of brass imported along the Guinea Coast, and its varied re-use in different areas to create sacred, symbolic or ceremonial items. People in the spiritual and community-centred environments of the Guinea Coast, were relating to brass as a medium for shared experience and for communion with the spirit world, or for expressing power and retaining control. It can be argued that brass production burgeoned due to technological progress and business acumen. However, the aquisitive motivation of individuals and the advantages of golden-coloured, rust-free, non-magnetic, spark-free brass were clearly strong factors too, attracting a huge market for brass that would further expand during the later 17th century.

For much of the world-wide brass industry, the early 1600s were a period of struggle to continue production in the face of hostilities, the main advantages falling to those who supplied arms, armaments and military buttons and helmets to the winning sides The other areas which benefited were those that lay, like Sweden and parts of India, slightly outside the main battle areas. Many were undoubtedly hoping for more peaceful future years.

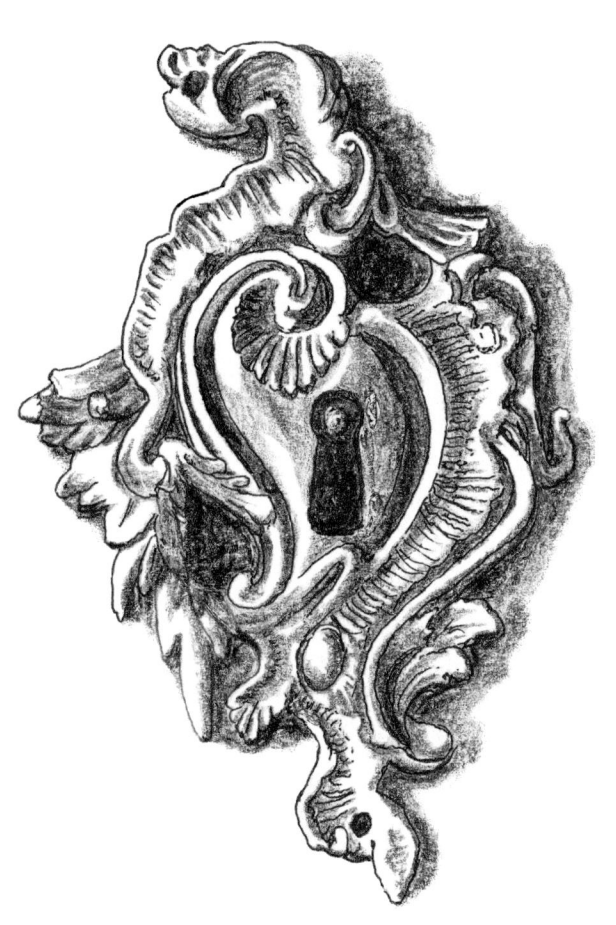

Tobacco box, 1762, made in Iserlohn, a source of zinc carbonate ore. The lid shows the conquest of Martinique in 1762, the base bears the initials of King George III of England. © Jan R. Schäfer, courtesy of Iserlohn Stadtmuseum

Chapter 6

Continuity and conflict in Europe
c.1650-1700

The brass industry was affected as several parts of the world emerged from years of conflict, including the Thirty Years War in central Europe, the Civil War in England and the conquest of China in 1644-7 by the Manchu from outside the Great Wall, founders of the Qing dynasty (1644-1912). Under the Qing dynasty, the wider-scale production of zinc metal began in China. Sites along the winding Yangtze valley distilled quantities of zinc metal to satisfy the increasing coinage needs of China's growing population, but the best brass was reported, in 1667, to come from Yunnan province,[1] where early zinc-distillation sites existed.[2] Chinese zinc distillation would have far-reaching global effects.

Chinese zinc distillation took place in vertical retorts with external condensers added at the top (as mentioned in chapter 4). In furnaces used in Guizhou province, south-western China, vertical retorts stood on a metal grid within the combustion chamber. The fuel consisted of briquettes moulded from measured proportions of coal-dust, yellow clay and furnace ashes. The retorts were propped up by briquettes and held slightly apart in the furnace by pieces of slag, to let the heat circulate evenly. When distillation was complete, the zinc was scraped from the retort lid, removed from the pocket, gently re-melted in a wok and stirred. Impurities floated to the surface and zinc-oxide residues were burned off. Ingots were cast from the distilled zinc metal.[3]

Yangtze River metallic zinc was at first intended primarily for Chinese coinage containing ever-increasing zinc levels. However, from 1644-57 some Chinese domestic implements were also made of brass. In the later 17th century, both copper and zinc, the two major ingredients of brass, became global commodities. Japan, for example, became a major and expanding seventeenth-century copper supplier to Europe, India and China, though Japan placed a embargo on copper export from 1638-41, declaring it a war commodity.[4] Any surplus Chinese zinc metal, after coin-minting, was sold at Canton (Guangzhou, Guangdong province) and at its trading harbour, Macao (Macau), where the Dutch learned from the Portuguese to choose zinc metal, rather than sugar, as ballast to steady their ships sailing from China to South-East Asia and India.[5]

---

[1] Needham 1974: 208, translating the author of a Chinese Geography
[2] Am Chuan Fan, personal communication
[3] Xu Li 1998 : 105-116
[4] Glamann 1977 : 281-4, 287
[5] Souza 1991: 297, 300

China generally excluded foreigners, but a few Jesuit missionaries were invited to the imperial court for their knowledge of astronomy and related arts. Flemish-born Father Ferdinand Verbiest (1623-88) was sent to China in 1659, but when the regime changed shortly after his arrival, he was imprisoned and tortured. In 1669, with the accession of the Kangxi Emperor, Verbiest returned to favour and was given charge of the Beijing observatory, pleasing the young emperor, who had himself depicted amidst dragons, a celestial globe and surveying instruments.[6] His observatory was 'fitted with great instruments, made by Verbiest, a Celestial Globe of brass six Foot Diameter being reputed the best'.[7] Other instruments included an ecliptic and an equatorial armillary sphere,[8] azimuth, quadrant and sextant.

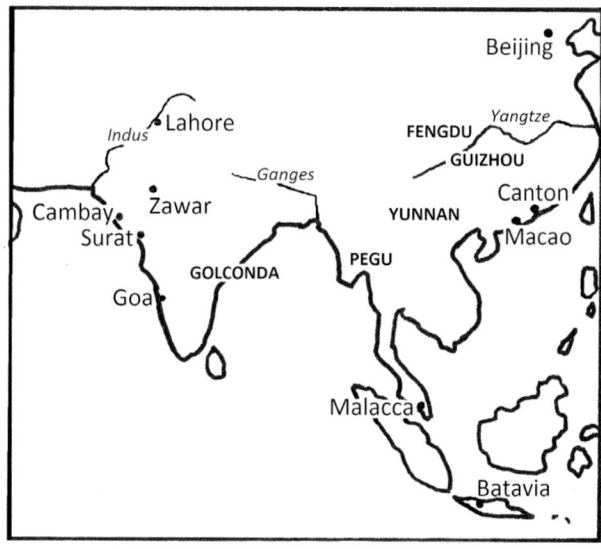

Figure 56. Map of Asia, showing some sites mentioned in the text

An armillary sphere – a spherical framework of metal bands – was centred on a small globe representing the earth. One fixed hoop represented the plane of the equator and another represented the meridian, linking the poles. Armillary spheres could carry increasing numbers of hoops, representing various planes of movement of objects in the sky relative to factors like earth tilt, elliptical orbit and seasons. An azimuth helped with astronomy and in finding bearings by measuring the angle between

Figure 57. Chinese imperial palace, equatorial sphere: diameter c.1m. 1669-1688

---

[6] Du Halde 1735, volume 1: 1
[7] *Philosophical.Transactions* 1697: 586
[8] based on Kepler's 1604 ecliptic theory for correcting distortion when using optical devices

three points, the horizon, due north and a star in the sky. Eight large imperial observatory instruments dating to the Qing dynasty were made from a relatively high-zinc cementation brass (27.6% zinc) that could be engraved with the required numerals and images, and a little added lead helped the metal to flow into the mould.[9] The instruments, embellished with imperial dragons, were described as well-cast. The astronomy introduced by the Jesuits ran counter to the Chinese calendar, which naturally angered the emperor's court followers.[10] The Chinese made brass temple images and guns, and exported quantities of brass basins and cauldrons.[11]

The Dutch actively exported Chinese zinc metal from the 1650s onwards. Dutch East India Company chronicler, François Valentijn, wrote that in 1650 the Coromandel Coast of eastern India imported *spelter* (zinc metal) from the Far East and still demanded it in 1678.[12] Evidently the eastern Indians, rather than producing metallic zinc, imported it from China, both to make use of it themselves and to trade it on to Europe. In the 1660s, a French traveller, Jean de Thévenot (1633-67), described metal he saw in Delhi as 'a certain Metal called Tutunac, that looks like Tin, but is much more lovely and fine, and is often taken for Silver; that Metal is brought from China'.[13] De Thévenot was describing either zinc metal, or a pale, high-nickel ternary form of brass alloy, which may also have been exported from China in small quanties by this time, and was sometimes mistakenly reported as tutunag by people confusing it with zinc metal.[14]

Curiously, zinc metal exports are not recorded from Surat or Cambay, the two ports most convenient for Zawar. Radio-carbon dating suggests that major industrial production of metallic zinc at Zawar, Rajasthan, started in the 14th century and finished early in the 19th century.[15] Distilled zinc from Zawar still seems to have been used in what were then northern India and Persia, where some brass astrolabes made in Lahore up to the later 17th century contained well over 30% zinc.[16] So why was zinc metal from China, rather than from Zawar, exported to Europe in the later 17th century, and why did much of India itself buy Chinese distilled zinc?

There may have been political reasons. In 1680, the Mughal emperor Aurangzeb (1618-1707) captured the central southern inland region of Golconda (today's Indian state of Telangana) for its diamond mines, and was engaged in warfare against the Hindu Maharatha, central Indian chieftains who resented both Mughal imperial rule over India and Mughal collusion with the aggressive Portuguese and Dutch traders

---

[9] Zhou Wenli 2012: 49
[10] Du Halde 1735, volume 3: 286-287
[11] Du Halde 1735, volume 2: 172
[12] Valentijn 1724-6, volume 5, part 1: 9, 33, 40, 44; volume 4, part 2: 12
[13] Thévenot 1683, 1687 edition, 46
[14] Gilmour and Worrall, 1995: 19
[15] Craddock, Freestone *et al* 1998: 42-46
[16] Newbury *et al* 2003: 360, 367-368

along its coasts. The Maharatha controlled Gujarat and the ports of Surat and Cambay, the two obvious outlet ports for distilled zinc from Zawar, which may explain why no zinc metal is recorded as leaving those ports for Europe.

India, in the 17th to 18th centuries, imported metallic zinc from China or Pegu (Myanmar).[17] Under Aurangzeb, the Mughal court flourished, delighting in smoking tobacco through the huqqa, its base often of *bidri* with brass decoration. Indian craftsmen made a variety of superb brass artefacts for their domestic market, including lamps, incense-burners, narrow-necked flasks, ewers, *bidri* basins delicately inlaid with brass and silver, and *pandans* (globular boxes), many of them inlaid in silver and brass with abstract design and stylised flowers.[18] India also exported brass artefacts to Europe, including a marquetry cabinet (1650-1700) with decorative lock-plates of pierced brass, thought to be from Goa.[19] India was in fact both a major consumer of distilled zinc and an exporter of brass artefacts.

Figure 58. Bidri pandan inlaid with brass leaves and silver flowers, Bidar, 17 century

Aurungzeb encouraged Islamic imagery like the openwork brass finials which topped religious standards held in Shi'a shrines. One surviving dragon-head finial features cut-out calligraphic words from the Qu'ran, which ingeniously form its head, crest and teeth.[20] In Shi'a Islamic areas of India, brass or bronze 'alams, or processional banners, were central to festivals and mourning rituals. Much of the inland eastern Golconda region of the Deccan maintained the Shi'a religion, and some other surviving 'alams from Hyderabad are flanked with dragon-heads bearing Arabic script. 'Alams often took the form of a brass tear-shaped medallion with projections at the top, or a protective five-fingered hand symbolising the Prophet and his four significant relatives. These elegant decorative banners were housed in shrines and houses of mourning. Originating from Iran, the standards were carried in royal processions and brought out annually to commemorate the martyrdom of the Prophet's grandson, Husain. The 'alams represented stylised battle standards carried by Husain and his troops, and formed part of a mourning ritual, accompanied by dirges.[21]

---

[17] Valentijn 1721-6, volume 5, part 1: 9 and 40
[18] Ashmolean Museum, University of Oxford: EA 1993.392
[19] The National Trust, Chastleton House: F/1
[20] Ashmolean Museum, University of Oxford: EA 1994.45
[21] Zebrowski 1997: 321-322

Figure 59. Islamic brass; *background*, eighteenth-century brass alams held in the royal Shi'a house of mourning, Hyderabad; *foreground*, cast-brass calligraphic dragon's head finial, 1650-1750 (10.7cm wide)

Emperor Aurangzeb's invasion of the Maharatha stronghold of Golconda, provoked a revolt that triggered the Deccan Wars (1680-1707). In the Deccan, Aurangzeb subdued the Maharathas with superior fighting power, extending the Mughal Indian Empire to its greatest size ever. Indian artillery, including cannons and 'small field pieces each mounted on a well-made handsomely painted carriage', were often confusingly described as brass by contemporary travellers such as François Bernier[22]. This artillery, was, however, cast from bronze, including the 4.3-metre-long bronze 'great gun of Agra', weighing over fifty tons, which perhaps supported Aurangzeb's 1681 seizure of Gujarat from the Maharathas. At this period, evidence from surviving smaller guns and swords show that fittings, banding and inlaid decoration were often of true brass, perhaps made using zinc mined at Zawar, then controlled by Aurangzeb. The craftsmen of south and east India, however, probably used Chinese distilled zinc, which arrived at southern Indian ports in large quantities.

Late seventeenth-century trade in metallic zinc was complex. The Dutch, who had a pre-1701 port in China, traded Chinese zinc metal to Japan, then re-exported it to Batavia (now Jakarta, Indonesia) for sale to India and the West.[23] The English East India Company, lacking access to Chinese ports, bought Chinese metallic zinc from Dutch traders in Batavia or India.[24] In 1678, the East India Company auctioned two

---

[22] Baber 1996: 67
[23] Valentijn 1724-1746: volume 5, part 2, 77
[24] British Library: MS IOR G/12/6, folio 830 (1700-1702)

tons of Chinese zinc metal in their London saleroom (in 250-kilo – 5-hundredweight – lots costing just over £16 a ton).[25] In both 1683 and 1693, London merchants, evidently familiar with zinc metal trading, asked for a price reduction on certain flawed consignments.[26] In January 1691, Sir Joseph Herne, owner of Redbrook copper-smelting works (south of Tintern, Wales), bought 537 slabs of Chinese metallic zinc in London at about £11 a ton, paying a similar price two years later.[27] English vessels managed to call in at Amoy (Xiamen, Fujian province), China, even before they were granted a trading post there. In 1698 an English frigate took on ten tons of metallic zinc and copper at Amoy,[28] and in July 1699, a thousand zinc ingots sold in London.[29]

All this long-distance trade required fleets of sea-going vessels, so brass, being rust-free and virtually non-magnetic, was much in demand for equipping them. Brass compass-parts, especially compass-bowls (or 'boxes'), were often ordered. Brass containing distilled metallic zinc was the best for compasses because distillation removes iron impurities. Brass made from distilled zinc is therefore non-magnetic, having no iron to disturb the accuracy of the compass needle, but compass-makers may not have understood this. Cementation brass can contain about 2% of iron, and even brass made with sublimed metallic zinc can include some iron. Earlier compass bowls were wooden,[30] but in 1669 an azimuth compass was made with a brass bowl and thick brass dial.[31] Foreign-bound British Royal Navy ships began to carry brass compass-bowls – in 1696, a brass box compass was ordered for the fleet to Newfoundland. From 1696 onwards, British Admiralty equipment orders for Deptford and Woolwich gunpowder depots included spark-free brass scale pans and sets of brass weights of 12.7 kilos downwards.[32]

Brasses containing particular percentages of zinc (by weight) have specific advantages. Low-zinc brass is generally more brittle and may crack if hammered, but is suitable for casting. Brass containing 20-30% zinc, being more ductile and workable, is better for hammering, wire-drawing or rolling into fine sheets. Higher-zinc brasses are better for incising, engraving or soldering, and they can be worked by deformation – which avoids repeated cycles of forging and annealing. Deformation-working involves heating the brass to a temperature lower than its melting point but hot enough to shape it. At this heat the molecular structure loosens and the metal can be shaped without cracking.

[25] British Library: MS IOR B/35 folios 58-9 (1670s)
[26] British Library: MS IOR G/36/5 (1683-1708)
[27] British Library: MS IOR B/40 folios 45 and 191 (1691-95)
[28] British Library: MS IOR G/12/5, folio 610 (1695-99)
[29] British Library: MS IOR B/41, folio 245 (1695-99)
[30] May 1973: 75, 84
[31] National Maritime Museum, Greenwich: NAV 0383 (1776)
[32] The National Archive, London: Admiralty papers, MSS, ADM 106/483/66; 106/482/118; 106/482/260 (1696); 106/496/249 (1696); 106/947/229 (1741)

In Europe, seventeenth-century braziers used imported metallic zinc for soldering. In 1671, the English writer John Webster (1610-82) noticed 'all this imported spelter' but the only users he could discover were braziers, who mixed it with copper 'to make their solder'. Webster bought metallic zinc from braziers, but only at 'very dear rates'.[33] By adding distilled metallic zinc, braziers could produce high-zinc brass. They added measured amounts of zinc metal to existing brass to make high-zinc brass solder which would melt and be malleable at low temperatures for joining or repairing brass parts. In 1692, in order to 'Braze or Solder a piece of work so thin or small that it will not endure Welding', the English brazier took small pieces of soldering-brass, laid them in place and heated them to seal the seams, as done in Amsterdam fifty years earlier.[34] Braziers also cut shapes from sheet metal, hammered them into shape, then soldered or riveted the pieces together. They could embellish this creative work with piercing, fretwork and repoussé, or create relief designs by layering and soldering one piece of brass over another.[35] They could also use rivets and solder to attach handles. These skills could be applied to brass-making for an emerging new society.

Following the tedious decades of the Thirty Years War and English Civil War, European society was ready for some luxury. The trend was led by the Sun King, Louis XIV of France (1643-1715) and his Versailles court, where skilled artists and craftsmen executed his extravagant concepts for palaces adorned with brass chandeliers, clocks and door furnishings. French taste soon permeated throughout Europe, with aristocratic society copying the luxurious style of the Sun King in their clothing, salons, boudoirs and model farms. The marquetry of Charles Boulle - inlaid brass and turtle-shell – embellished fine French furnishings. From 1660, the love of lavish living was echoed at the English court of King Charles II, and the flamboyant lifestyle of the nobility permeated to the rising European merchant and middle classes.

The diary of Giles Moore, an English country parson (1656-1679) shows his enjoyment of luxury. His listed brass included candlesticks, andirons, stirrups, sundials, a clock, fire dogs, fire-shovel and tongs, and a toasting-fork. For his kitchen he had two sizes of cauldron (6½ and 13 kilos), a chafing dish, skillet, skimmer and ladle and, finally, a warming-pan for his bed.[36] For the more sophisticated city buyer, two London braziers, from about 1645-1690, produced the so-called 'Surrey enamels', brass buttons, candlesticks, sword-hilts, spurs, firedogs and stirrups, all richly decorated in coloured enamels.[37]

English probate inventories show that braziers (brass-workers) worked in small and large towns all over the country. Analysis of French, Flemish and Spanish scientific

---

[33] Webster 1671: 247
[34] Houghton 1692, 1728 edition, volume 1: 13
[35] Eveleigh 1995: 15
[36] Green and Butler 2011: 22
[37] Blair and Pattison 2000: 10-18

instruments suggests that certain European brass alloys contained added metallic zinc from at least 1650, and probably well before, as suggested by analyses of monumental brasses, showing an average of over 32% between 1600 and 1625,[38] and samples from Taynton brass works.[39] Two brass tokens minted in Bridgewater in 1654 contained over 34% zinc, and from 1675 a zinc content of 33% was regularly exceeded in western Europe.[40] Seventeenth-century English braziers called metallic zinc '*spelter*', so they probably first bought it from Dutch merchants. The English East India Company, which called it *tutenag*, had entered the trade later, and John Houghton in the late 1690s, reported 'a sort of tin' brought from India. He added 'I think they call it toothenaag; whereof our tinners think it in their interest to keep tolerably low'.[41] In other words, braziers kept quiet about their source of imported zinc metal, and were content to leave others thinking it was tin.

Brass varies in colour, depending on the zinc content. Up to 10% weight of zinc to 90% copper gives a slightly reddish brass. The bright warm gold colour appears at around 10-12% zinc. With 15% zinc, the alloy takes on a tinge closer to real gold. 17%-zinc brass has a bright golden colour, suitable for gilding, and brasses containing about 22-28% zinc take on a bright but cooler yellow-gold colour. As the zinc content increases between 15% and 30%, the redness of the gold colour reduces and the greener yellowness increases. A ratio of 30% zinc to 70% copper gives a duller greyish appearance, and brasses containing over 35% zinc (made using metallic zinc) take on a paler colour, turning from yellowish to white with increasing amounts of zinc.[42]

In Nuremberg, Johann Rudolf Glauber (1604-68) made laboratory experiments, and in 1656 published a method for adding Chinese metallic zinc to molten brass to increase its zinc content[43]. Prince Rupert of the Rhine (an experimental scientist and leader of the Royalist army in the English Civil War, 1642-46), studied in Glauber's Nuremberg laboratory, and subsequently used a similar formula for his own 'Prince's Metal'. In 1668, Prince Rupert hired Temple Mills, Hackney Marsh, on the River Lea north-east of London, as a brass works.

Temple Mills, Hackney, like most later seventeenth-century British brass works, was near the major English port of London. Temple Mills lay astride a mill-channel by the River Lea[44] (just south of today's Lea Interchange). In 1668 the Samyne family acquired these mills, with their dams, floodgates and land[45]. In August 1668, Prince Rupert,

[38] Clare Calver, 1990. folios 68 ff
[39] Morton 2009: 274, analyses by Chris Salter; Dungworth *et al* 2010: 6
[40] Pollard 1996: 216, 218, 223-224
[41] Houghton 1693, 1728 edition: 175
[42] Fang and McDonnell 2011: 57
[43] Glauber 1656b: 6-7
[44] National Grid reference TQ 3761 8542
[45] Hackney Archive: MS, M 795 ii, 1668

CONTINUITY AND CONFLICT IN EUROPE

Figure 60. Map of England and Wales, showing some sites mentioned in the text

backed by several wealthy shareholders, leased them[46], equipping them to make his prince's metal (about 34% zinc by weight)[47]. The brass was to be used for his invented method of boring gun barrels[48], though Princes Metal is better known as a gilding brass. By 1671 the 'Temple Brass Works on the River Lea had become 'celebrated for the production of prince's metal, brass guns etc.'.[49]

From the late 17th century, the Temple Mills brass-works at Hackney was run by Colonel John Shorey and his son John Shorey, who employed men who lived in their home, the old Golden Cock Inn, Basinghall Street, City of London, where they cast and finished metal in the cellars.[50] In the City of London and in Tottenham, the Shoreys had other warehouses where they sold finished brass, export goods for the Guinea

---

[46] Hackney Archive: MS, M 795 i, 1668
[47] Pollard and Heron 1996: 219
[48] Granger 1675: 138
[49] Aitken 1866: 239
[50] Clifford 1990: 130

trade to Africa,[51] and brass household items including spittoons and 'all manner of small boxes'. John Shorey's own 'room over the Parlour' boasted a brass hearth fender with tongs, fire shovel and poker.[52]

In 1690 the Shoreys of Hackney brass works decided to start a new brass factory at Bisham[53] on the River Thames near Marlow in Berkshire,[54] naming it Temple Mills after their Hackney mill. Two London investors, Samuel Clarke and a partner, saw the profit to be made from 'the working and making of Brass kettles and Brass wire and other Manufactures of Brass'. In 1693, they leased part of Temple Mill, Bisham, with its Thames islands and decayed weirs and locks. Their twelve shareholders agreed to back them, providing that Gerlach Becks, a Stolberg brass craftsman of high repute, managed the mill and recruited foreign workmen. Gerlach Becks spent £244.10s on recruiting sixteen skilled continental brass-workers, returning with them in January 1694.[55]

The foreign Temple Mill workers included 'Mr van Oyle' and Matthius Thiller, whose uncle lived at Aachen (Aix-la-Chapelle) 'from whence the foreign battery comes'.[56] Matthius had previously worked at Esher brass mills, and his two brothers, Christian and Adam Thiller later worked at Baptist Mills brass works, Bristol.[57] The Temple Mill accounts for 1694 include payments to 'the women where the men lodge'. The London partners favoured an expense-account lifestyle, putting in claims for business meetings at Vernon's Coffee-house, and the Kings Arms, Swan and Nags Head taverns, sometimes together with the manager Gerlach Becks. By 1695, they had built more mills, workshops and warehouses at Bisham..

Sheet brass was cast between granite casting-stones, but there was a problem with obtaining stones of suitable granite from its source on the north-west coast of Normandy. William III of Orange had formed an anti-French alliance, making it impossible for Temple Mills to import Normandy granite casting-slabs through Saint Malo. However, the enterprising Gerlach Becks procured several pairs through a Dutch contact, in time to cast brass sheet in September 1695. By 1697 the workmen were growing angry and rebellious, having received no pay, because some shareholders had refused to pay up for their shares. Eighteen shares were therefore re-allocated, and Samuel Clarke and five new shareholder partners took on the Temple Mills lease, so

---

[51] The National Archive: MS, C 104/105 part 1, item 28
[52] The National Archive: MSS, C 104/105/1; C/104/106/2
[53] National Grid reference SU 8406 8436;
[54] The National Archive: MS, C 5/354/3
[55] The National Archive: MS, C 5/151/70
[56] *Journal of the House of Commons* 1711-14: 163
[57] Kungliga Biblioteket, Stockholm: MS, BS M2491, folio 410 (c.1725)

the men were presumably paid.[58] The Shoreys' London warehouse, meanwhile, still traded goods to the Guinea Coast of Africa.

In 1689, under William III, British enterprise in brass-making was at last encouraged by the passing of a Mines Royal Act, ending the Society of Mineral and Battery Works' royal monopoly on brass-making.[59] William III also removed the Guinea Coast trade monopoly, granted to the Royal African Company by Charles II in 1674. This encouraged the city of Bristol to petition in 1690 for a larger slave quota to supply its American and Caribbean plantations (to which they already sent brass) Bristol merchants were given permission to trade to West Africa and the Caribbean,[60] and in October 1699, the ships *Beginning* and *Wakeing Lyon* left Bristol for Guinea.[61] This sad trade created a demand for Bristol-produced brass.

During the 1690s, the London-based Royal African Company traded brass mainly to the Gold Coast, where demand was highest,[62] and to Sierra Leone, the River Gambia, Arda (Alada) and Whydah (Ouida), both in southern Benin, and occasionally to Annamaboe (Anamabo, Ghana) and the Windward Coast.[63] The Saharan Songhai Empire had collapsed during the 17th century, causing the migration of Islamic horseback traders southwards and eastwards into forested areas. Local Asante (Ashanti) groups united to resist the intruders,[64] which led to Islamic wars (1699-1701), during which many captives were bartered as slaves. The Portuguese and Dutch not only bought or captured slaves from coastal areas, but organised coastal groups to venture inland and buy them slaves from independent merchants.[65]

Most European countries, including Sweden and Denmark, were now drawn into the lucrative slave trade, with brass among the commodities exchanged for slaves along the Guinea coast, who were then transported to America and the Caribbean – the triangular slave trade. The French traded mainly to Senegal, but the Dutch still led the field. In truth, almost all European brass works must have become implicated in the slave trade, either directly or indirectly. At this period, Dutch slave ships traded vigorously and chose their cargoes with care. In 1680, the brass goods they sent to the Guinea Coast included locks, trumpets, rings and bells,[66] but the factors (local agents) at the new English slave trading stations were fast learning the trade.

---

[58] The National Archive: MS, C 5/151/70 (1697)
[59] Statutes of Parliament 1688, volume 3: 412
[60] Dresser 2001: 22, note 89; Latimer 1903: 178,
[61] The National Archive, London: MSS E 190/1156/1 (1698-9), 6 and 24 October
[62] Herbert 1984: 133
[63] The National Archive, London: MS T 70/22 (1705-1719) and 23 (1719-1724),
[64] Ade Ajayi and Crowder 1985: section 35
[65] Evans and Richardson 1995: 675-678
[66] Donnan 1931, volume 2: 291

The English East India Company, whose ships called in for African gold to trade in India, bartered about a ton and a half of brass battery goods on the West African coast in 1658, and three and a half tons in 1659, together with a thousand muskets, probably furnished with brass components.[67] For the period 1678-81, the local Guinea agent for the Royal African Company of England, ordered 'Black brewing kettles of the largest sort' ('black' possibly meaning that they were made from copper or low-zinc brass), 'Manelloes, bright' (manillas [bracelets or anklets] perhaps of purer brass), and 'Wyer bound kettles of the largest sort'. The factor at Arda (Alada, southern Benin) ordered 'brass basins of all sorts', including those used for washing oneself. In 1679-80, the Sherbro River factor requested brass basins and wire-bound cauldrons; and the Cape Corso factor, in 1681, asked for brass basins and 144 manillas.[68] A few undated manillas survive at Bristol, two found in today's south Nigeria, and two in the Bristol area.[69] Another manilla was excavated on Charlestown main street, Nevis (Caribbean), opposite the presumed site of the old slave market,[70] but few manillas of this date have been analysed to discover the grade of brass.

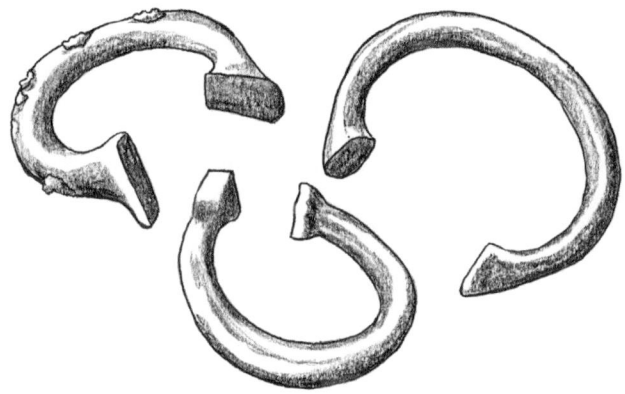

Figure 61. Manillas from the three corners of the slave trade. *left*: manilla excavated at King Street, Bristol; *centre*: manilla excavated beside the former slave market site, Nevis, Caribbean; *right*: manillas from the kingdom of Benin

International rivalry and envy were rife in the slave trade. In December, 1679, James Nightingale, English factor at Cape Corso, wrote to protest at the poor quality of goods sent out. With better quality, Nightingale claimed that he could have got more for the goods, but that the Dutch earned more money 'by their tricks and cheats than wee who venture our lives for your honors at this place'. On July 10, 1680, Nightingale wrote that 'Trade is very bad upon the Gold Coast ... for want of Goodes but it is not soe with the Dutch for they gett fine Ships with good Goods'.[71]

After the English Civil War (1642-6), British brass works gradually managed to meet the demands of English agents in West Africa. One of them was Esher Mill,[72] located at

---

[67] British Library: MS, IOR E/3/85, ff.67, 105, 128 (1658-9), East India Company voyages to Guinea
[68] The National Archive: MS T 70/20 (1678-81) Royal Africa Company agents writing to London
[69] Bristol Museum: E 12096
[70] Roger Leech, archaeologist, personal communication
[71] The National Archive: MS, T 70/20 ((1678-1681)
[72] National Grid reference TQ 1315 6575

a division of the River Mole in Surrey, started in 1649 for brass wire-making by Jacob Momma (from an Aachen brass-making family), Daniel Demetrius from Dordrecht and his partner Peter Hoote.[73] The last two were Dutch merchant partners operating in London, an astute choice by Momma, because the London brass and copper trade was controlled by a small coterie of powerful Dutch merchants. Momma imported most of his copper from Sweden. He obtained his calamine, initially, from Aachen,[74] having no wish to pay money to the Society of Mineral and Battery works for a 'privilege' for using English Mendip ore.

When, in 1655, the Society accused him of failing to pay them 'privilege' money for making brass, Momma pleaded ignorance as a foreigner, and until 1663 he continuously evaded payments. Powerful Dutch merchants in London constantly tried to overthrow his business by dumping cheaper continental brass on London and trying to block him from buying Swedish copper. From 1663, the Society of Mineral and Battery Works gradually lost its authoritarian grip on brass makers, and by 1664 Jacob Momma was prosperous, possessing a fine mansion in Esher with twenty fireplaces.[75] The sister Company of Mines Royal discovered that, for ten years, the Earl of Devonshire had mined copper ore on his own land at Ecton, Staffordshire, without paying them for a 'privilege'.[76] The Ecton copper mine carried on nevertheless, and Jacob Momma leased it in 1665. Under the management of his son, also Jacob Momma, Ecton copper ore was blasted out with gunpowder as happened in Germany (the Mommas' Esher mill being conveniently near to gunpowder works). Ecton mine ran for two and a half more years before the meagre accessible copper seams gave out.[77]

Momma died at Esher in 1668, but in 1670 his son, the junior Jacob Momma, spent eight thousand pounds on building new wire-drawing mills. His improvements prompted the Society to charge him more for the remaining sixteen years of his brass-making 'privilege'.[78] In 1691, the Esher mill lease was taken by an active entrepreneur, William Dockwra, inventor of the 1683 British Penny Post, who was now Comptroller of the Post and owner of Upper Redbrook copper-smelting works. By 1693, Esher brass mill imported refined copper from Sweden,[79] and later from Upper Hungary, Morocco, Algeria and certain small German mines.[80] Dockwra took on two partners, a brazier and a London banker named John Coggs, who raised £12,000 of public stock to fund Dockwra's new Esher buildings and skilled foreign workforce. By mid-1696, the

---

[73] Pettus 1686: BR
[74] Riksarkivet, Stockholm: MS E III 5, f. 612 (1691-1693) Odelstierna
[75] British Library: typescript YD 2005 b.234, Greenwood, G. B., 1980, 4
[76] British Library: MS, Loan 16/2, folios 131 and 140v
[77] Norton and Robey 1985: 195; Plot 1686: 185
[78] Surrey History Centre: MS, K 61/5/15 (1760)
[79] Houghton 1693, 1728 edition: 2, 188
[80] Liverpool University Archive: MS, typescript 7.1 (24), f.3, (1691-3)

capital was drained and profits were poor, so a new manager, George Ball, was hastily appointed to set things right.[81]

In 1697, Esher brass crucibles were charged with two parts refined copper, four parts calamine, one part scrap brass, and charcoal. Scrap brass melted faster, speeding up the process, which made it cheaper. The heat was maintained for up to twelve hours in furnaces with air ducts beneath them, then poured into one larger vessel, where workers skimmed off the dross and left it to settle. Analysis of solid brass droplets (prills) from an Esher crucible[82] revealed an all-purpose brass containing about 26% zinc (suitable for casting, wire-drawing or battery-work)[83]. The molten metal was poured between two granite casting-stones, each weighing a ton or more, tilted up at one end until the metal filled the whole cavity, then set horizontal to cool as a sheet or (thicker) plate. Since the current French wars prevented delivery of granite casting stones from Normandy, George Ball negotiated with the English government to exchange French prisoners for 'four pairs of grist stones to be brought from St Malo, which Mr Dockwra desired for the Esher brass works'[84].

Brass plates weighing almost forty kilos were cast between these stones. Each plate was passed several times through a Swedish water-powered rolling mill (described below) until thin enough to cut into seven or eight strips for wire-making. Rolling strained the grains within the brass, distorting them by elongation in the direction of working.[85] Between each rolling, the brass was annealed to a temperature at which the distorted molecular structure re-crystallised and the metal regained pliability. For wire, the long strips were spread on twenty-four water-powered wire-drawing benches, and were drawn repeatedly through 'many holes in irons', to the thicknesses desired, and the wire was coiled into 12.7kg 'rings' for dispatch to the pin-makers.[86] About fifty tons of brass wire was produced annually at Esher in the 1690s (eighty rings a week). The manager, George Ball, had been hired in haste with promises of a large salary, annuities, tax-free housing, commissions on sale and free beer and coal. Finally, in 1698, Coggs, the banker partner, dismissed, him, leading to a debilitating fifteen-year lawsuit, which eventually ruined Coggs.[87]

Ember Mill, Thames Ditton, Surrey, had, in 1638 been converted to brass making by James Ledges. After 1649 Jacob Momma the elder and Daniel Demetrius had taken it over, just as they had taken over Esher Mill, but the disruptions of the English Civil War made Ember Mill difficult to keep going. In 1653, Momma was arrested

---

[81] Brown.1761: 1, 296
[82] National Grid reference, findspot: TQ 1320 6570
[83] Morton 2009: 274, analysis by Chris Salter
[84] Calendar of State Papers Domestic 1696: 382
[85] Barclay, 1993, 35, figure 8
[86] Houghton 1693, 1728 edition: 2, 190
[87] *Journal of the House of Lords* 1711-1714: 54

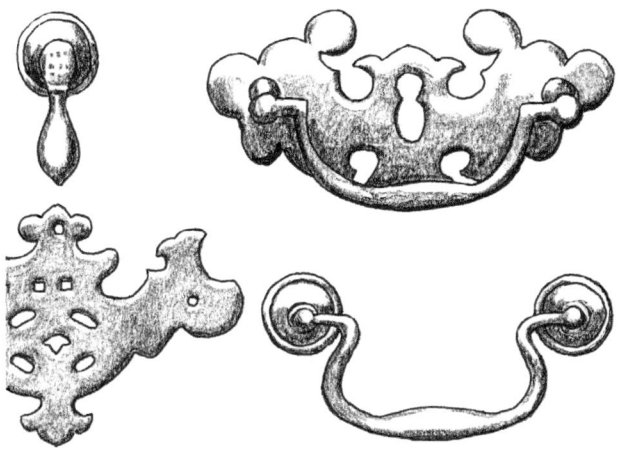

Figure 62. Drawer handles, *top left* tear-drop handle with key-plate c.1690; *top right* bat-wing plate c.1720; *lower left* bat-wing style, later mid-1700s; *bottom right* loop handle with disc plates

on suspicion of colluding with the enemy, but was released after five days.[88] In February 1656 the Society of Mineral and Battery Works discovered that Jacob Momma persistently obtained Mendip calamine, alloyed brass and cast it at Ember without paying for 'privileges'. Momma claimed to their court that, as a 'stranger and foreign born', he knew nothing of the Society's privileges, but he succumbed, and paid them £10 a year.[89]

In 1656, Momma, as an English importer of Swedish copper, protested strongly against the duties that Oliver Cromwell's Parliament imposed on imports of foreign copper ore.[90] In February 1663, when Momma's manufacture was foundering, his partners implored the Society to petition the government against copper import tariffs. They also protested against the fact that Dutch merchants in London supplied London pin-makers with cheaper foreign brass wire, forcing British brass makers out of their home market.[91] After 1663, in a period of high fashion, brass wire was produced in varying lengths and thicknesses for rustproof haberdashery pins and needles for tailors and seamstresses. Most pins had spirally-wound globular or, in the later 17th century, conical, heads.[92] Pinning was also the only means of holding loose papers together[93]

A wider range of seventeenth-century household products appeared, including cast brass spoons.[94] Brass handles appeared on English furniture from about 1660. A walnut chest, dated c.1690, at Chastleton House, Oxfordshire (where much of the furniture is original to the house) has teardrop handles backed by round plates,[95] a style popular until about 1725. Heavier drawers had drop handles curving inwards at the ends, backed by very thin brass plates with elaborately sculpted edges and stamped with

---

[88] *Calendar of State Papers Domestic* 1652-1653: 168, 175
[89] British Library: MS, Loan 16/2 folio 131
[90] *Calendar of State Papers Domestic* 1655-1656: 318
[91] British Library: MS, Loan 16/2 folios 132v, 142, 142v
[92] Tylecote 1972: 187
[93] Colonial Williamsburg Foundation, ER 13232-19B; and ER 987C-19B (pins)
[94] Coleman Smith 2002: figure 30, 1680-1720
[95] National Trust, Chastleton: CHS F/80

Figure 63. Ember brass mill interior, 1689-91, after a drawing by Eric Odelstierna

leafy designs. From c.1675-1700, rim locks for doors were enclosed in brass cases (replacing wood), some of them pierced and engraved, including one at Dyrham Park, Gloucestershire, which was cast in Birmingham to a Moorish design.[96]

By 1672, the Momma family had given up Ember mill, by then in a bad state of repair,[97] but by 1691 it was producing brass again, under John Stapleton, who had been granted a brass-making 'privilege' for Ember mill (available after the failure of Byron's Nottinghamshire venture). Such 'privileges' were in effect a royal tax on industry, bringing revenues direct to the king as opposed to parliament. In 1691, John Stapleton, who had studied brass-making in Nuremberg, took out a British patent for high-zinc 'white' and 'yellow' metal, which could be beaten or rolled into fine foil or leaf for gilding.[98] Chinese zinc metal, known as 'Nuremberg Metal' had been readily obtainable

---

[96] National Trust, Dyrham Park: DYR M.35
[97] British Library YD 2005 b.234, typescript, Greenwood, G. B., 1980, 8
[98] Patent Specification 285, 1691

by 1678 from East India Companies in London or Amsterdam. In 1683 and 1693, London traders, evidently familiar with this commodity, recognised certain flawed zinc metal deliveries, on which they asked for an 'Abatement', which suggests that it had been regularly imported already.[99]

This was the start of the era of Swedish observers, or, to put it more bluntly, industrial spies. A succession of experienced Swedish mines inspectors and brass-makers were sent to England – which they had begun to see as a brass-making competitor. These seventeenth-century 'observers', Odelius (alias Odelstierna), Cletscher and Kalmeter provide valuable eye-witness glimpses into the late seventeenth-century British brass industry. In 1691, Eric Odelius, who made a sketch of an annealing furnace and battery hammer at Ember brass mill, observed a reconstructed British brass industry, now greatly envied by foreign competitors. He noted that Stapleton's company produced brass that included 'white metal' (high-zinc brass). Odelius noted that the company had 'very long experience of the West Indies, France, the Netherlands and Germany and of every kind of metallurgical and chemical work, with active production over a long and continuous period'.[100]

In 1694, John Stapleton, in order to renew his Ember mill lease, took on financial partners including John Hitchcock, a London merchant and Quaker, They planned to replace the mill, then 'greatly out of repair', with a substantial new brick-built mill at a cost of £1,800,[101] but in December of that year, Stapleton sold up, leaving Hitchcock the main owner.[102] Two years later, Thomas Cletscher from Sweden commented that Ember brass works had recently declined for lack of skilled workers, but was now set up properly by two Germans. They installed continental equipment and eight furnaces, two of which were already producing brass from the more forgeable Swedish copper and English calamine from Somerset, Cornwall or Nottinghamshire (perhaps Derbyshire), which they found as good as calamine from Poland or Aachen, and cheaper.[103] Hitchcock had a small copper refinery at West Ditton, near Ember Mill, to refine annually 40 tons of Devon copper, first smelted at Cuckolds' Point, Rotherhithe.[104] Efficient copper smelting and refining was vital to the brass industry, but England was running short of timber, now in heavy demand for shipbuilding, as well as fuel.

In the early 17th century, to solve the lack of timber fuel, iron smelting was revolutionised by the adoption of the reverberatory furnace which could use coal

---

[99] British Library: MS, IOR G/36/5 (1683-1708)
[100] Riksarkivet, Stockholm: MS, RAS E III:5, 611-12 (1691-1693), Odelstierna
[101] Brown 1761: 1, 522
[102] Morton 1985: 252
[103] Liverpool University Archive: MS, typescript 7.1 (21) folio 20 (1696), Cletscher translation
[104] Liverpool University Archive: MS typescript 7.1 (23) folios 56-57 (1725), Kalmeter translation

fuel instead of charcoal or wood.[105]. Described by Biringuccio, such furnaces had been used for cannon production in Hungary around 1500,[106] and were known in England since before 1611.[107] By the later 17th century, the scarcity of English timber made charcoal more expensive than coal, so in about 1685,[108] a Bristol worker, John Coster, smelted copper in a reverberatory furnace, using coal fuel.[109] Swedish industrial spies keenly described English reverberatory furnaces, but deplored the quality of their copper product compared to Swedish wood-smelted copper.[110] Forest timber was still plentiful in Sweden and Germany, so coal-fired reverberatory furnaces were not introduced there until later.

Most brass mills produced sheet brass and wire, but sold it on to specialist brass-founders and smiths for creating objects. Flemish brass founders around Liège were famed for casting decorative stocks and locks for guns and muskets.[111] Already in the 16th century, musical instruments like trumpets and flutes were fashioned from sheet brass, which was more malleable and easier to work in sheet form than bronze. A Nuremberg trombone, dated 1612, has already been mentioned, and Thomas Beale, English seventeenth-century state trumpeter to both Cromwell and Charles II, was both player and maker. However, analysis of one of his trumpets, dated 1667,[112] showed that only the bell, a replacement, was of true brass,[113] so the earliest 'brass' instruments may not all have been brass. Harpsichord hinges and virginal strings were produced in brass, and in 1694, England imported about 100 kilos of virginal wire from Germany.[114] Wire-makers like the Shoreys and the Hallens, though well known for their brass wares, often kept their methods secret. The antiquary John Aubrey (1626-97) wrote in 1718 that 'at Wandsworth is a Manufacture of Brass Plates for Kettles, Skellets, Frying-Pans &c by Dutch Men, who keep it as a Mystery'.[115]

In 1671 the Wandsworth 'Frying pan Houses' lay east of the River Wandle,[116] whereas in 1729 they were to the west,[117] so workshops existed at different times on both banks, while the hammers were powered by a mill straddling the river.[118] Having established

---

[105] Barker 1966: 33-4
[106] Belenyesy 2018: 158
[107] Neri 1611 (1662 edition): 239
[108] Liverpool University Archive: MS, typescript 7.1 (23) folio 12 (1725), Kalmeter translation
[109] Rees 1968: 496
[110] Liverpool University Archive: MS, typescript 7.1 (21) folio 11 (1696) Cletscher translation
[111] Gentle and Field 1975: 31
[112] Bate Collection of Musical Instruments, University of Oxford: x 78
[113] Barclay, L. 2002: Examination of a trumpet by Simon Beale, London 1667; and Bacon, L., 2001, Preliminary findings on the technical analysis of the different parts of a trumpet dated 1667, 1, (unpublished reports for the Bate Collection, Faculty of Music, University of Oxford)
[114] Houghton 1697, vol. 2: 184
[115] Evelyn, 1676, 1718 edition: 14
[116] Aubrey, John, 1671, 1975 edition: 14 containing John Ogilby, map of Surrey
[117] Senex, 1729: Map of Surrey
[118] British Library: MSS, Add 7184 A; Add 7184 J; Add 7184 V, Althorp Papers

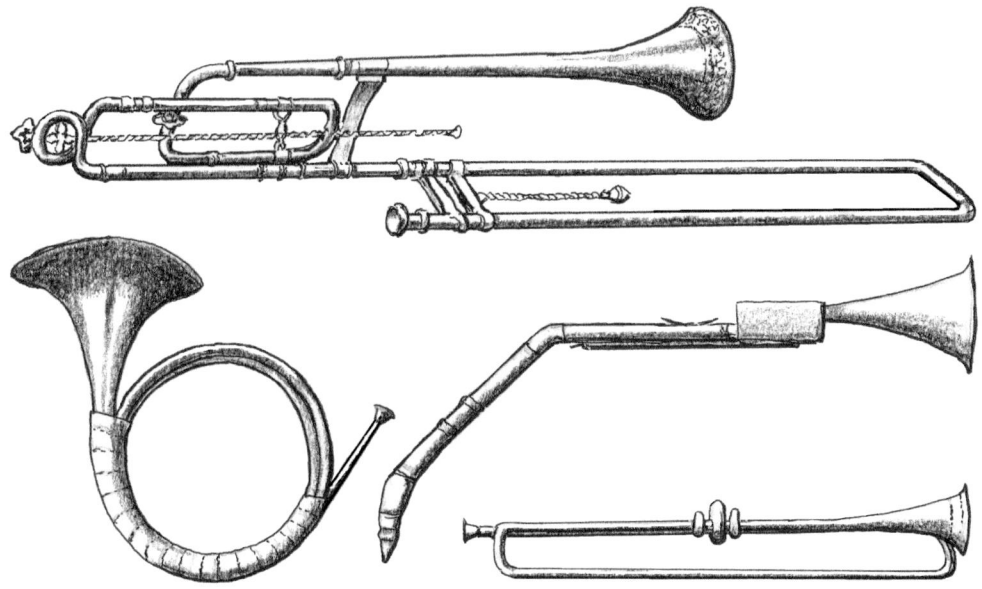

Figure 64. Musical instruments; author's drawings, *top*, bass trombone, 1612, Isaac Ehe, Nuremberg; *below*, *left* French horn in F, 1700-25, Christian Bennet, London; *upper right* basset horn, late 18 century, Johan Heinrich Grendel, Dresden; *lower right* natural trumpet, 1667, Simon Beale, London

his Wandsworth brass-works, Cornelius van Hallen 'shifted his sons and grandsons about England as he judged most advantageous to their common interests'.[119] Some of them cast iron, brass and copper at Keele, Newcastle-under-Lyme, Coalbrookdale and Stourbridge in the West Midlands'. In 1686, Dr Robert Plot, a chemist with an enquiring mind, wrote that the secret of making frying-pans, or skillets, lay in regulating the furnace heat. He found the Hallens' process 'so ingenious and wonderful' that he was moved to describe John Hallen's pan-making sequence at Keele and Newcastle-under-Lyme.

The process started by beating out flat, circular plates under a battery hammer weighing about a quarter of a ton. The flat plates were first beaten singly, then two, three or four together, until broad and thin enough to make a nest of nine plates. To work them into shape, the nine frying-pan plates were 'laid upon one another, the largest at the bottom and the smallest at the top of the pile', and were clasped together by turning up four flanges fixed to the lower plate. The battery-man could then shape all nine by revolving the whole pile under the fall of the rapidly beating trip-hammer. This saved time and energy because the nine juxtaposed pans retained a constant heat long enough for them all to be forged simultaneously, whereas a single pan cooled faster and had to be re-heated at least nine times before it was fully forged. The heat was kept low enough to prevent the pile of pans from sticking together

---

[119] Hallen 1885: 23

during forging, and up to twenty different-shaped battery-hammer heads were used to achieve the final shape and finish.[120] Cornelius Hallen had brought the process with him when fleeing from his native Liège (Lüttich).

Great quantities of brass wares from European foundries and workshops were in demand for the cargoes of Dutch West India Company ships heading for the Guinea Coast or America. By the 1640s, Dutch settlers in America played brass mouth-organs and traded brass kettles and cauldrons to Native Americans in exchange for furs, and in 1656 the steeple of the Dutch church at Albany (200 km west of Boston) had a brass cockerel (rooster) weathervane crowning its spire. A brass hairpin was excavated nearby,[121] and both were probably crafted in the Netherlands from Aachen or Stolberg brass.

Sweden, now a leading brass-making nation, pitched into the Guinea Coast slave trade, and set up a trading post on the Ghanaian coast in 1650, followed by the Danes in 1652. Sweden's brass industry was controlled by very few people, most of them closely related to merchants in the Netherlands (for a map of Swedish sites see Chapter 4). In order to obtain calamine and sell brass goods, Sweden's merchants and brass-makers maintained close contact with their Aachen or Amsterdam relatives. Sweden's seventeenth-century brass industry, however, was still dominated by its monarchs. Favoured businessmen, if raised in social status by the monarch, dropped their previous name and assumed a new one. However, anyone first mentioned here under one name will continue to be mentioned by that name, to avoid confusion.

By the mid-17th century, Isaac Cronström employed twenty brass workers at his Skultuna brass works about a hundred kilometres west of Stockholm. He added two new cauldron workshops and a chandelier foundry,[122] and by 1670, a rolling-mill. One of Skultuna's most skilled workers, Hans Honafwer, was a celebrated chandelier mould-maker, whose name lives on. Mould-makers, as the key designers of brass objects, gained much respect. In Sweden, with its long dark winters, chandeliers, candlesticks, wall-sconces and flame-snuffers were highly prized. Skultuna was a very successful producer of beautiful brass lighting equipment, but independent chandelier- and candlestick-makers also worked in foundries around the country. Brass, besides being strong and fireproof, was an excellent material for lighting, because it gave off a rich, warm glow under candlelight. Brass lanterns, some with horn windows that protected candles from wind, and brass chandeliers, already a prominent feature in churches and cathedrals, also began to light wealthy homes.

Northern European late seventeenth-century candlesticks often had plain, dome-like bases and slightly tapered, squat round stems. Some Dutch and Flemish candlesticks

---

[120] Plot 1686: 335-336
[121] Huey 2009: 18, 24
[122] Forsgren 2010: 16

had a small hole drilled sideways into the holder, through which a point was inserted to eject candle-ends. Two-piece moulds, used from the late 17th century, left a vertical seam and, around 1670, baluster candlesticks often had one knop (stem bulge), resembling an inverted acorn.[123] Although most seventeenth-century chandeliers in London were imported from the Netherlands, by the later 17th century some were cast in London and by 1699 they were copied in Bristol and the west of England.[124] In 1697 a delivery of forty-three brass lamps arrived in England from 'Germany'.[125]

Swedish chandeliers, cauldrons and other brass wares for the domestic market were sold either direct from factory to customer or by travelling peddlers. The earlier cauldrons or pots often had ears riveted to the sides and a rolled-over border, but a few had rounded bases or a lid. A few urn-shaped pots with ring-handles and rolled borders were designed to hang from beams. In the 17th century, thousands of brass cauldrons, bowls and basins were exported from Sweden to other Scandinavian countries, western Europe and the Guinea Coast of Africa, but particularly to Russia, whose merchants sent fleets of ships to purchase hundreds of chandeliers, cauldrons and other finished Skultuna brass wares.[126]

In 1670, Sweden's seven brass works had a total of eighty-three furnaces between them. Norrköping had twenty-one furnaces, Gusum twenty, Nyköping eighteen, Skultuna nine, Bjurfors six, Vällinge six and Nacka three. Their main exports were sheet brass and strips for wire-drawing, whereas hand-crafted everyday utensils and ornamental goods were produced for the Swedish domestic market.[127] Most of the brass was turned into wire, but small, independent workshops hammered brass into cauldrons and basins, or cast chandeliers and everyday household items and equipment. Immigrant brass hand-workers (*gyrtler*, girdle- or belt-makers) originally made small parts for belts and horse harness, but had long expanded their repertoire to include sleigh fittings, dog collars, buttons, sword-belts, sword-hilts, flagpole-tips, clasps, buckles, candle-sticks, snuffers and brass smoothing-irons. In the 17th century, a smoothing-iron made of brass was a symbol of wealth. Another particularly Swedish seventeenth-century local product was the elaborate brass bridal crown. In 1681, a Gothenburg journeyman girdler advertised coffin furniture, signs, candlesticks and plates. Girdlers cast items in iron, brass or copper, and undertook silvering or gilding.[128]

In 1666 Skultuna brass works proprietor, Isaac Cronström, started a second brass works at Bjurfors, Västmanland, twenty kilometres further north.[129] The site provided

---

[123] Noel-Hume 1969: 94
[124] Sherlock 2002: 1
[125] Houghton 1693, 1728 edition, part 2: 184
[126] Erixon 1969: 298
[127] Forsgren 2010: 18
[128] Erixon 1978: 19
[129] Riksarkivet, Stockholm: typescript Register, Bjurfors 1666, folio 115

energy from the Bjurfors stream, forests full of timber, and experienced charcoal-burners and metal-workers available from a recently closed ironworks. A disadvantage was the difficult forty-kilometre overland transport for zinc ore and finished goods through forest and lake country to and from Vasterås, for shipment to Stockholm. One overriding advantage, however, was that Bjurfors lay only three kilometres from Europe's greatest copper refinery at Avesta. From 1663, Isaac Cronström's father (Isaac Kock) introduced new processes for refining copper from the huge Swedish Falun mine (100 kilometres further north), making it cheaper for brass-mills to buy refined copper from Avesta than to refine copper themselves.

Avesta, besides the copper refinery, housed the royal mint, where Cronström's father introduced an innovative rolling mill which replaced hand minting (Leonardo da Vinci illustrated a rolling mill in 1488). At Bjurfors, in 1667, Cronström laid out an entire new brass works, and in 1671 he installed a similar high-quality labour-saving rolling mill.[130] The mill straddled the border between two provinces, Västmanland and Dalarna. A 1698 map shows a channel leading through the rolling mill, stone remains of which are visible east of today's Bjurfors manor house and lake.

Rolling was used to plastically deform metal by passing it between rollers – called rolls – supported on a mill-stand with a downwards screw (as on old clothes mangles). The brass piece to be rolled passed between two rolls, running at slightly different speeds to get the desired thickness, the screw bearing down on them to adjust the gap between the rolls. Cold rolling at room temperature, below the recrystallization temperature of the metal, reduced the thickness of flat cast sheets. It simultaneously increased the hardness of the metal by strain-hardening it, and improved the surface finish. It thus produced thinner, harder, flat sheets from cast brass plate. Once launched, Cronström's rolling-mill was a sensational improvement, speeding up the production of brass strips to be drawn out into wire, and reducing the number of workers from nine to two. Bjurfors also had a calamine-crushing mill, two furnaces, battery hammers and workshops for wire-drawing, punch-work, annealing and the finishing of hollow vessels like cauldrons. In the late 17th century, Cronström's Bjurfors and Skultuna mills between them produced 1,088 tons of brass a year, a third of Sweden's brass output.

Isaac Cronström, besides having technical expertise, worked fast with financial insight, and could push a hard bargain. He invested his own money in forest land to secure charcoal for the works. In 1672, in a joint letter to the Board of Mines he mobilised other brass mill-owners[131] to argue against brass export tariff increases[132]. Karl X1 (r.

---

[130] Forsgren 2010: 123
[131] Erixon 1957, 17-18
[132] Forsgren 2010: 18, *Deduktion om Mässings Brucken, dheras opkombst, och huruwida dhe äre nödige och nyttige till Kongl Maij:ts och Fädernes Landsens tiänst* (not read in the original, due to problems with Old Swedish script)

1660-1697), proposed imposing high tariffs on brass exports to make Swedish copper expensive, for two reasons: firstly, to stop a glut of cheap copper on the Netherlands market, and secondly to help pay off ransom for the return of Älvsborg, Sweden's only fortified North Sea port, lost to Denmark in the 1560s[133]. (In a 1571 treaty, the Danes had promised to return Älvsborg if Sweden paid a huge 150,000-daler ransom, still outstanding a century later). The brass mill owners, however, argued that tariffs on brass exports hurt not only the brass mills but the population at large, and that the king should not impose heavy taxes on the very brass-workers who raised the national income by boosting copper trade and brass exports. Sweden's seven brass mills, they claimed, used 1,020 tons of raw copper annually. If exported, this amount of copper would raise 270,000 daler of income for the state, but the same amount of copper, converted into 1,360 tons of extra brass, would raise a massive 520,000 daler. The mill-owners' plea was successful and tariffs were not increased.[134]

Bjurfors brass mill was threatened in 1688, when Karl XI of Sweden introduced his national 'Reduction' scheme, to seize back all landed estates (previously granted by the Crown to various aristocrats), in the hope that this would help pay off the royal debts. Despite her husband's death in 1679, Kristina Cronström amanaged to hold on to the Bjurfors brass mill estate and keep it going. In 1695, her mill had six brass furnaces, twelve wire-drawing benches and a foundry for cauldrons and 'all sorts of shiny brass work'. Unfortunately, by 1699, the headstrong young King Karl XII (1697-1718) was fighting in Sweden's first skirmishes of the Northern War (1700-1720), in which Russian influence over north-eastern European territories was pitted against Swedish control. The expenses of war caused Sweden harsh economic deprivation which was far from counterbalanced by the increased demand for military brass buttons.[135] The fate of individual Swedish brass works shows the depth of national economic suffering in the late 17th century, causing Sweden to lose much of its flourishing brass industry and trade to other European nations.

Back in more prosperous times, Nyköping brass mill was revitalised from 1646 onwards by Mathias Römer and Willem Momma, who also leased a brass mill at Vällinge, south of Stockholm. Financed by the son of the Netherlands-born financial giant, Louis de Geer (1587-1682),[136] Nyköping had a 'great mill', sawmill, homesteads, charcoal-burning rights and three trip-hammers. After 1651, when Sweden's economy was stable, Willem Momma added 18 furnaces, 15 hammers and 30 wire-drawing benches at Nyköping brass works. Momma enjoyed tax exemptions on imports of calamine and casting-stones, purchases of coal and copper, and brass manufacture. Queen Kristina of Sweden negotiated favourable contracts and treaties with foreign powers,

---

[133] Hekscher, 1932, 8
[134] Forsgren 2010: 18-19
[135] Forsgren 2010: 124-126
[136] Forsgren 2010: 54-55

so Nyköping exported brass to the Netherlands, France and England.[137] In 1665, just when Willem Momma planned to introduce a rolling mill, the city burnt down, leaving the brass mill in ruins. Undaunted, Momma, pressed ahead, completely rebuilding the Nyköping factory and establishing a foundry and wire mill upstream.

At this point, however, the Swedish economy was crippled by expenditure on the second Northern War (1655-60) of King Karl Gustav X (r. 1630-1660), in Poland, Lithuania and, latterly, Denmark. This war, accompanied by threatened increases in brass tariffs, severely cut brass production. Momma's financial position was at risk, but he managed to attract investors, including his son-in-law, artist David Ehrenstrahl, who took over the Nyköping brass mill when Willem Momma died in 1681. Ehrenstrahl part-illustrator of *Suecia Antiqua*, a now-famous work promoting Sweden, had become court artist to Karl XI in 1661. With Nyköping brass mill, Ehrenstrahl inherited disputes – with the family over inheritance, with landowners over charcoal rights, with suppliers over import charges (on calamine and casting-slabs), and with Avesta refinery over copper quality. As court artist, Ehrenstrahl maintained close ties to the king, who passed him occasional cash windfalls from customs revenues. Ehrenstrahl's business sense kept the factory operating smoothly (while busy decorating Drottningholm Palace in 1673, he negotiated a new ore source for Nyköping brass works). When Ehrenstrahl died in 1698, the mill had eighteen furnaces in six furnace-houses, three wire-drawing workshops and five foundries, and sold brass wire, sheet, cauldrons and basins,[138] but in 1705, Swedish industry was on such hard times that the mill was sold.

In 1666, Jakob Momma (c.1625-78) had purchased the much larger Norrköping brass works,[139] fifty kilometres further west, which by then had twenty-one brass furnaces, a calamine mill, twenty-six wire-drawing benches, two lathe shops, a copper warehouse, cauldron workshop, rolling mill and four further workshops. The water-wheels of the brass mill on Mill Island (*Kvarnholmen*), Norrköping, were driven by very deep, powerful falls of water on the River Motala.[140] Aachen-born Jakob Momma, a powerful Stockholm merchant, was supported by influential relatives. Norrköping mill lacked timber, but Jakob Momma, who owned ships and managed Swedish Crown estates, could obtain timber, so the brass works flourished. Supported by Dutch capital, he exported brass wire, plate brass and cauldrons, probably to supply the lucrative Dutch slave trade. He penetrated the English and Scottish markets and had permanent agents in several European cities. Production remained high, and Norrköping under Jakob Momma became Sweden's largest brass works.[141]

---

[137] Riksarkivet Stockholm: MS Anteckningen till Riks (Drottning Kristina) 1660-1661
[138] Forsgren 2010: 56-59
[139] Helmfrid 1971, volume 3: 102
[140] Molbech 1820: 68
[141] Forsgren 2010: 88-89

The Swedish economic crisis, however, caused arrears on land-rents owed to Jakob Momma, whilst the Danes seized a valuable cargo from one of his ships, and related legal processes devoured vast sums. In the 1670s the Franco-Dutch War (1672-8) hampered brass exports and, in 1677, flooding damaged the mill-dams, water-wheels and workshops. After Jakob Momma's death in 1678, his son Abel Reenstierna (1665-1723), took over at Norrköping, but he was less financially competent than his father, and conditions were against him. As a last straw, just as Abel Reenstierna gained European markets, the rival English brass industry revived, largely thanks to the quite different Jacob Momma whose father (yet another Jacob Momma) had started brass-works in England in 1649. In 1687 a major cave-in and further collapse at the Falun copper mine caused a Swedish copper shortage. The Swedish economy was failing due to war breaking out against Europe, and by 1695 local timber was so sparse,[142] that Norrköping brass mill had to buy charcoal from more distant parishes.[143]

In 1661, while the Swedish economy was still prospering, Henrik de Try (c.1601-1669), from Aachen, founded a brass mill at Gusum 150 km south west of Stockholm, where he built himself a substantial country house.[144] Gusum mill used English, Polish or Upper Hungarian zinc ores. Brass made from English zinc ore was considered slightly brittle, whereas Polish ore gave tougher brass and Upper Hungary produced copper for making brass wire. From 1661 onwards, coils of brass wire of different weights became Gusum's sole export. A brass wall candle sconce (decorative wall bracket), donated by Henrik de Try, survives in nearby Tryserum church. De Try, formerly manager at Norrköping brass works, was financed at Gusum by his wife's uncle, Claude Hägerstierna, a leading French Huguenot merchant in Stockholm. Hägerstierna's business ethics were far less sound than de Try's, and he appropriated money earmarked for paying the mill's customs dues. In April 1663 he felt personally entitled to remove 85 tons of brass products from Gusum. This forced de Try to seek new sources of funding, but Sweden was just entering a serious credit crisis connected to an issue of worthless bank notes. Confusion reigned at the mill, and one workshop burned down in 1665. It was soon rebuilt, but de Try was struggling to repay his generous creditors. The de Try family withdrew from the mill, and Henrik de Try, who had never learned Swedish, spent a dismal last few years, depending for friendship on his main creditor, Jakob Momma, who brought herrings, peas and money to help de Try's family.[145]

After de Try's death in 1669, tenants, including Abel Reenstierna, leased Gusum Mill, then the third largest Swedish brass works, importing copper, calamine and casting-stones through the mill's Baltic harbour at Valdemarsvik, five kilometres south of Gusum. The mill had 12 brass furnaces, 26 wire-drawing benches, a foundry and a calamine mill. The Gusum rolling-mill, built in the mid-17th century, was burnt down

---

[142] Riksarkivet, Stockholm: typescript Register, folios 553 and 2395
[143] Riksarkivet, Stockholm: typescript Register, folio 560
[144] Forsberg 1993: 16-18
[145] Forsgren 2010: 98-100

in 1689, together with its materials, water-wheels, gear-wheels, axles and workshops full of tools, all destroyed. A huge weigh-house caught fire, with bowls, weights and about eighteen tons of brass.

The brass mill at Vällinge, ninety kilometres south-west of Stockholm, after a shaky start, was taken over by Henrik Thun (1624-76), a leading Stockholm copper and brass exporter, and a Dutch cousin. The partners' broad economic view and mutual interests in commercial and shipping companies linked them to the vital Dutch credit market. By 1670, Henrik Thun could turn more attention to Vällinge, where over seventy people, including his family, lived around the mill. Thun planned an elegant chapel there, designed by Nicodemus Tessin the Elder (1615-81). It was completed in 1679, after his death. Three 37 cm-square brass-casting stones remain set into the chapel floor. The mill had six furnaces, and produced wire, pins, chandeliers, pans and cauldrons. In 1693, two very wealthy bankers (owners of ships, ironworks and a sugar factory) leased Vällinge mill, but a few years later business declined and the workshops urgently needed repair. As the economy slumped further, brass production ceased, and the mill officially closed in 1712.[146]

Nacka brass mill, nearer Stockholm, was leased in 1673 by one Mauritz Ernst. In 1682 a rolling mill was introduced and by 1695, water from the Stockholm shipway was diverted into the Nacka stream over a weir which controlled a five-hundred-metre-long flow of water, dropping eighteen metres down towards Lake Järle (Järlesjön) below. A further weir channelled water to a rolling-mill, with lathe and annealing oven.[147] A succession of buildings downstream included a mansion house with a grand pillared entrance, an adjacent brass furnace house, wire-drawing workshops (with thirteen wire-drawing benches and five shears) and a brass battery-works with three hammers, culminating with a calamine crushing mill at the lower, Järlesjön, end. In 1682 a rolling mill was introduced. However, by November 1694, production had sunk and the workers complained of severe poverty, whilst the owner craftily removed all the raw materials before his unpaid lease expired. The workers implored the College of Mines to find them skilled work elsewhere, before they and their families perished of hunger in the approaching cold winter. Selling the works was difficult because the Nacka furnaces had denuded the local forests, leaving no timber for future brass-making, so in 1699 the Crown had a valuation map made, in a failed attempt to find a buyer.[148]

After 1688, when King Karl XI of Sweden had tried to reduce his debts by seizing back former Crown estates, Skultuna mill was leased by Jacob Momma's son, Abel Reenstierna, Although a member of the Momma family and a favoured courtier, Abel Reenstierna was now also bankrupt, immersed in bank loans. Legally, lease-holders of

---

[146] Forsgren 2010: 68-70
[147] Riksarkevet, Stockholm: MS Vide proto, folio 422
[148] Forsgren 2010: 75-6

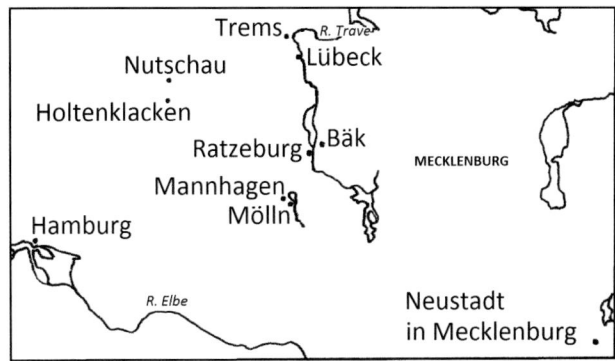

Figure 65. Part of Holstein, map showing some sites mentioned in the text

commercial mills took a drop in salary when production declined, so, when he fell further into debt, Reenstierna was held personally responsible and had to remain in post at Skultuna till every creditor was paid off. The State Bank (*Riksbank*) had lent huge sums to brass works in the good years, so now in the harsh late seventeenth-century Swedish economic climate, the bank claimed all bankrupt assets. In 1695, after the financial collapse of Skultuna and Nacka brass works, the bankrupt Abel Reenstierna left his grand chateau for a bare prison cell.[149] The Swedish producers who had flourished in the mid-17th century collapsed with the national economy at its close.

Scattered across mainland continental Europe many brass works emerged or restarted after the Thirty Years War (1618-48). Several were set up around the independent Baltic trading city of Lübeck. In 1664 the Küsel family, for example, built new large brass and copper mills at Trems, four kilometres north of Lübeck, on a mill-stream leading from the River Trave into a lake (*Tremser Teich*). Maintaining sluices and dams against the strong flow of the River Trave was costly, so in 1682, to get extra cash, the mill proprietors re-stocked their mill-pond with fish and let out the fishing rights. Direct river access to the Baltic Sea and to the nearby city of Lübeck made it easy to import raw materials and export brass, so the brass-works kept going.[150]

Resulting from the 1648 Peace of Westphalia at the end of the Thirty Years War, the richly forested prince-bishopric of Ratzeburg, in the lake district south of Lübeck, was controlled by the Danish Dukes of Holstein-Gotorp, whereas, from 1653, the region to its east, including Lake Mechow, came under Mecklenburg rule. As a result, Ratzeburg, on its beautiful lake peninsula, became a focus for fifty years of conflicting inheritance claims and aggression. Despite the disputes, Matthius Leers, who already a had brass mill on the steep Bäk (stream), across the water from Ratzeburg, in 1661 built a second one, to be operated continuously by his family for 150 years (see map in previous chapter). He made brass cauldrons and other hollow wares, and employed scrapers, burnishers and polishers to finish his products. In 1675, after a few short

---

[149] Forsgren 2010: 90, 199
[150] Warncke 1949: 211-213

tenancies, his widow, Cathaleina Leers took over the brass mill, and leased farmland just upstream near Lake Mechow, which lay in Mecklenburg.[151]

In 1692, the Duke of Mecklenburg laid claim to Ratzeburg, prompting King Christian V of Denmark, Duke of Holstein-Gotorp, to bombard Ratzeburg to the ground, in an attempt to wrest back what he saw as his rightful land. The brass mills at Bäk, located in woodland across the lake from the town, survived his onslaught, so Cathaleina Leers held on to them. In 1695, a Hamburg agreement granted Ratzeburg to Mecklenburg. Because Cathaleina Leers happened to lease farmland from the Mecklenburg authorities in Mechow, she obtained a 'privilege' not only to make brass, but to sell brass and brass cauldrons and bowls throughout Mecklenburg province and even beyond. In 1696 Cathaleina's son, Johann Jacob Leers, with two partners, leased the brass mills together with his mother's manufacturing and trading privileges.[152] Mill-owners had broken out of the medieval feudal form of land control, and, like Cathaleina Leers, who knew her rights, were informed enough to defy a powerful overlord who confronted them. Her lake-district brass mills, at Bāk near Ratzeburg, illustrate the central European problem of trying to maintain a business in the face of aristocratic families laying claim to pieces of interlocking territory.

In 1653 Mannhagen brass mill, also in the lake district south of Lübeck, likewise fell under the rule of the Duke of Mecklenburg. Two years later, a second locally-managed Mannhagen brass mill was started, and in 1695, a Lübeck citizen, Christian Meyer, built a new brass-mill and battery-works on a series of descending dammed mill-ponds nearby. He sold them the following year to another Lübeck merchant, Hermann Fock, who apparently prospered, because battery mills (probably no longer brass) still existed at Mannhagen in the 19th century.[153] For much of the last quarter of the 17th century the Duke of Mecklenburg was also overlord of a brass mill at Neustadt (in Mecklenburg) alongside existing copper works.

In today's central and north-eastern Germany, smaller locally-supported brass-works were operating in quite remote valleys, an industrial model still alive in Germany today. One brass factory founded in the late 17th century was the Hegermühle (Heger mill), at Finow in Brandenburg, about 60 kilometres north of Berlin. During the Thirty Years War (1618-48) the canal connecting the Oder and Havel rivers was destroyed, leaving only the river Finow navigable.[154] To encourage skilled workers back after the war, the Hohenzollern ruler of Brandenburg issued a royal edict in 1685, authorising Huguenot refugees to enter Brandenburg. In 1698, he ordered his court jester and dancing-master, Gottlieb Schutz, a skilled brass-maker,[155] to set up the Hegermühle

---

[151] Kock 1981: 83
[152] Kock 1981: 84-85, 89, 93
[153] Lisch 1842: 84-85
[154] Schillig, C., 2008: www.monumente-online.de/06/01/01_Finowtal.php?media=print, 2
[155] Woiwode 1996: 22

brass works beside the strongly flowing River Finow. Extensive local forests provided wood and charcoal for fuel, calamine came from Gleiwitz (Glewice) in Upper Silesia and copper from the Harz,[156] two hundred kilometres to the south-west.

The Harz mountain area, besides supplying copper, still had brass-works of its own. Oker brass-works, near Wernigerode in the Harz, was built beside a liquation plant where copper and silver were separated from lead-rich ores. Whilst the ore was being heated (in order to separate both lead and silver from copper), zinc from the ore sublimed as vapour and solidified to form zinc oxide and zinc metal accretions on the walls of the liquation furnace flues. This zinc by-product was scraped off and collected, so in 1647 Oker brass-works (modelled on Bundheim brass works) was erected to exploit it. By 1648 Oker employed seventeen brass workers, and by 1671 they alloyed the zinc and copper to produce approximately 21 tons of brass annually, worth over ten thousand silver *taler*. A 1689 inventory mentions that this brass-works had four furnaces surmounted by two great chimneys. A water-wheel with iron axle-shaft powered the battery-hammers, supported by annealing furnaces and a large chimney. A second mill drove three grindstone crushers to pulverise calamine for storage in a tiled warehouse. The foundry contained two pairs of stone casting slabs, secured with iron bands and equipped with screws, rings, hooks, ropes and winches. The wire-drawing workshop, about eighteen metres long, had apartments above. A hedge surrounded the entire works and a bridge across the Oker from the kettle workshop and manager's house led to copper and charcoal sheds.[157]

Aachen and Stolberg had long relied on Harz copper, but after a city fire at Aachen in 1656,[158] the Aachen brass industry further collapsed and more brass makers left for Stolberg. By 1667, thirteen Stolberg brass-works operated more than 90 furnaces, the Beck family holding five brass-works, the Mommas four, and the remaining families two or three each. These prosperous mill-owners had splendid fortified houses and workshops, built around courtyards. In the year 1698, Stolberg furnaces produced about 1,400 tons of malleable and ductile brass,[159] suitable for brass plate, sheet, wire and 'semi-prepared' (unfinished), vessels, a proportion of which would have been destined for the slave trade on the Guinea Coast of Africa.[160] Semi-prepared brass goods were often finished (etched, lathe-polished and given handles, feet or lugs) in the Netherlands or in Flemish cities like Liège (Lüttich) in today's Belgium. Many seventeenth-century Dutch paintings show brass cauldrons and bowls glinting from shadowy corners. Some of these products were exported to America and the Caribbean islands, to pay for new transatlantic products like tobacco and sugar that were arriving in Europe.

---

[156] Herr Roland Gabsch, personal communication, April 2008
[157] Priesner 1997: 25
[158] Schleicher 1974: 18
[159] Roderburg 1927: 13-15
[160] Schleicher, 1974: 34-35, 50

Novel commodities like tobacco and sugar required new domestic equipment. Sugar, for example, might be served from a brass shaker of 'lighthouse' design, with a decorative perforated lid,[161] and imported coffee beans were ground in brass mortars. Dutch imports of tobacco from their American slave plantations introduced Europe to tobacco-smoking and snuff-taking. From the early 17th century, Netherlands craftsmen produced small oval tobacco-boxes, changing around the middle of the century to twelve-centimetre-long rectangular brass tobacco boxes with rounded ends. Iserlohn, whose collapsed calamine mine-shafts were restored in the 1680s, besides producing fine wire, soon made tobacco boxes and Sweden did so slightly later. From 1660, brass tobacco-tampers for pipe-smokers were made in the Netherlands for pressing down tobacco in the pipe-bowls. Many tamper handles were decorated with small figures.[162] Narrow early pipe bowls needed narrow tampers but later ones were wider. Snuff (tobacco sniffed in powdered form) was popular among both sexes, and required the sniffer to possess a snuff-box.

Previously unknown sugar and tobacco imports were not the only innovations. The later 17th century saw enthusiasm for new scientific discoveries in mathematics and astronomy, which gradually introduced the Age of Enlightenment, and a new interest in empirical knowledge, which made frequent reference to the world of the Ancients. The religious fervour which had inspired and controlled peoples for centuries, when the unexplained was interpreted as the work of the gods or God, now met challenges. Educated people became fascinated by fields such as the motion of planets, blood circulation, the properties of gases, gravity or the speed of light – ultimately bringing into question received views based primarily on religion. The Royal Society in London, whose erudite members had met informally since the 1640s, was founded in 1660, and in Paris, scientists, who had been meeting informally from 1666, became the Académie Royale des Sciences in 1699. These learned societies built observatories and botanical gardens, and older myths about the universe began to be replaced by natural laws deduced by reasoning from newly discovered knowledge. The belief arose that people could improve society by taking responsibility for their own destiny and their own institutions.[163] The movement remained largely the domain of educated and wealthy people, but the world had entered a mercantile age, so the ideas it engendered began to diffuse further.

Also in later seventeenth-century Europe and America, the written word was all-important,[164] the resulting paperwork held together with rust-proof brass pins. The ones for paper started in the 19th century The communal religious and spiritual dimension of the European medieval and early baroque era now gave way to a

---

[161] Crawford 2006: 2
[162] A collection is displayed in the Rockefeller Museum, Colonial Williamsburg, Virginia
[163] Alcock 1998: 157-159
[164] Moreland 2001: 57-58

compartmentalised social order, with the individual beginning to take precedence over the community.[165] Popular imagination had been seized by individual self-awareness. The household started observing regulated time schedules (requiring clocks and sundials, for which brass was extensively used) and spatial rules (needing survey instruments), and entertained according to the rules of polite society. People no longer saw every phenomenon as divinely ordained, but started to question and rationalise. This, the Age of Enlightenment, was attended by world-wide interest in astronomy, time-keeping, measurement, planning, and musical and scientific pursuits, all of which demanded brass equipment.

In 1656, a Dutch astronomer, Christiaan Huygens, applied the pendulum to a brass clock movement, a design made and marketed first in the Netherlands, then in England. Clock-making became a trade, and brass was used for clock gear wheels,[166] front and back plates, engraved dials, decorative spandrels (corner-pieces) and many tiny parts. By about 1660, brass-faced table clocks called bracket clocks appeared.[167] Nuremberg craftsmen still produced superb timepieces, precision brass compasses, dividers and other instruments for surveyors and draughtsmen, as well as everyday brass objects like stopcocks and syringes, cast in moulds of fine local white clay.[168]

From 1656 to 1670, in outer Nuremberg, the brothers Lang ran the Hammer brass mill at Laufamholz. An inventory dated 1681 lists plentiful accommodation, including the owner's house (*Herrenhaus*),[169] a dwelling house with washroom and bathroom, another rented house, thirty apartments, an office and stables. There were facilities for alloying, forging, annealing, ore grinding and crushing, brass sawing, wire drawing and crucible making, more than twenty buildings altogether. In about 1680, an inn had been built, together with a new sandstone clock-house with decorative half-timbered gables, to lodge eight workmen with larger families. Paternalism towards the work force made financial sense. Each worker's rent-free apartment at Hammer had a living-room with an oven, a small unheated room, kitchen and attic. Fuel was delivered free, and the living room was white-washed and the oven renovated free once a year. With his apartment, each worker could have a rent-free allotment and potato-shed, and some had pig-sties or goat-sheds. Those unable to work due to illness, disability or extreme age, and widows and children, had rent-free accommodation as needed.[170]

Nuremberg merchant, Johann Cristoph Volkamer (1644-1726), proprietor of Hammer Mill, was a garden enthusiast, ardent botanist, plant-collector and early specialist in

---

[165] Leone 1988: 235-61
[166] Colonial Williamsburg Foundation, Virginia: CWF 8-18-13A, clock gear-wheel,
[167] Noel-Hume 1969: 151-152
[168] Nuremberg Stadtarchiv: MSS, Landauer 1, Amb.279, folios 133-4; Mendel II, Amb.317b, folios 101-2, 179-80
[169] Nuremberg Stadtarchiv: MS, 455/2. 1687
[170] Wittek 1984: 121-122, 124

citrus trees, who travelled to and from Bombay (Mumbai), Calcutta (Kolkata) and the Near East. In 1690, Volkamer passed on the running of Hammer brass mill to his heirs, but continued his voyages. In the process he established invaluable brass-trading contacts with the Near East and India, through the port of Venice.

Volkamer's garden not far from Hammer brass mill, Laufamholz, was a fine example of the geometrically laid out gardens currently fashionable throughout Europe. In 1693, on returning from Bombay and Calcutta, he ordered an obelisk from Constantinople for his formal garden.[171] As a member of the rising merchant class, it was appropriate for him to be fascinated by geometry, astronomy, the classical world and natural history. The material possessions associated with such pursuits sent out messages, presenting the owner in a desirable light and distinguishing his or her status as an individual.[172] Possession of interesting plant species or curious scientific instruments helped to position their owner in society.

To meet the new fashion demands, the skilled craftsmen of the Meuse Valley varied their product, adding secular items such as wine coolers, candlelight reflectors, lanterns, warming-pans and decorative candlesticks each with a tall point on which to spear the candle. However they still furnished churches with traditional brass offertory plates, alms boxes, holy-water buckets, altar candlesticks, crucifixes, carillon bells and sanctuary lamps, often with cut-away and elaborate repoussé or chiselled decoration.[173]

Fashion is embedded in social behaviour, and society was changing rapidly at the end of the 17th century. In Europe and America, as self-awareness of the individual increased, the rising merchant class was socially emulating or even overtaking the landed aristocracy. People related to the showy, high-fashion brilliance of brass as an indicator of individuality and status, and the extravagant trends of the Sun King, Louis XIV of France, lingered on in fashion products like brass buttons, buckles and sword-hilts, designed to attract and impress. A surviving brass naval sword-hilt of about 1690, from the wreck of the English vessel *Stirling Castle*, was richly embellished with cast designs.[174] John Webster (1610-82), physician and chemist, writing in 1671, mentioned contemporary uses for different brasses. Hard brass that 'will abide the hammer in some measure' could be beaten into plates or leaves, and was suitable for fire tongs, fire shovels, snuffers and mathematical instruments. Other more brittle brass was suitable for pots, pans, chafing-dishes or candlesticks.[175]

---

[171] Mertel 2010: 11
[172] Burke 1994: 149-150
[173] Toussaint 2005, many figures
[174] Capes and Chamberlain 1998: 134
[175] Webster 1671: 250

The interest in fashion and luxury was met by brass-founders in European centres like Birmingham, which between about 1640 and 1670 produced a constant and varied flow of small brass objects known as 'toys', including brass shoe and garter buckles, buttons, locks, miniature guns, lockets and brooches. Such 'toys' were calculated to be popular in France, one rising into fashion as the last was rejected.[176] Birmingham's population almost trebled between 1650 and 1700, boosted by the influx of refugee craftsmen, who experimented with decorative brass inlay on guns, swords and knives, often using 'Prince's metal', containing metallic zinc, to counterfeit gold. People spent extravagant sums on decorative brass buttons and showy buckles. Belt-buckles dating from the 1650s have been excavated in the Aldgate business area of London.[177]

The aristocracy or wealthy elite still collected precision instruments related to astronomy and measurement. In 1679, one collector, twenty-five-year-old Count Carl of Hesse, who had a deep fascination for scientific instruments, built a brass-works complex well outside the Kassel city walls, at Bettenhausen. He intended to alloy brass using local copper ore, in order to produce brass objects, including scientific equipment, exploiting copper ore from his own region of Hesse. He realised that it was no longer worth exporting his copper ores cheaply whilst importing expensive brass wares. To profit from his Hesse copper, he stopped the export and import of raw materials and decreed that anything mined in Hesse should be worked within Hesse, at Bettenhausen.[178]

The River Losse powered the Bettenhausen mill-wheels with a strong and constant year-round flow. Completed in 1680, two long buildings (50m x 10m and 14m high) with baroque gable-ends were erected along opposite sides of a great courtyard. High walls with arched entrances completed the ends of the rectangular courtyard. The ends and one wing are still standing. The calamine-crushing mill and battery hammers were driven by four undershot water-wheels, two of which powered seven battery hammers for shaping brass plates into pans and cauldrons. The wheels were housed in the wing nearest to the constant strong flow of the mill-stream. The foundry, wire-drawing workshops and crucible-pottery were in the opposite wing.

The foundry, with its five furnaces in large circular holes in the ground, and a huge surviving vaulted flue above, produced plate and sheet brass of varying thicknesses. Brass sheets were slit to be drawn into wire of different diameters, for producing articles, including pins, chains and hob-nails. A counting-house, weighing-rooms and delivery storage area were located in a half-timbered building at the town end of the courtyard.[179] Following the 1685 Edict of Nantes, in which Louis XIV of France expelled

---

[176] Hamilton 1926, 1967 edition: 129-130
[177] Thompson *et al* 1984: small finds ALD 173/1262; ALD 150/1222; ALD 425/1133
[178] Gronau 2011: 18-19
[179] Schaeffer, Bernd, 2013: www.errinnerungern-im-netz.de/aw/Testseiten/~Der_Messinghof_Wiege_des_Herkules, 1-2

Figure 66. Bettenhausen brass works, west façade, 1679, 50 metres wide

Protestant pastors but refused escape to lay Huguenots, about two hundred thousand workers fled France at risk of their lives, Count Carl of Hesse offered skilled Huguenot craftsmen profitable jobs at the brass works, with the chance to settle in Kassel.[180] In 1697 the spacious Bettenhausen site lay amongst a rich timber source of deep forest, and it had a subsidiary copper works, a crucible pottery and good workers' housing.

The earlier factory, at Kaufungen slightly further up-river, had used Goslar copper, but the new Bettenhausen mill used Hesse copper from the Richelsdorf hills, fifty kilometres south of Kassel, or from Frankenberg-on-Eder, fifty kilometres south-west. Hesse had no calamine, so it was transported 450 kilometres from Silesia, but later 60 kilometres from Brilon.[181] Not only charcoal, but coal, was used to reduce oxygen during the brass-making process. The crucibles were charged with copper and calamine, to which was added a half quantity of charcoal. Once it had burned through, powdered coal was introduced, then, after eight hours, further charcoal was added.[182] The Kassel-Leipzig trade route ran alongside Bettenhausen brass-works mill-stream, giving it an obvious transport advantage, but the site lacked a navigable waterway. This meant that, in bad weather, horse transport over hilly forest tracks might be

---

[180] Heppe 1996: 4
[181] Seib 1977: 170
[182] Hachenberg and Ullwer, 2013: 181, quoting Marcus Fulda MS, 1717, folio 52

delayed or halted by mud, snow or cart damage,[183] not to mention shortage of fodder for the draught animals.

Two hundred kilometres south-east of Kassel, in the Auertal, a valley in Saxony (near today's Czech border), two partners Melchior Haugk and Jacob Kôrber, gained permission to re-start the Ellefeld brass mill which had been largely destroyed during the Thirty Years War (1618-48). A nucleus of skilled workers, who remained in the area, restored the dilapidated workshops, mills and water channels, furnace-house, and workshops. The earlier mill had used Brilon calamine, now nearly exhausted, so a fresh source was found in Upper Silesia. A polisher and scraper were hired. The men were paid by their plate brass output, and by the year 1666, when both Körber and Haugk died, the mill was producing again.[184] Their sons Johann Haugk and Christian Körber (aged only fifteen) took over and their brass output in 1670 was 38-40 tons a year. In the late 1670s, Christian Körber lured Ellefeld workers to a new brass-works he had started further north, at Grunau in Thuringen, on the Loquitz stream five kilometres south of Leutenberg. He placed it under the same management as brass-works at Hockerode, and Arnstadt, respectively five kilometres north and forty kilometres north-west of Leutenberg.

In 1695, Martin Albert, Mayor of Freiberg, took over Rodewisch brass-works, downriver from Ellefeld. He formed a small 'brass-trading society', which linked the proprietor to the business and regulated the status of its four shareholders. In 1697 Rodewisch produced two and a half tons of brass and its first coarse brass wire. In 1710 the number of shareholders increased, to keep an eye on the business. A factory agent handled orders, sales, deliveries and markets, and judged disputes, backed by shift foremen to oversee and police the craftsmen. The major drawback was that the agent was obliged to contact a shareholder for a ruling on every detail of pay structures, prices, ordering, building, repair, new equipment or dispatch of brass to the Leipzig trade fairs. This involved lengthy written correspondence which, as economic historian Hans Otto Gericke points out, made the Rodewisch agent's job hopelessly impracticable, particularly as the proprietor turned up only twice a year to discuss proposals.[185]

In 1666, a rival brass-mill was established only twenty-five kilometres away across the mountains at Stříbrná in Bohemia (today's Czech Republic). From 1675 onwards, the brass mill at Stříbrná competed with Ellefeld and Rodewisch, which lay over the border in Saxony. Former managers, Ficker and Körber, quit the Ellefeld firm, fearing serious competition, leaving Johann Haugk in sole charge. In 1679 he recruited a skilled battery-worker from Oldesloe, Holstein, to hammer out plate brass into semi-prepared pans, cauldrons, and brewery vessels, but Haugk lacked leadership and over-

---

[183] Gronau 2011: 19-20
[184] Gericke 2008: 83, 59-61
[185] Gericke 2008: 73, 76-7

paid himself from factory funds. Ellefeld workers complained in court and Haugk was ordered to manage the mill better. He failed, and from Easter 1690 the mill lay idle, so most of its skilled brass workers left, one for Nuremberg, some to Finow-Eberswalde north of Berlin, one to Jakobswalde in Silesia and many to Štřibrná itself, just across the mountains.[186]

Štřibrná, now a hamlet of 25 houses, lay up a steep forested valley east of Kraslice (*Graslitz*) in the western Bohemian Krušné Hory (Erzgebirge) mountains. In 1666 the Habsburg Counter-Reformation declared Bohemia a Catholic country, and the Bohemian Duke of Nostitz bought the lordship of Štříbrná, forcing brass workers to become Catholic or flee. Many fled, so the Kraslice copper mines and smelting-works declined. Balthasar Rössler, a Bohemian-born metal-working expert with intimate knowledge of Štříbrná and Kraslice (where he worked from 1634-49) left for Saxony in 1649 and wrote down the methods he knew (published fifty years later by his grandson).[187]

Štřibrná had three battery works with many battery hammers, sheet-slitting workshops and at least four large furnace-houses, where 75 lbs (34 kg) copper smelted with calamine ($ZnCO_3$) would produce 100 lbs (45 kg) brass,[188] which would have contained rather less than 25% zinc, useful for both casting and forging, and probably golden in colour. Kraslice copper-smelting ovens were heated by bellows and copper was refined by re-smelting, to produce the highly refined copper needed for brass making. Štřibrná brass works probably exploited zinc metal droplets and zinc oxide gathered from nearby lead- and silver-smelting furnace-flues. Rössler mentions the distillation of arsenic and mercury but not zinc.[189] During its most productive years – between 1660 and 1680 – Štříbrná delivered brass to Leipzig, Dresden, Prague, Nuremberg and Vienna.[190]

Around 1655, the small independent state of Salzburg, which lay further south-west – about 200 km west of Vienna – restored and enlarged its two dilapidated brass works at Oberalm and Ebenau, which had lain idle for almost fifty years. They were directly controlled by the court of the prince-bishop of Salzburg, and their brass sheet, wire and ingots were aimed mainly at the Salzburg home market, which, however, was too small to consume great amounts of brass. In 1661, the two mills produced 68 tons annually, rising gradually to 100 tons in 1671, and reaching a steady 109 tons or more from 1676. The best years of the Salzburg-owned brass works were from 1660 to 1700, before entering a gradual decline.[191]

---

[186] Gericke 2008: 67-70
[187] Rössler 1700: unnumbered fourth page of Forward
[188] Rössler 1700: 19, paragraph 43
[189] Rössler 1700: 19, 141, 146-147, 159
[190] Jaroslav Zaplatal personal commumication
[191] Priesner 1997: 59

Achenrain, a mountain valley brass works in the Austrian Tyrol, stood about 100 km south-west of Salzburg, and two kilometres north of Brixlegg in the alpine Inn valley (for Austrian sites, see the map in Chapter 4). Copper and silver had been mined and produced nearby since 1463 (when it had been part of Bavaria). Initially, small-scale brass-alloying occurred high up on the mountain slopes, but by 1500 this proved increasingly difficult, so the brass-makers moved down to Achenrain. For fuel and charcoal, they used timber floated down the River Ache from the nearby Brandenberg mountain.[192] A surviving half-timbered former managers' house at Achenrain bears the date 1571. In 1645, Carl Aschauer, a customs man, married into a leading family of tin-founders and became interested in making brass.[193] He spent two years in Sweden (1645-7), learning the latest techniques of their brass manufacture. Aschauer returned to Achenrain, 'rich in knowledge' and 'brought back every type of brass, but above all he brought understanding, science and technology, so that every brass worker could use this expertise to bring it to perfection'. In Sweden he evidently learned the 'Netherlands' method of embedding entire crucibles in charcoal within an inner chamber of the furnace, because, at Achenrain, this charcoal was renewed after seven hours of cementation.[194] He also hoped to 'improve the quality of battery brass by using expensive foreign methods and ingredients' – which included using purer imported calamine.[195]

In 1648, the enterprising Carl Aschauer with a merchant partner, Andreas Pranger (his wife's relative), bought Achenrain brass mill, with its foundry, furnace, house, orchard, stables, meadow, river channel and forge.[196] Aschauer introduced wire production, whilst Pranger aimed to underbid other enterprises until all competition fell away. Huge sums were invested, and good-quality raw materials were organised to arrive promptly on site.

Achenrain brass pins, wire and sheet-brass were exported to Germany, Switzerland and Italy, and became recognised for their quality. Through the Bolzano and Venice markets they sold in Verona, Florence, Siena, Urbino and Milan.[197] The armourers of Milan were renowned, and a fine breast-plate decorated with brass studs (1690) survives in the Sforza castle, Milan.[198] When, in 1655, the efficiency of the Achenrain plant reached the ears of Emperor Ferdinand III (1637-1657), he personally promised food, wine and clothes for the brass workers; and exempted brass-maker entrepreneurs from petty court processes. They were authorised to train apprentices to become journeymen free to travel anywhere within the Holy Roman Empire. To stave off

---

[192] Kôfler 1972: 367
[193] Urbanner 1972: 47-48, 52
[194] Hachenberg and Ullwer 2013: 181
[195] Aschauer 1953: 317, quoting Franz Riesner
[196] Tiroler Landesarchiv: Schwatzer Schatzarchiv 200, inserted in documents 10-11, dated 1670
[197] Köfler 1972: 370
[198] Museo del castello sforzesco, Milan: inventory number. 125.198

competition, the emperor forbade the erection of other brass-works in the Tyrol, except at Lienz.[199] A third small brass-works, at Nassereith in the Inn Valley, had been burnt down in 1651 and was not revived.

In 1653, after the Thirty Years War, the Lienz brass factory in the alpine south Tyrol was re-started by Andree von Winkelhofen. In 1679 Carl Aschauer from Achenrain bought the Lienz brass mill, together with its Jauken calamine mines in the Drau valley, Upper Carinthia (*Kärnten*). The Lienz mill continued production and was soon ordering extra copper supplies from Upper Hungary and from Agordo in Italy, but Aschauer was more interested in the calamine supply than the brass-works, and left its workers poorly paid.[200]

Achenrain now had an assured calamine supply, delivered from Carinthia, Nassereith and Auronzo.[201] Copper derived from high-quality ores mined from the mountain slopes of the Inn valley, was delivered from Brixlegg refinery.[202] Further supplies came from Schwach, Prettau (now Predoi, Italy, north-west of Bolzano) and Defreggen (75 kilometres south-east of Innsbrück).[203] Before 1660, barely sixty tons of Brixlegg copper had been used annually for brass-making at Achenrain but this soon doubled.[204] After acquiring the Lienz brass-works, Aschauer and his partner gained a total monopoly over brass production in the Tyrol. To increase efficiency, the partners acquired an iron-smelting works on the Pillersee to supply iron equipment, and made a water-supply agreement with the monastery up-river to ensure a good water flow for the battery hammers. A dispute over timber-felling rights, provoked by shortages of charcoal timber, forced Aschauer to return illegally-felled tree-trunks to the Brixlegg copper-refinery.

In this prosperous period under Carl Aschauer, the Achenrain works cast fine artistic brass figures, including a rider on horseback, dating to 1650-80.[205] The brass-works earned 15,000 gulder a year from sales to the Tyrol, Germany and Switzerland, and to Italy through the Bolzano (Bozen) market. To expand production capacity, the partners took on growth capital, protected by taking out credit. The relatively high interest rate of up to eight per cent might have presented no problem had the partners not been extravagant with company profits, leading, in 1685, to unmanageable debts. Compromise with the creditors prevented immediate collapse but the firm never fully recovered, because of mismanagement and disputes with Tyrolean regional officials,

---

[199] Priesner 1997: 53
[200] Ucik 2002: 172
[201] Amman 1990: 333
[202] Mutschlechner 1990: 231, 236
[203] Mutschlechner 1990: 20
[204] Priesner 1997: 53
[205] Tiroler Landesmuseum Ferdinandeum, Innsbrück: on permanent display

who were astonished at the brass-works debacle.[206] Formerly faithful creditors furiously demanded their money back and confiscated any brass wares they could find.

As the Achenrain works fell silent and the customers disappeared, the government stepped in. Merchant partners ran the site, but they had to pay cash for raw materials. After Carl Aschauer's death in 1693, his ill-qualified son Oswald inherited the brass-works. An Innsbruck merchant, Paul Pauletti, was appointed as junior manager, but could not cope with Oswald Aschauer's disputes and debts. Oswald Aschauer, however, suddenly died and the brass-works failed to prosper for many years.[207] Achenrain illustrates two age-old problems that bedevil otherwise successful production, firstly, inheritance by a weaker son and secondly, embezzlement of factory profits for the manager's own extravagant ends.

Figure 67. Horse and rider, cast brass, Achenrain brass works c.1650-60, 43.5 cm. high

In 1647, a brass-works at Möllbrücke, south-eastwards across the Alps in Carinthia (Kärnten), was bought by the Hendl brothers, two Venetian merchants who lived extravagantly in a château in Obervellach, eighteen kilometres further up the River Mühldorf. In 1664, the brothers fell into debt, so Duke Johann Ferdinand Portia acquired the whole works, improved the site, raised the output, and sold it in 1689 to Carl Aschauer (already owner of the Achenrain and Lienz brass-works) and his brother-in-law Karl Anton Wagner, a senior Tyrol mines official. Through Carl Aschauer, Achenrain's management problems now affected Möllbrücke, so in 1693 Wagner took Aschauer's share of the failing Möllbrücke brass-works, and in 1696, the Viennese court exchequer took responsiblity.[208]

In 1654, Reichraming brass works in Carinthia, was taken over, after Hans Eggar died, by Matthius Rieser von Riesenfelz, but he was inexperienced so the brass was poor, and in 1658 the Steyr authorities took back the mill. Reichraming had recently been earning two to three thousand florins a year from export through Venice to Turkey,

---

[206] Priesner 1997: 53-54
[207] Köfler 1972: 372-373
[208] Ucik 2002: 175-176

but Salzburg now seized this trade. In 1660, only sixty tons of Reichraming brass was forwarded from Vienna to Venice. Rieser's son Franz took over but did little better. In 1669, Vienna sales were down by about a third since the previous year, and in 1671, Steyr craftsmen bought only one ton of Reichraming brass, having previously bought 24-30 tons annually. Only the Venice trade was still viable, though payment was unreliable, and strong competition came from Salzburg and the Tyrol. Goods carried to Vienna on river rafts were subject to excise duties so in 1676 Rieser tried (unsuccessfuly) to be exempted from them.[209] Even Franz Rieser's remote forest timber was stolen by foreign producers of pitch for Venetian boat-builders.[210] Franz Rieser, disheartened, gave up Reichraming brass-works in 1705.

The late seventeenth-century brass industry reflects not only the destructive power of military hostilities over national economies. It also shows the effects of business competition by strong entrepreneurs and managers with technical and business sense and interest in the workforce, as opposed to the alternative weaker or less experienced heirs interested primarily in personal profit. This was a time when Chinese export of distilled zinc affected world brass technology, and when America became a significant market, increasing both transatlantic trade and the slave trade. Brass also took a more important place in extravagant fashion. This period leading up to the early 18th century opened the way for more complex and vigorous global contact and trade.

---

[209] Stiftsarchiv Seitenstetten: MS, Reichraming Messingwerk, 53/V
[210] Brunnthaler 2000: 122

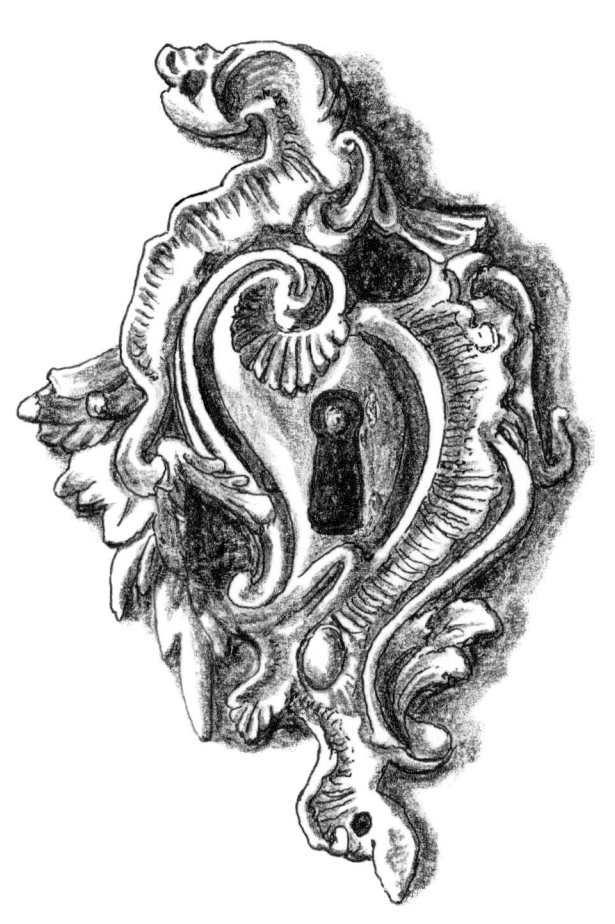

Cast thimble and bell, excavated from a mid-eighteenth-century domestic site in the colonial settlement at Williamsburg, Virginia. photo © author, courtesy of Collections of Colonial Williamsburg Foundation's Department of Archaeology. Inv. thimble: ER1339.19B,00030-19BB, bell: ER1346.19B,00024-19BB

Chapter 7

Trade and technology

c. 1700-1750

Eighteenth-century evidence provides more detail about individual brass users, makers and technical processes. The trade opportunities opened by the age of the great discoveries meant that thousands of ships now sailed the oceans. In the early- to mid-18th century, the American colonies imported quantities of brass objects from Europe, particularly from the Netherlands and Britain, and the slave trade was moving European-made brass around the West African coast. Copper arrived in the West from many corners of the world, including Japan, South America and the Cameroon. Chinese metallic zinc was exported to South-East Asia, India and Europe, and Indian brass artefacts were entering Europe.

Even in the 18th century, few people in Europe understood where metallic zinc came from, and there was good reason for this confusion. Most ships returning from the Far East called in at Indian ports, sold goods to local agents and returned directly to China. Goods re-exported from India to Europe carried a record of the port of origin of the ship, without noting where its cargo components came from. Zinc metal often arrived with goods from several ports, in, for example, a Dutch East India Company cargo arriving at Batavia (Jakarta) from Japan, carrying textiles from China and adding tin at Malacca, all of which were first forwarded to Bengal and then to Surat for further onward distribution.[1]

English ships began trading directly with China, having gained a trading post off Amoy, southern China, where, in 1701, they loaded about 60 tons of zinc metal. In 1703, they took on 570 tons from the ports of Chusan (Zhoushan island, Zhejiang), Amoy (Xiamen, Fujian) and Canton (Guangzhou, Guangdong). Chinese exporters told European agents that the price of Chinese zinc metal had risen sharply because 'the Stones [ores] are farmed'.[2] 'Farming' of ores meant processing or manufacturing, so English merchants began to realise, perhaps for the first time, that Chinese zinc metal was not just mined, but also processed (distilled). From 1704-26, Dutch ships delivered metallic zinc and other Chinese goods, from Japan to Batavia (Jakarta),[3] where ownership of brass objects denoted high status. Greater numbers of European ships now carried Chinese metallic zinc,[4] making good profit from the trade[5] – in 1745 the Swedish East Indiaman *Götheborg*,

---

[1] Valentijn 1726, volume 4, part 1: 257
[2] British Library: India Office MS, IOR G/12/6, folio 830, 915; IOR G/12/7, folios 964, 972, 1017-18, 1303
[3] Valentijn 1726, volume 4, part 1: 257
[4] British Library: India Office MS, IOR G/12/7, folio 1094
[5] Du Halde 1735, volume 2 : 172

arriving from Canton, sank in Gothenburg harbour with a 133-ton cargo of 6,052 distilled-zinc ingots of 98.99 per cent purity.[6]

Under the Qing dynasty (1644-1912), China expanded its empire west across the Himalayas and increasingly exported metallic zinc and brass to India and Europe. In 1703, 300 Chinese thick brass plates and 1,410 fine brass sheets were sold in London.[7] Chinese brass locks, hinges and ornamentation also arrived on the thousands of fashionable lacquered chests and cabinets imported to Europe from the Far East. Chastleton House, Oxfordshire, for example, has a Chinese lacquered cabinet dated to c.1720-50, with a huge elaborately-pierced brass lock-plate and hinge clasps.[8] In 1703, English ship's captain, Alexander Hamilton, saw what he described as a five-ton (five tonnes) brass cannon in Canton.[9] Brass ordnance was more common in India, China and South-east Asia, than in the West. The Qing emperor Yongzheng (1722-35) encouraged craftsmanship and technology, and after 1736 a high-zinc content was normal in Chinese brass coins, as the content of copper reduced and the distilled zinc content increased. After 1641, the government had small amounts of lead added to facilitate cutting and improve surface lustre, and tin to harden the brass and improve the wear of the coins.[10]

Brass coins of a very different sort simultaneously reached inland forested Canada. They took the form of medallions, excavated from the early eighteenth-century King's Stores, Quebec City, and across Canadian forest areas hunted by indigenous peoples. From the early 18th century, French traders bartered brass medallions bearing images of Christian saints, the Sacred Heart and Immaculate Conception, in return for furs. Local trappers used these Counter-Reformation medallions as burial goods to accompany them to the after-life,[11] prizing them highly though not in the way envisaged by their European producers. By the 1740s, native Canadian fur trappers were closely associated with the French, who opposed the English not only in Europe, but in America and at Indian coastal trading-posts.

It was in India, in 1727, that English sea-captain Alexander Hamilton noted the devotion that inspired Hindu and Jain pilgrims, young and old, to travel 400 miles from Surat to Benares (Varanasi), where priests filled 'Brass and Copper Pots made in the shape of short-necked Bottles, with water from the sacred River Ganges'. The priests consecrated these flagons, containing about 18 litres, sealed them up, and delivered them all over India.[12] Tall-necked brass pilgrim flasks, with rings for attaching ropes to string them from long poles, had been used for at least two centuries, some taking sensuously

---

[6] Needham 1974, volume 5, part 2 : 212
[7] British Library: MS, IOR B/44, folio 139; folios 75-111
[8] National Trust, Chastleton: CHS F/5
[9] Hamilton 1727, volume 2, 296
[10] Dai Zhiqiang and Zhou Weirong, 1992, 53
[11] Marsette 2003, 32-33
[12] Hamilton 1727, volume 2: 23

curvaceous forms.¹³ From Rajasthan, a surviving contemporary female image of Dīpalakshmi is of high-zinc brass.¹⁴

In 1701, at the siege of Panhala fort, twenty kilometres north of Kolhapur, Maharashtra, eastern India, the Mughal artillery officer was impressed by twelve field guns of brass (of unknown purity) brought by the British ambassador, who, during a court audience, felt obliged to present them to Aurangzeb, together with six gunners to manage them during the siege¹⁵. From 1707, the Mughal dynasty, rulers over most of the Indian subcontinent for a century, began to decline in the face of now-rebellious Hindu Maharatha and Afghans. In 1717, a weakened Mughal emperor granted the British East India Company duty-free trade in Bengal, but local Bengali Muslim-appointed nobles seized control to deter British ships from landing. In 1727 and 1739, Hindu Maharatha troops entered and plundered Delhi, defeating the Mughals at Bhopal, and harrying Bengal. From 1746-8, English and French merchants again attacked each other's Indian trading posts.

Off the south-eastern Indian coast, Sri Lanka had long undergone Portuguese violence and religious oppression, with the destruction of Buddhist temples and their contents. By the 18th century, the Dutch occupied the island, and only the forested mountainous interior kingdom of Kandy (Maha Nuwara) remained independent. A few Buddhist brass figures and vessels survive from the isolated Kandy uplands,¹⁶ to which the artistic indigenous Sinhalese people increasingly resorted in the face of hostility. The main brass-making centres were located just south-west of the higher uplands, at Philimalawa, Daulagala, Elugoda and Pamunuwa Temple, roughly 100 kilometres north-east of Colombo. The brass workers decorated their wares with motifs such as lotus petals, cockerels and peacocks, casting by the lost-wax process, cutting out shapes from the metal with chisels, and hammering out reverse designs from the back using blunt chisels – repoussé work. For the Tamil (originally south Indian) populations, brass was also integral to everyday Hindu household culture, and included fittings such as door and window locks and hinges. By tradition, brass lamps on pedestals, water pots, betel servers, meal servers and drinking-water pots were polished at New Year, and certain brass items were acceptable as dowry payment.¹⁷ A fine inlaid eighteenth-century Tamil brass drinking-pot (*tembu*) survives in Sri Lanka but most Hindu objects were of bronze, not brass, and showed south Indian influence.¹⁸

By this time, most Europeans realised that India was not the source of the metallic zinc being imported. In 1728, the astronomer Edmond Halley (1656-1742) wrote that 'Tutunague' (metallic zinc) was 'peculiar to China only. 'Tis found in Malacca a port for

---

[13] Zebrowski 1997: 198, figure 301
[14] Biswas 1993: 325
[15] Richards (1993): 289
[16] Mr Senarath Wickramasinghe, curator, Colombo National Museum, personal communication
[17] Kent 2004: 269
[18] Colombo National Museum collections

zinc metal from Pegu (Myanmar) and Sumatra, but the finest is in China'. He added that it was 'a species of Tin, not so bright as either Tin or Lead, but closer and firmer, and finer than Lead, and will work very thin beyond either Tin or Lead'.[19] Distillation was already practised in Europe, but usually for isolating other substances such as saltpetre,[20] arsenic or mercury,[21] not zinc.

Sublimed zinc oxide and zinc metal also reached Europe in small quantities from the Arabian Peninsula, for medicinal purposes. In 1686, Sir John Pettus mentions *Lapis Tutij* as a compound made of calamine, good for sore eyes.[22] London sales of English East India Company goods from Mocha included, in 1699, two chests of *Lapis Tutia*,[23] and in 1706, 4469 pieces of '*Lapis Tutier*', together with coffee, alloes and myrrh.[24] Metallic zinc seems not to have been used along the coasts of West Africa in the 18th century, but imported brass was re-cast for local uses, which lowered the original zinc content.

Figure 68. Oba Ewuakpe with attendants, Benin

An altar-group portrait of the ruler of Benin, Oba Ewuakpe (1700-1712), with attendants indicates that brass was still used to express power.[25] Edo peoples, worshippers of Osun who exerted magical healing powers through forest potions, cast a dramatic surviving Benin brass head.[26] Snakes emerge from the eyes and nostrils; a crown of birds represents powers of prophesy and protection, and axes in the forehead

---

[19] Halley 1728: 216
[20] Glauber 1656: 16-17, 62
[21] Rössler 1700: 159
[22] Pettus 1686: page LE
[23] British Library: MS, IOR B/43, 1699, folio 49
[24] British Library: MS, IOR B/45, 1704-8, folio 226
[25] Ethnologisches Museum, Berlin: C 8165
[26] National Commission for Museums and Monuments, Nigeria: NMM 54.19.1

symbolise the destructive force of lightning.[27] Eighteenth-century kings of Benin used brass heads to represent their own mystical powers. Ornate brass staff heads, symbols of authority, have been excavated at Onitsha, east of Benin.[28] Brass altar animal-heads, including the highly-prized leopard's head,[29] were carried in processions to offer magical support to the king during rituals of renewal.

Orders from harassed European trading agents along the West African coast suggest that brass demand fluctuated wildly. In 1705, for example, the British agent at Whydah implored London to send no more brass neptunes (broad flattish salt-pans with wide rims) till further notice. However, later in 1706 he ordered 'about 150 kilos of Brass Neptunes for each 100 Negroes – to be sizeable' but weighing no more than 2-3 kilos each. That year, the Cape Corso agents had to stockpile their 'Brass Diglins' (cooking-pots, often with two lugs or handles), 'Brass Pannes', and 'Brass Kettles' but up the River Gambia brass basins were still acceptable. In November 1708, the agent at Cape Corso (in today's Ghana) sent home a sample style of basin required by the Ashanti, who also wanted brass or iron cauldrons of standard weight but as light as possible, and with lugs or hinged handles.[30]

Dutch merchants could trade Stolberg brass (converted into artefacts in Flanders or the Netherlands) direct to the Guinea Coast and to the Royal African Company in London for their West African trade. In 1712, London merchants representing Dutch and German interests refused to export British-made brass to the Guinea coast, arguing that their Dutch producers provided 'all such Brass Wares as are vendible there'. They also reported supplying 'all the American Islands, and continent' with 'Copper and Brass Wares, as Kettles, Pots, Pans and all other Conveniences, and Utensils such as Coppers [large water boilers], Furnaces, and Stills for the Sugar Plantations &c'.[31] Dutch ships naturally had the first choice of these home-produced brass goods.

In March, 1714, the English agent at Sherbro (an island off Sierra Leone), ordered a large, continuous supply of even lighter-weight brass basins and cauldrons than before. Weight was clearly important, and it was probably the repeated requests by Guinea-coast agents for very light-weight cauldrons and pans that motivated Abraham Darby of Bristol, and later of Coalbrookdale, to invent technology for making them lighter (the Bristol brass industry is discussed later in this chapter). When, in 1720, the Dutch in West Africa sold off cheap surplus brass goods, undercutting their competitors, the English agents at Cape Corso begged the Royal African Company to stop sending them brass cauldrons. The agents' orders hint at the vast bulk of brass being traded. In 1721, they ordered almost half a ton of brass 'Diglins', specifying that they must weigh from

---

[27] British Museum website, 2007: www.thebritishmuseum.ac.uk/explore/highlights/highlight_objects
[28] Museum für Volkerkunde: Vienna, 98.162
[29] Museum für Volkerkunde: Dresden, 13627
[30] The National Archive: MS, T 70/22
[31] British Library: pamphlet, 8245 a 18, (1712), folio 4

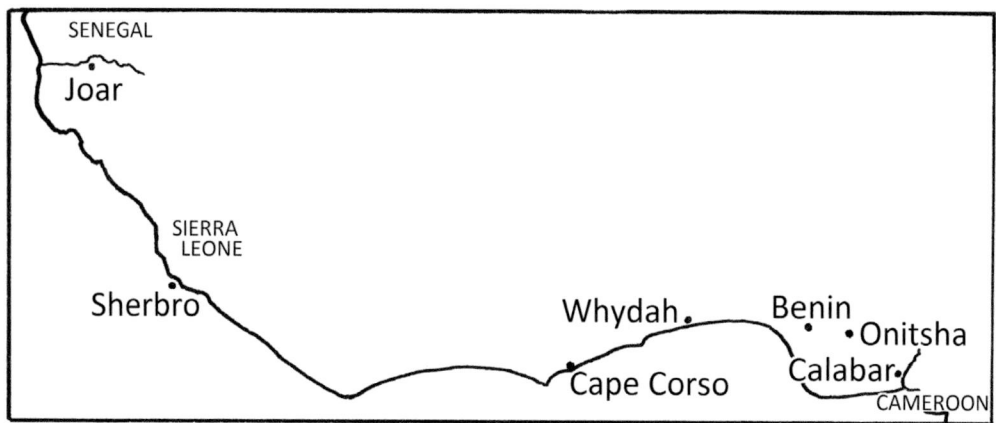

Figure 69. West Africa, map showing some sites mentioned in the text

0.7 to 1.4 kilos each; 10 kilos of brass pans (0.7 kilos each); 20 kilos of brass 'kettles' (2.7 kilos maximum); 3,500 plain stock English guns with round locks and square barrels, and 100 buccaneer guns.[32] European-made guns in the early 1720s normally had brass moving parts and decoration on the barrel or stock.[33]

In 1721, ship's doctor John Atkins described a 'Guinea cargo' that included 73 brass 'kettles' (cauldrons), 251 Guinea pans each weighing about a pound (half a kilogram) and 216 Guinea basins weighing two, three and four pounds (one to two kilograms).[34] In October 1723, an English Royal African Company agent bought 3,000 brass cauldrons in Amsterdam, plus 9,000 'kettles' and 500 brass neptunes (wide salt pans). The French made three sizes of straight-sided 'Guinea pan' weighing half, three-quarters and one kilogram respectively, ranging from 5.5 cm to 8 cm deep and from 22 cm up to 31 cm in diameter. In Senegal, sales of these were very slow in the late 1740s but rose to almost 20,000 pans in 1754-5.[35]

The months that elapsed between order and delivery rendered goods from Europe out of fashion before they arrived, so agents tried to stock popular types.[36] Between 1723 and 1732, for example, goods for the African western (windward) coast included brass cauldrons with handles, large, wide, brass neptunes, guns, pistols, manillas with bells, and horse and hawk bells. In 1724 the exasperated Cape Corso agent complained that an item, while 'a mere Drug' one month, might be the 'only commanding Commodity' the next. That July, brass pans sold, but not brass neptunes, and by November even the Dutch were selling off brass cheap.[37] In the 1730s, Benin imported bright brass rings

---

[32] The National Archive, MS, T 70/23
[33] Schloss Gottorf Museum, Schleswig: 1937/341, flintlock pistol, c.1720
[34] Atkins 1737: 159-160
[35] Curtin 1975: 315
[36] Daaku 1970: 38
[37] The National Archive: MS, T 70/23 and T 70/25

(made from thick-drawn wire), whereas people up the Cameroon River accepted only hammered brass pots and kettles. At Calabar (Cameroon delta), people bought brass arm or leg rings, but the agent complained that they 'often turn over a whole cask before they find two to please their fancy'.[38]

Through such fastidious choice of goods, local buyers reduced the power of their European suppliers, thus cannily controlling the brass trade. The main brass buyers were local inland leaders who captured and sold slaves. From 1730-34, Francis Moore, factor at Fort James, Joar (now Jawahar), on the north bank of the River Gambia, described them as 'much like the Arabs; which language most of them speak, being to them as the Latin is in Europe; for it is taught in Schools'. They followed Koranic law, spoke both Arabic and Fulani and accepted brass pans in exchange for slaves[39] destined for the Caribbean, Brazilian and northern American plantations.

In the late 17th and early 18th centuries, sheet or ingot European brass and quantities of brass objects were exported across the Atlantic, including wool cards, horse harness and haberdashery.[40] Most American men or women without land just subsisted, but earning, even at this level, strengthened their consumer voice, which in turn boosted sales until, by 1700, many could own a brass cooking pot or skillet.[41] The British forced their American colonists to buy brass from Britain, by prohibiting them from buying the raw materials to alloy brass themselves. The ban on production, however, did not prevent casting recycled brass. By 1701, a Bristol brass founder, William Smith, was working in Philadelphia.[42]

A German immigrant to America, Caspar Wistar (originally Wüster), was born in 1696, the son of a royal forester-huntsman from Hilsbach in the deeply forested hills inside a bend in the river Neckar just west of Heidelberg. The family were loyal to the Protestant Elector, but in 1688 a Catholic succession resulted in the area being overrun and laid waste by French troops. Life grew hard, so young Caspar Wilstar left for America in 1717, arriving in Philadelphia in 1720. By 1725 he had opened a brass button factory in the High Street and joined the elite Quaker community. This combination soon brought him great wealth, and was still very profitable when he died in 1752.[43] American settlers had a growing home market, so they began to produce better-quality goods and diversify. By 1738, for example, Wistar also made glass, though he always described himself as a button-maker and brass founder (*messing gieser*).[44] Francis Richardson, brass founder in Philadelphia from 1716-37, produced

---

[38] Barbot 1732, 1746 edition, volume 5: 361, 382, 383, 386
[39] Moore 1739: 21, 30-31
[40] The National Archive: MS, E 190/1160/5; E 190 1163/4, E 190 1165/4
[41] de Vries 1994: 112-113
[42] Whisker 1924: 176
[43] Pierce 1992: 92-108
[44] History Society, Philadelphia: MS Wistar family papers, 10 November 1733,

Figure 70. Kitchen wares: *top left* three-footed English cauldron, *top right* skimmer; *centre* domestic pan; *lower left* 'frying pan' or skillet; *lower right* pestle and mortar

'very neat clocks and brass jacks, handles and escutcheons, tea tables, bolts and desk hinges'[45]. Numerous mid-eighteenth-century Philadelphia brass-founders are mentioned in tax documents[46] – one brazier offered ready money for old copper and brass to re-use,[47] several paid their apprentices' passages from the Netherlands[48] and others made firearms or were recruited to the militia to cast and repair brass gun and harness parts.[49] From 1725-50, clock parts were also cast in Philadelphia, and from the 1740s both American and European brass clock-spandrels featured lively cherubs.[50]

Around 1738, in the colonial town of Williamsburg, Virginia, the Geddy family established a foundry, from which archaeologists have excavated the rough casting of a brass clock-face. After 1740, consumption of European goods in America grew faster than the population, and clothing formed a far larger part of the eighteenth-

[45] *Pennsylvania Gazette*, 16 September 1736
[46] Beiler 2011: 101
[47] *Pennsylvania Gazette*, 3 August 1738
[48] Whisker 1924: 81, 126
[49] Whisker 1924: 48, 51, 61, 79, 111, 150, 197
[50] Museum of the History of Science, Oxford: 39438, *c.*1775

century economy than ceramics or cutlery. Delaware farmers, for example, wore elaborate brass cuff-links and buttons and by 1740 their decorative cast brass buckles sometimes cost more than their shoes.[51] Ordinary American people found that they could create a more exciting new identity for themselves through fashion and its accessories. Moralising rebukes were voiced in the pulpit and broadcast by the Boston and Philadelphia news-sheets, highlighting the pitfalls of succumbing to luxury and to thoughts above one's rank. Women – often blamed for the excesses – eagerly acquired the new-style brass items, to show that an individual of any sex or class could make an informed personal choice.[52]

The main early eighteenth-century exporters to America, the Caribbean and West Africa were still the Dutch, who could rely on Stolberg brass. Brass is occasionally recovered from ships like the *Stirling Castle*, wrecked off the English Kent coast in 1703, which yielded brass finds, possibly Dutch, including an ornate sword-hilt, parts of a brass-barrelled musketoon, a candlestick (with a finger bone still attached to it), candle-snuffers with accompanying table-holder, dividers and a set of four small book-clasps.[53] England, which came into brass-making late compared to the rest of Europe, enjoyed a period of relative peace in the first half of the 18th century. This allowed the English brass industry to grow particularly strong. British society had also changed during the late 17th century, giving its industry an added freedom. Its brass-works now belonged outright to merchants, tradesmen and their shareholders (as opposed to the mainland European tradition, where overlords, mainly aristocrats or monarchs, could still control brass works).

At the beginning of the century, several British brass-works still traded through the Port of London. From 1703, John Shorey, pewterer, owner and leading partner at Temple Mill, Bisham,[54] on the River Thames, west of London, became something of a property tycoon, with leases on houses and warehouses in London and several counties.[55] From 1708, his son, also John Shorey, was his business partner. Their 1708 inventory of brass-works equipment includes a pair of brass scales, a 'brass whitning Pan' (used in the etching process, before polishing) and two 'brass Ingot moulds'. For 1710 the list includes 'Brass Shruf' (scrap battery brass) and various battery hammers, including an 'outside hammer' and 'bourges hammers'. Also listed are burnishers, shears, furnace tongs, a brass pestle and mortar, casting pans, shears, moulds, blocks, ladles, anvil blocks and a 'fier shovel'. John Shorey and his son already manufactured brass kettles and plates at their Hackney brass-works, selling them from warehouses at Tottenham and in the City of London.[56] In 1712-13, their London warehouse inventory

---

[51] Bedell 2001: 97-99
[52] Breen 1994: 251-257
[53] Capes and Chamberlain 1998, 134-136
[54] The National Archive: MS, C 5/354/3
[55] The National Archive: MS, C 104/106/ 2
[56] The National Archive: London, MS, C 104/105/ 1

included barbers' basins, Guinea jugs and basins,[57] showing that they made brass for West Africa. The Shoreys employed men to cast and finish metal objects in the cellars of the family home, the former Golden Cock Inn in Basinghall Street.[58]

In his 1708-20 cash ledger, Shorey's products included military items for the Artillery Company and the East India Company.[59] Small weapons required brass parts,[60] and ships' compasses needed non-magnetic brass casings. After a disastrous shipwreck off the Scilly Isles (west Cornwall) in 1707, the wooden compass-bowl was blamed, and from 1708 Royal Navy ships were commanded to carry compasses with brass bowls,[61] leading to many Admiralty orders.[62] Brass gimbles – pairs of supporting rings, each pivoting at right angles to the other – maintained the compass bowl and dial at a steady horizontal level while the ship dipped and rolled.[63] Military engineers and navigators used brass instruments for measuring and charting; quadrants became essential for telling the time and making calculations for surveying and astronomy. Brass sextants were available from around 1730, and octants of that period had brass index arms.[64] They were cast and engraved in Nuremberg, London and other centres. Brass made using metallic zinc was hugely advantageous for maritime instruments like compasses, being fully non-magnetic, and its use in this context was a major innovation.

In 1713, the Shoreys imported 51 tons of 'Spellter' – a huge amount of distilled zinc – indicating that at Temple Mills they were mixing distilled metallic zinc directly into brass (for a map of British sites, see Chapter 8). In 1716-18, the Shoreys' new Temple Mills Brass Company at Bisham had a copper store house, a furnace house containing four furnaces; twenty battery hammers driven by five water-wheels; a rolling mill with three pairs of rollers; a calamine crushing mill, and fifteen workers' houses.[65] A works for sand-casting brass thimbles was run next door by John Loftus from Islington, who employed a Netherlands expert.[66] In 1719, the Shoreys' success brought them an offer of 35 shares in the now prosperous Bristol Brass and Copper Company – an attempt by Bristol to bring them into a consortium – but things were about to go badly wrong.

The 'South Sea Bubble' was about to burst. The French had inherited a contract to supply slaves to the Spanish West Indies but in 1714 a Utrecht Treaty transferred the

---

[57] The National Archive: MS, C 104/105/1
[58] Clifford 1990: 130
[59] The National Archive: MS, C 104/105/1
[60] The National Archive: MSS, ADM 106/490/57 (1696); ADM 106/1087/245 (1750)
[61] May 1973: 75-76
[62] The National Archive: MSS, ADM 106/483/66; ADM 106/498/5 (1696); ADM 106/944/151 (1741); ADM 106/1043/10 (1747); ADM 106/1090/228; ADM 106/235 (1751); and ADM 106 /1160/131 (1767)
[63] National Maritime Museum, Greenwich: MSS, NAV 0289 and 0409 (c.1780)
[64] National Maritime Museum, Greenwich: numerous artefacts on public display
[65] Defoe 1725, 2006 edition: 158
[66] Houghton 1693, 1728 edition, volume 2: 195-196

contract to Britain, decreeing that a London company must supply an annual 4,800 slaves, via Jamaica, to the Spanish colonies. London's South Sea Company therefore needed to find brass and other goods to exchange in Africa for this slave quota for the Spanish plantations. Spanish demand for slaves, however, far exceeded Spanish capacity to pay for them, so the scheme was already doomed. Most London merchants, including John Shorey, had invested heavily in the South Sea Company, so the crash of 1719 left them reeling, just when insurance charges for London shipping soared due to the current French war. The Bristol and Liverpool merchants seized the occasion and stole the trade advantage from London.[67]

John Shorey, senior, had invested heavily in stock against the value of his non-audited company. In March 1720, his son protested at this inefficiency, but too late, because shares in the hitherto well-funded Temple Mills were decimated by the 'South Sea Bubble' which burst over the London stock market. The father-son partnership dissolved[68] and the stock was divided.[69] In 1720 the works were described as largely fallen into decay. By January 1721, however, John Shorey, cushioned by his property investments, had repaid fifteen per cent of the £3,000 he had borrowed against the value of shares in Temple Mills[70] at Bisham, and the mill continued to produce brass kettles and pans of all sorts.[71]

Immigrant workers were the essential life-blood of emergent brass industries. Wherever brass-making started, the skilled, experienced key workers were immigrants, often refugees. It was they who could transmit technology in the immediate and competent way that made brass-making viable enough to justify investment in new processes. From 1725-40, for example, brass tableware, including spoons, forks, cream jugs, tankards and beer-jugs, was often silvered over. In this painstaking process, the immigrant silversmith heated the brass item to 'blue heat' before applying and burnishing silver leaves, layering them up to sixty leaves thick for a hard-wearing surface. The new product, known as French Plate, was widely sold.[72] Temple Mills, Bisham, on the Thames, employed about forty men, including the immigrant manager Gerlach Bechs and workers 'Mr van Oyle' and Matthius Thiller whose uncle lived in Aix-la-Chapelle (Aachen).[73] Matthius, who had earlier worked at Esher brass mills, had brothers working at Bristol brass works.[74]

---

[67] Parkinson 1952: 91
[68] Clifford 1990: 131
[69] The National Archive: MS, C 11/51/46
[70] The National Archive: MS, C 104/106/2
[71] Liverpool University Library: MS, typescript 7.1 (23) folio 44, Kahlmeter
[72] Cameron 2013: 3-4
[73] *Journal of the House of Commons* 1711-1714: 163
[74] Kungliga Biblioteket, Stockholm: MS, BS, M 2491, folio 410

Another London-centred brass mill operated from 1717 at Ember, on a Thames tributary west of the capital. John Hitchcock and partners re-introduced brass-making at Ember, after a short spell making iron barrel-hoops.[75] Analysis of solidified droplets from an Ember crucible has revealed a brass containing 29 per cent zinc,[76] good for forging and wire-drawing. The partners added a small copper refinery at nearby West Ditton, processing annually up to forty tons of copper, mined near Tavistock, Devon and smelted at Cuckolds' Point, Rotherhithe.[77] Refining was transferred from West Ditton to Rotherhithe in October 1725[78]. In 1728, Ember mill closed and reverted to an iron mill.

Also trading through London, Edmund Brydges,[79] in 1694, converted Byfleet Mill, Surrey,[80] to brass production. Taken over by John Hitchcock, Byfleet Mill produced brass wire for sieves, used in bakeries, paper works[81] and gunpowder mills – because brass was, of course, an invaluable spark-free metal, which prevented explosions. In 1703, Thomas Wethered, with a three-quarter share in Ember Mill, leased Byfleet brass mill and spent about £850 on rebuilding it from planks of elm, oak and deal, laths, tiles, lime, clay, cast iron and nails. He bought two great water-wheel shafts and stocked up with copper, brass, tools and millstones, paying wages to a carpenter, bricklayer, millwright and smith. The new mill was equipped to produce copper and brass plate, brass wire for sieves, and iron barrel-hoops.[82]

Having overspent on rebuilding Byfleet Mill, Thomas Wethered decided to extort 'great summes of money' from a wealthy local man named Walter Kent but, by 1706, Kent realised that Wethered was not repaying the promised dividend. So, in 1710, Kent dissolved the partnership, ousted Wethered and sold the mill for four thousand pounds to John Hitchcock, who remained a partner but immediately sold it back to Kent.[83] This cunning strategy gave sole control to Kent. In 1717, Hitchcock also became a partner of the Bristol Company, which purchased Byfleet Mill, owning it wholly by 1723. After this, Byfleet brass was delivered to Bristol rather than London,[84] reflecting a trade shift away from London and towards the British west coastal ports. A Surrey map dated 1729 shows Byfleet Brass Mills with three water-wheels,[85] but by 1762 it had become an iron mill.[86]

---

[75] British Library: typescript YD 2005 b.234, Greenwood, G.B., 1980, The Elmbridge Water Mills, 12
[76] Morton, 2009, 274, Chris Salter analysis
[77] Liverpool University Library: MS, 7.1 (23) 7, folios 44, 56-57
[78] Kungliga Biblioteket, Stockholm: MS, BS, M.2491, folio 408
[79] Surrey History Centre: MS, G 1/47/2
[80] National Grid reference TQ 0727 6068
[81] Berg and Berg 2001: 323
[82] The National Archive: MS, C 5/125/14
[83] The National Archive: MS, C 5/125/14
[84] Morton 1985, 254
[85] Senex map, 1729, Surrey
[86] Roque map, 1731, 1762 edition, Surrey

The main early eighteenth-century English brass works were close to Bristol. In 1700, five Quaker merchants had tried to start brass-making in central Bristol but were thwarted by guild restrictions, so they chose Baptist Mills two kilometres north of Bristol, on the River Frome.[87] Like Hammer brass mill, just east of Nuremberg, Baptist Mills lay outside the city walls, and therefore outside guild restrictions, but just inside a city parish. The brass works, set up in 1702,[88] was managed by Abraham Darby, described as 'of good favour and orderly behaviour' but an 'ingenious scheming man', 'small of stature, very strong and active ... his eyes were black and very bright, his complexion dark'.[89] Abraham Darby used reverberatory furnaces fuelled with coke derived from local 'caking coal', because Bristol was short of wood fuel, which was heavily used for shipbuilding. Caking coal is the only type suitable for turning into coke, because it is reactive and high in bitumen, and when heated it becomes plastic, softens, swells and then sets as hard, strong, cellular coke.[90] The advantage of coking is that it removes harmful sulphur from coal. In 1698, diarist Celia Fiennes, riding to Bristol through Kingswood Forest, described many horses 'returning loaden with coals dug just thereabouts ... this is the cakeing coal'.[91] The Kingswood Great Seam of good caking coal was worked from about 1670-1740 at Clay Hill, two kilometres east of Baptist Mills,[92] where drainage allowed coal mining to deeper levels.[93]

By 1703 Abraham Darby had recruited skilled Stolberg brass makers, battery-workers and wire-drawers,[94] including the Thiller brothers,[95] and others who had Huguenot names. Bishop Secker of Gloucester (1693-1708) noted that local parishioners included 'a number of foreigners and their families who were sent for by the Quakers to their brass works at Baptist Mills ... there are twenty or thirty men of these foreigners'.[96] In 1754 they were described as originating from the Duchy of Julich,[97] which included Stolberg. The immigrant battery workers made brass vats for soap or tallow, sugar-stills, brewery or dye vats, and many items for the Guinea Coast of Africa, including an annual 80 to 90 tons of large pans – 30-160 cm in diameter.

Baptist Mills brass works exploited calamine from the nearby Mendip Hills and copper shipped in from Cornwall and South Wales. Abraham Darby managed to obtain suitable granite for casting-plates from Cornwall instead of Brittany,[98] a huge

---

[87] National Grid Reference 6029 7427
[88] British Library: MS, Add 22675, folio 36
[89] Ironbridge Gorge Museum Trust: MS, Lab/ADBI/2/4; and undated notice, text by Lady Labouchère
[90] *British Standard* 1016, 1980: 5
[91] Fiennes 1698, 1982 edition: 192
[92] Bristol Coalmining Archive, Kingswood Great Vein coal analyses, 1884, numbers 7225 and 1904
[93] Ellacombe 1881: 183
[94] Labouchère 1988: 21
[95] Kungliga Biblioteket, Stockholm: MS, M 249: 1, folio 410
[96] Ralph 1985: 45
[97] Berg and Berg 2001: 145
[98] Cornish granite source of a casting-stone fragment verified by University of Oxford Museum of Natural

advantage during current hostilities with France. Two large Cornish granite casting-slabs, ground flat, measured roughly 100 cm by 40 cm, with a surface coated with clay or loam and organic material to facilitate the release of the cast brass plate from the mould. Three iron bars were inserted between lid and base, to adjust the thickness of the brass plate, then the two granite slabs were 'bound together very firm by a stout leaver [sic] across and screwed at one end'. The massive slabs were tilted by means of a pulley and tackle to allow the hot metal to flow down evenly, or to open the hinged upper stone 'lid'.[99]

By 1705, Abraham Darby was experimenting with casting the extremely lightweight brass hollow-wares demanded at the time by local slave-trading agents on the Guinea Coast.[100] From 1698 free traders from Bristol had won 13 years of trade to West Africa, breaking the London monopoly of the 'Royal Africa Company'. Bristol ships appear to have copied the 386 private Liverpool ships that traded openly with the Guinea Coast by 1701-7, often evading the ten per cent duties on cargoes for Africa by falsely declaring to the Customs House that they were bound for intermediate or ultimate ports, such as Ireland, Madeira or America. From 1714-17 both Liverpool and Bristol free traders overtly declared large brass exports to Ireland and America, but less to their lucrative West African markets (where brass was in fact paying for 25,000 slaves annually).[101] The customs declarations only reversed in the 1720s.

Sea transport was both highly competitive and hazardous (due to the French harrying English and Dutch merchant vessels), so British free traders to Africa lost many sailing ships. They also protested that the more experienced Dutch acquired 'Gold and other Affrican commoditys 50 per cent cheaper', and paid no taxes on cargoes or fitting out ships, whilst carrying cheaper, more marketable, cargoes and paying their seamen less.[102] When Bristol's thirteen years of free trade lapsed in 1713, Bristol City Council petitioned Parliament against restoring London's trade monopoly, arguing that Bristol depended on manufacturing brass for export 'to the coast of Africa for the buying of negroes'.[103] In 1725, 63 Bristol vessels traded to the Guinea Coast,[104] even though the London monopoly continued until 1730.

From 1711-12 foreign (German or Dutch) traders in London hurled parliamentary petitions and pamphlets at the Bristol brass manufacturers,[105] who responded with equal vigour.[106] However, many English brass craftsmen making finished items like

History
[99] National Library of Wales: MS, Ynysfor 7/5/3 (1748)
[100] The National Archive: MS, T 70/22, 17 November 1708 and 10 March, 1714
[101] Parkinson 1952: 86-89
[102] The National Archive: MS, CO 388/12/197v
[103] Bristol Archive: MS, M/BCC/CCP/1/9, folio 306
[104] Richardson 1986, volume 1: multiple pages
[105] British Library: MSS, 8245 a 18 1-6
[106] *Journal of the House of Commons* 1711-14: 160-164

cauldrons, pots and fish kettles, surprisingly supported foreign London merchants by arguing that English brass quality was inadequate. In 1712, Baptist Mills brass works produced 259 tons of satisfactory brass, which local braziers pronounced 'equal in all respects to Dutch battery',[107] implying that levels of impurities like sulphur were now low enough to produce malleable brass.

British brass exports increased markedly between 1711 and 1714,[108] and the domestic market rose sharply. Excavation at Oyster Street, Portsmouth, for example, revealed copper-alloy buttons, and cuff-links decorated with a raised sailing ship motif (1710-1720).[109] Zinc leaches out over time from the surfaces of small buried brass objects like buttons. George Woodruff, a Birmingham button-maker, wrote that Birmingham used fifty to sixty tons of English brass annually for button-making, though many still thought Stolberg brass 'softer' and better.[110]

This was a time of mergers, takeovers and cartels. As early as 1709, Baptist Mills took over the brass mill at Esher in Surrey, which by 1707, already produced about 142 tons of brass wire a year, double its output of ten years earlier. In 1728, Kalmeter noted that, at the first heating, a ratio of 40 lbs of copper to 60 lbs of zinc ore produced 56 lbs of brass. Esher crucibles contained 28 *pund* or pounds (12.7 kg) of refined and unrefined copper, 14 pounds (6.4 kg) scrap brass and 56 pounds (22.7 kg) of Derbyshire calamine.[111] Esher joined Bristol to gain access to its trans-Atlantic markets, after being weakened by the personal financial troubles of its banker partner, Coggs.[112] Brass wire production at Baptist Mills and Esher reflected a strong American export market for wool cards (which later reduced, as calico grew more popular than wool). The combined Bristol and Esher factories produced about £80,000-worth of brass annually,[113] but the English brass makers' monopolistic business mergers enraged the German and Dutch importers of brass goods in London. They saw their trade threatened by the '[few [Bristol] mercenary Men, who have acquired prodigious Stocks of Money by their Oppression' ... 'engrossing Trade and Business into their own hands'.[114]

Mendip calamine was now mined more for the domestic market than for export. In 1720, after John Hitchcock of Esher Mill became a Baptist Mills partner, the South Sea Bubble brought financial instability. For security, shares in Baptist Mills and Esher brass works were split into 80 parts, held by 14 shareholders.[115] In 1720 the

[107] British Library: MS, 8245 a 28, folio 7
[108] The National Archive, MSS, E/190/1160/3; E/190/1160/5; E/190/1164/2; E/190/1168/2; E/190/1175/2; E/190/1180/1
[109] Fox and Barton 1986: 110, 150, 236; excavated finds 16L41 and 161.49
[110] *Journal of the House of Commons* 1711-1714: 61
[111] Kungliga Biblioteket, Stockholm M.2691, folio 409 (1766) Kalmeter Dagbok
[112] *Acts of the Privy Council* 1710: 49
[113] British Library: MS Add 22675, f.35v (1720)
[114] British Library: MS 8245 a 18 (1712), f. 2
[115] British Library: MS, Add 22675, folios 36-37

brass works consortium amalgamated with a Bristol copper works. Bristol Quaker merchant, Nehemiah Champion, was the power behind the Bristol brass works, with its 36 furnaces in six furnace-houses.[116] In 1722 he further merged Baptist Mills with the Upper Redbrook Copper Works (south-east Wales) and with Byfleet brass mill in Surrey.[117] A vignette illustration dated 1734 shows Baptist Mills with several substantial buildings and tall, smoking, conical outer furnace chimneys.[118]

In 1723 Nehemiah Champion patented a reverberatory annealing oven,[119] to separate fuel-generated sulphur fumes from finished brass items. Between each shaping under the battery hammer, part-finished brass cauldrons were stacked tightly inside one another by size, within two protective containers on a trolley that was wheeled into the annealing oven on rails. The vaulted, metre-long annealing oven had outer side-walls which the Swedish observer, Kalmeter, estimated at roughly one and a half feet (45 cm) thick. Coal fuel was shovelled into spaces running inside the length of each. Two adjustable draught holes above the hearth drew the flames up towards the chimney. Once inside, all apertures were closed, the joints sealed with clay, and the annealing heat held constant for two to three hours.[120]

By 'granulating' copper before alloying cementation brass, Nehemiah Champion helped copper to absorb zinc vapour more efficiently – small globular or granular shapes offer greater surface area relative to volume, so more is absorbed and less is lost through evaporation.[121] Molten copper, first poured through holes pierced in an iron sieve, dropped into cold water to cool as solid spherical droplets (rather than as broken-up copper fragments). The sieve was lightly coated with refractory (fireproof) Stourbridge clay, reducing the size of the holes to produce copper droplets so tiny that they solidified before reaching the bottom of the water tank.[122] As the hot granules hit the water surface, boiling water sprayed up dangerously, simultaneously heating the water. To keep the tank cool, Champion pumped in cold water at one side and drained out warm on the other. In 1725, Kalmeter, the Swedish observer or industrial spy, reported that the weight of copper needed to produce 16 pounds of brass before granulation, now produced at least 20 pounds. Baptist Mills now made more brass (300 tons a year) containing a rather higher percentage of zinc (not accurately calculable solely from the claims Kalmeter heard), and wasted less copper.[123] *Arco, archal* or *ercol* referred to the first, unrefined, cementation brass process. The zinc content for brass

---

[116] Morton 1985: 259
[117] Kungliga Biblioteket, Stockholm: MS, M 249:1, folios 410, 529 (1725), Kalmeter, *Dagbok 1*
[118] Buck, C. and N., 1734: Print, border vignette, 'North-West Prospect of the City of Bristol, drawn from Brandon Hill'
[119] *Patent Specification* 454, 1723
[120] Riksarkivet Stockholm: MS, SAB E 111 10, Kalmeter, 1725, folios 292-294
[121] *Patent Specification* 454, 1722
[122] Swedenborg 1734: 352
[123] Liverpool University Library: MS, 7.1 (23), 1725, folio 68, Kalmeter

excavated from Baptist Mills varied from 15-28 per cent, so they were supplying brass for a range of uses.[124]

In 1745, the Bristol-Esher consortium prevented their copper suppliers from selling to their competitors by forming further cartels,[125] but by 1750 Esher had stopped making brass.[126] In 1748, Baptist Mills sourced calamine from the Mendip Hills as well as from Flintshire in North Wales (then considered better). Each furnace house contained eight circular 30 cm-diameter furnace-holes, their top edges level with the stone-paved floor, and with air ducts to introduce a draught 'from a back yard'.[127] A 1750 plan of Baptist Mills brass works shows a large area of stables, yards and calamine storage bays along the river.[128] Some calamine buddling (washing and sieving) or calcining (slow roasting) occurred at the mine-face, but in 1754 the brass works itself had calcining kilns, two calamine-grinding mills, a new brick windmill under construction and a horse gin for calamine-crushing. There were workshops for scraping and repairing cauldrons and other wares, and large formal riverside gardens. A row of small workers' dwellings had individual gardens for growing vegetables.

The narrow River Frome lacked the power to meet Baptist Mills' high demand for brass processing, so five satellite mills were established along the larger River Avon and its tributaries. Abraham Darby, manager at Baptist Mills from 1702-9, started the first three. The first, Avon Mill, Keynsham, at the mouth of the River Chew, begun in 1705, had 12 battery hammers, rolling, slitting and wire-drawing equipment, to meet the heavy demand for brass Guinea pans, kettles (cauldrons) and manillas (arm or leg bands) to provide part payment for slaves involved in the triangular slave trade from Bristol to the Guinea Coast and on to the Caribbean or America. Chew Mill, the second, built just upstream in 1706, had battery hammers and an annealing oven.[129] By 1736, the third, Woodborough Mill,[130] at Pensford, Somerset, further up the River Chew, had three brass battery mills for cauldrons.[131]

Erected in 1711, Weston battery mill just west of Bath, was a Baptist Mills satellite on the River Avon, made fully navigable with lock-gates from 1727.[132] Bristol imported rum and sugar from the Caribbean, so Weston Mill produced brass parts for the sugar stills, pans and utensils need for rum-making. By 1727 barges could transport

---

[124] Morton, 2009: 274, Chris Salter analysis
[125] British Library: MS, Add 22675, folios 35v-37
[126] Surrey History Centre: MS 546/2
[127] National Library of Wales: MS, Ynysfor 7/5/3, Joseph Harris journal
[128] Bristol Archive: SMV/6/3/4/7, de Wilstar, plan of Baptist Mills
[129] Berg and Berg 2001: 139
[130] National Grid Reference ST 6360 6412
[131] Somersetshire Archive: MS POP 69 (1736)
[132] National Grid Reference ST 7245 6486

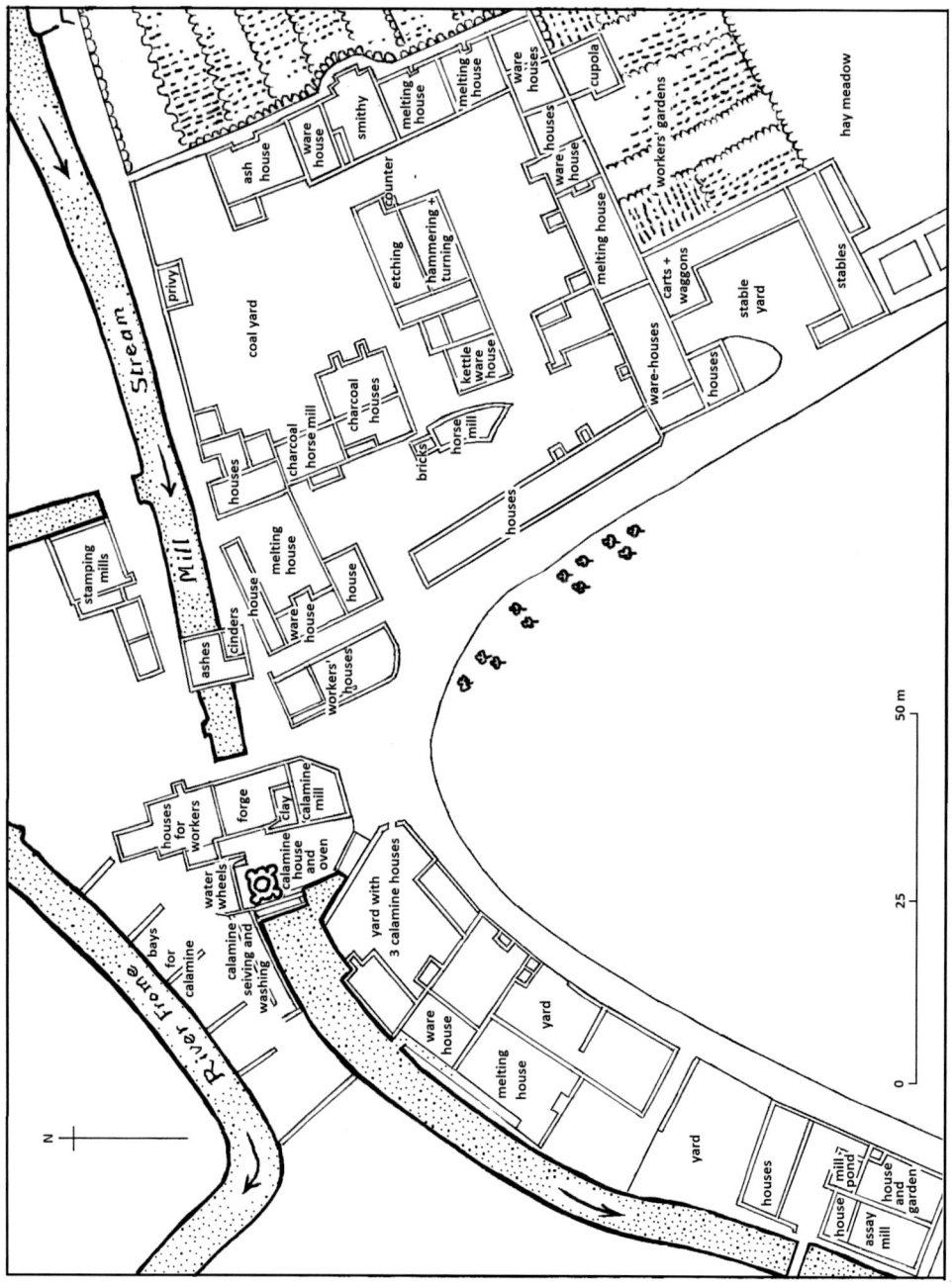

Figure 71. Plan of Baptist Mills, based on a plan by de Wilstar, 1750

brass and other materials between Bath and Bristol.[133] Saltford Mill,[134] a Baptist Mills satellite built in 1721 on the River Avon, was a stone-built, pan-tiled brass mill with three battery hammers, four annealing ovens and a rolling mill.[135]

In 1721, after the South Sea Bubble, the English trade to America, the Caribbean and the Guinea Coast largely transferred to Bristol and Liverpool, away from the Port of London, so the Surrey brass works declined. Liverpool's and Bristol's brass exports to North-west Europe and West Africa were now almost equal. Because the main Dutch coastal trading-base was at Elmina (coastal south-west Ghana), Bristol merchants traded brass 200 km further east, beyond the River Volta, calling in at Sierra Leone and at ports between the River Gambia and the Ivory Coast.[136] The markets of both Bristol and Liverpool grew from 1731 onwards, their exports to Europe expanding rather faster than to Africa. Even before the 1743 Anglo-French war (which had started after the French invasion of the Spanish Netherlands in 1672), Bristol shipping was harried by French privateers. Bristol ships faced higher insurance charges because the Bristol Channel entrance was vulnerable to French attack, causing the brass trade to swing from Bristol to Liverpool.[137] Around 1730, London, Bristol and Liverpool each sent about seventy ships to Africa,[138] but in 1744 Bristol ships dropped to eight, and just sixteen sailed in 1745.[139] The number of vessels sent to Africa by London merchants decreased to less than ten a year, so the French successfully seized their trade to Whydah (Ouida, southern Benin) and much of the Gold Coast.[140] It was time for the European domestic market in brass to take over.

Brass sheet, wire and semi-prepared hollow wares were finished by founders, braziers and smiths up and down Europe, to produce items like tea kettles, curtain rings, handles and candlesticks.[141] The design of domestic items like candlesticks varied with the demands of fashion. By 1680, in the Netherlands, candlestick stems had tulip-like balusters, and around 1700, Swedish candlesticks often had a distinctive spiral column design.[142] After 1700, English plain octagonal candlestick bases slowly grew more decorative and curvilinear, and petal-based candlesticks appeared in England around 1730. Most brass door knobs were spherical from about 1700-1750,[143] often accompanied by a plain brass door-lock case, sometimes with two knobs, one to close the door and one to lock it. Lock escutcheons or plates had been diamond-shaped and

---

[133] Warner 1801: 215; British Library, map, K.Top.38.22a, undated, late 18th century
[134] National Grid Reference ST 6871 6702
[135] Gloucestershire Archive: MSS, D 1628, Sturges reports (1830), sale note (1852), auction notice (1862)
[136] Richardson 1986, volume 1: xxiv-xxv, xviii
[137] Bristol Reference Library: MS, Southwell Papers, volume 7, February 1741
[138] Postlethwayte 1746: 98
[139] Richardson 1986, volume 1: xiv
[140] Postlethwayte 1746: 98
[141] Eveleigh 1995: 11-12
[142] Gadd 2007: 2
[143] Colonial Williamsburg Foundation: CWF 12-24-DC, spherical knob,

Figure 72. Brass candlesticks, left to right, Netherlands type with tulip-shaped cup, 1680; Spanish type, c.1650; English type 1700-1720

engraved with rosette patterns in the later 17th century, but were plain brass by about 1740. Daniel Defoe, in 1728, commented that 'most of the Brass Locks of all the fine Palaces in France, if narrowly inspected, will be found to be English'.[144] For simple farmhouse doors, a brass latch-lifter could be poked between door-planks to raise the latch from outside. From about 1720-50, loop drawer-handles, outward-turned at the ends, were backed by heavier 'bat's-wing' plates.[145]

Flemish-type chandeliers were still being made until after 1735, still with large spherical components like great golden balls, but Germany and England soon became influenced by French court crystal chandeliers, which required a brass framework.[146] Equipment for places of worship included items like the 1725 silvered brass reservation dish for the family chapel at Erddig House, Flintshire.[147] In the 18th century, early weight-driven brass 'lantern' clocks gained long cases. Cast-brass cherub, filigree and foliate scrolls and chased gilt decoration appeared as spandrels inside the clock-face frame, for example on a London bracket clock by Richard Rooker (1694-1735).[148] Various complex elaborations on the sundial, dating to 1725-50, became popular with amateur gentleman scientists, but these skilfully made artefacts were produced by individual craftsmen, not in large brass factories like Baptist Mills.[149]

William (1709-89), son of Nehemiah Champion of Baptist Mills, Bristol, was Europe's first industrial-scale producer of distilled zinc metal. During the late 1720s, young William travelled 'into most parts of Europe' to observe 'Works and Manufactures'

---

[144] Defoe 1725, 1983 edition: 290-291
[145] National Trust, Chastleton: CHS F/59
[146] Haedecke 1970: 94-95, 100
[147] National Trust, Erddig
[148] National Trust, Chastleton: CHS H/5, London bracket clocks, 1720
[149] Museum of the History of Science, Oxford: 41625 and 37904

for 'Mineral Production and Metallick Operations'.[150] He was motivated by the rising price of Chinese metallic zinc in the early 18th century. In 1713, John Shorey of Temple Mills brass works paid about £381 a ton for 'spelter',[151] whereas in India, three years later, it cost an estimated £340 a ton and in London in 1723 it cost about £365 a ton.[152] By 1727, Chinese officials increased the zinc ratio in their own coinage to 50% (due to copper shortages) to meet the cash needs of their great population,[153] which reduced the export zinc available and must have boosted its price.

William Champion reportedly visited the laboratory of Jean-Frédéric Henckel (1678-1744), in Freiburg (Saxony) to watch zinc distillation experiments,[154] before returning from mainland Europe

Figure 73. Cherub spandrels, on an eighteenth-century long-case clock, Richard Rooker

in 1730. Then aged about 21, he began six years of experiments to distil Mendip calamine, initially in a laboratory at Baptist Mills.[155] In 1731, Nehemiah Champion hired Baber's Tower, a former gunpowder works off Old Market, Bristol,[156] where his son William attempted to develop a full-scale zinc distillation process.[157] It is noticeable however that in that year, Bristol merchants imported 538 tons of distilled metallic zinc. At least 24 tons arrived in the names of Nehemiah Champion and fellow Bristol merchant, William Cook (who traded with China) and 300 tons lack merchant names.[158] Nehemiah Champion imported more, off and on, until 1736, when only six

---

[150] Journal of the House of Commons 1750: 54
[151] The National Archive: MS, C 104/105/1, 1713
[152] British Library: MS, IOR G/12/8, folios 1094 and 1431
[153] Souza 1991: 305
[154] Henckel 1737: 309
[155] *Journal of the House of Commons* 1750: 54
[156] National Grid Reference ST 598 732
[157] Buchanan 2000: 140
[158] Bristol Archive MS: SMV/7/1/1/25-6 (1730-1731),

tons entered Bristol. In 1738, William Champion described zinc distillation for his British Patent in vague, even devious, terms.[159]

In 1742, a Swedish chemist, Anton von Swab, devised a way of isolating metallic zinc in his laboratory, but by that year noxious fumes from William Champion's Baber's Tower zinc distillation plant were provoking complaints to Bristol City Council, so he stopped to restore the damage. In June 1743, he informed the council that he had 'deffrayed the works at Baber's Tower complain'd of as a Common Nuisance'.[160] In 1644, following renewed public complaints about pollution, Nehemiah Champion bought William a deserted nearby iron-foundry building[161] for his zinc distillation experiments. By this time Chinese imperial politics had seriously reduced zinc metal exports.[162] Nehemiah Champion bought a working coal mine site at Warmley, east of Bristol – the coal seams near Baptist Mills being almost exhausted – and in 1746, his son William Champion built Warmley brass-works and zinc distillation factory there[163]. In that year, Andreas Marggraf, a Berlin chemist, distilled metallic zinc in his laboratory, using calamine from several sources including Holywell, North Wales.[164]

Metallic zinc, distilled for centuries in India and China, had been imported to Europe from the late 16th century, but William Champion was the first European manufacturer to distil zinc industrially. He boasted that he would soon distil enough zinc metal to supply Britain's entire domestic needs.[165] He distilled local Mendip calamine ore ($ZnCO_3$) in vertical retorts. Vapour rose up the retort and molten metallic zinc passed down a central tube into 1.8-metre-long condensing pipes, before dropping a further three metres into buckets below, losing valuable but noxious gases to the atmosphere as it fell. Swedish observer, Reinhold Angerstein, in 1754 criticised the inefficiency of this part of Champion's distillation process, suggesting that it would be healthier for brass-workers and less wasteful of zinc metal (which vaporised into the air) if it fell directly into steam and water. Angerstein's drawing of William Champion's Warmley zinc-distillation furnace shows steps leading down to an underground chamber, three metres deep, with vats to receive molten zinc.[166]

Bristol and Esher were the main early eighteenth-century English brass-alloying mills, but in 1709 Abraham Darby moved to Coalbrookdale on the River Severn in Shropshire, where he produced very lightweight iron pans and cauldrons. However, analysis of residues on two crucibles from Darby's Upper Forge, Coalbrookdale,

---

[159] *Patent Specification* 564, 1738
[160] Bristol Archive: MS, M/BCC/MCC/2/1/3/52 (1736-1747)
[161] Bristol Archive: MS, 4658 (6)a (1744)
[162] Souza 1991: 305-306
[163] National Grid Reference 670 728
[164] Marggraf 1746: 50-53
[165] *Journal of the House of Commons* 1750: 54
[166] Berg and Berg 2001: 142-144

suggests that some brass was also produced there.¹⁶⁷ The inventory for the 'Upper Air Furnace' includes some brass pot and kettle patterns, and Darby's accounts show that the Upper Forge and its 'copper house' received calamine, brass, brass-wares and brass-making supplies by river from Bristol.¹⁶⁸ Copper may have come from mines north of Shrewsbury.¹⁶⁹

At Coalbrookdale and nearby Madeley, a member of the Flemish Hallen family from Wandsworth already worked in brass and iron. In 1705, the will of a Shropshire yeoman included a brass pot, skillet, 'ffrying pan' and small brass kettle,¹⁷⁰ possibly cast at Lower Forge, Coalbrookdale by William Hallen of Madeley. He and his sons (born at Madeley) cast plate metal and forged battery ware.¹⁷¹ William, the eldest son, became a 'Panmaker' at Stourbridge, the second, Cornelius (1673-1731) was a plater and pan-maker at Lower Forge, Coalbrookdale,¹⁷² and the younger, John, made pans in Birmingham. Young Cornelius Hallen learned foundry work at the Hallen's expert Wandsworth works in London.¹⁷³

By 1707, Cornelius Hallen, aged 34, had returned from Wandsworth to work his father's Lower Forge, Coalbrookdale,¹⁷⁴ and in 1709 supplied Abraham Darby with brass flat-irons,¹⁷⁵ prestige possessions in Europe at the time. In 1710, Darby's lease for a different mill prohibited the alloying of brass 'as that work is being done at Coalbrookdale'.¹⁷⁶ 'In 1711 Cornelius Hallen leased Coalbrook Farm, and lived in its surviving late sixteenth-century half-timbered house, also leasing a Pan Shop behind the house, and farmland and a wharf on the River Severn.¹⁷⁷ His adjacent 'Lower or Hallen's forge' had a battery mill beside a dam below a long mill pool.¹⁷⁸ In the 'Pan Shop' he cast and finished brass, copper and iron skillets, flat-irons and other objects. In 1717 Cornelius Hallen still rented Lower Forge,¹⁷⁹ producing brass objects, perhaps those listed in a 1726 Shropshire inventory, which included a clock, seven candlesticks, four pots, three bottles, a mortar and pestle and a warming pan.¹⁸⁰ By 1754 the Hallens rented workshops, now owned by the Coalbrookdale Company, delivered sheet metal (referred to by this period as 'latten'), semi-prepared goods, pans, skillets, cover dishes

---

[167] Dungworth *et al.* 2010: 14-15
[168] The National Archive: MS, E 190/1257/11 folio 6 r/v
[169] Dungworth *et al.* 2010: 4
[170] Trinder and Cox 1980: 428
[171] Shropshire Archive, MSS, 6320/1(1711), and microfiche P 180/A/2/1 (1673)
[172] National Grid Reference SJ 6665 0375 (1711)
[173] Hallen 1885: 34
[174] Shropshire Archive: MSS 1987/19/3 (1705), Coalbrookdale
[175] Shropshire Archive: microfiche 6001/328 (1708-9), cash book 1
[176] Shropshire Archive: MS 112/11/15/66 (October 6 1737)
[177] Shropshire Archive: MSS, 6320/1-2 conveyance (1711) and indenture (1793)
[178] Ironbridge Gorge Museum Trust library: map, Slaughter, 1753
[179] Shropshire Archive: microfiche 6001/328 folios 16, 24, 28, 38 (1709), and P.316/1/11/30 (1717)
[180] Trinder and Cox 1980: 434

Brass from the Past

and plate warmers both locally and downriver to Bristol and Wales.[181] By 1786 the pan shops were disused.[182]

Around 1710, Abraham Darby and Thomas Harvey (his wife's relative), were seeking a rolling-mill site. At that time Thomas Harwood of Tern Hall, Shropshire (now Attingham Park) invited a friend to invest in a metal-slitting mill on the River Tern, near where it joined the River Severn.[183] The friend refused, so Thomas Harwood's solicitor prepared Darby a Tern Forge lease, which forbade the alloying of brass but allowed a 'Rowling Mill, to Rowl Brass plates and to make those up into Kettles and other goods'. Darby arranged for Thomas Harvey and six Bristol partners to lease four dilapidated corn mills, with 'Shattered Walls and Fottered Thatch Roofe', for a rolling mill and large battery hammer. A raised canal appears to have supplied water direct to Tern Forge from a dammed pool a kilometre upstream (as in the upper Leen Valley, Nottinghamshire), by-passing a shallow pond called Tern Mere. In 1728, Thomas Harvey's unreliable son, Benjamin, became manager but by 1731 his father-in-law, Joshua Gee, had taken charge. An inventory of Thomas Harwood's house goods at Tern Hall, includes, in the kitchen, '1 Brass Ladle, 2 brass Skymmers' and '4 brass candlesticks'.[184] An eighteenth-century drawing of Tern House shows mill buildings and workers' cottages[185] for about fifty workers, but Tern forge was turning over to iron.[186] The brass forge came to an end by 1760, but not before it had raised the age-old pollution dilemma.

Figure 74. Brass flat-iron, cast at Skultuna, with decorative dolphin features, 1726

'Where there's muck there's brass' – and *vice versa*. Pollution had always attended brass-making. As early as 1540, Vanocchio Biringuccio had warned that the penetrating fumes from brass-making were very harmful and, over time, became a deadly poison, causing asthma, stupidity, unconsciousness, paralysis or any number of diseases.[187] Another early documented example had been the contamination, in 1565, of the

---

[181] Berg and Berg 2001: 335
[182] Shropshire Archive: MS, 1681/179/1 (1786), plan, G. Young
[183] National Grid Reference SJ 551 099
[184] Shropshire Archive: MSS, 112/11/15/66 (1737); 112/12/26/30, /33 (nd, c.1750); 112/12/26 /67 (?1713)
[185] Bodleian Library, Oxford: MS, National Trust archive, Attingham, drawing 35
[186] Shropshire Archive: MS, 112/11/15/67 (?1713)
[187] Biringuccio, 1540, 1959 edition: 75-76, translation by Gnudi and Smith

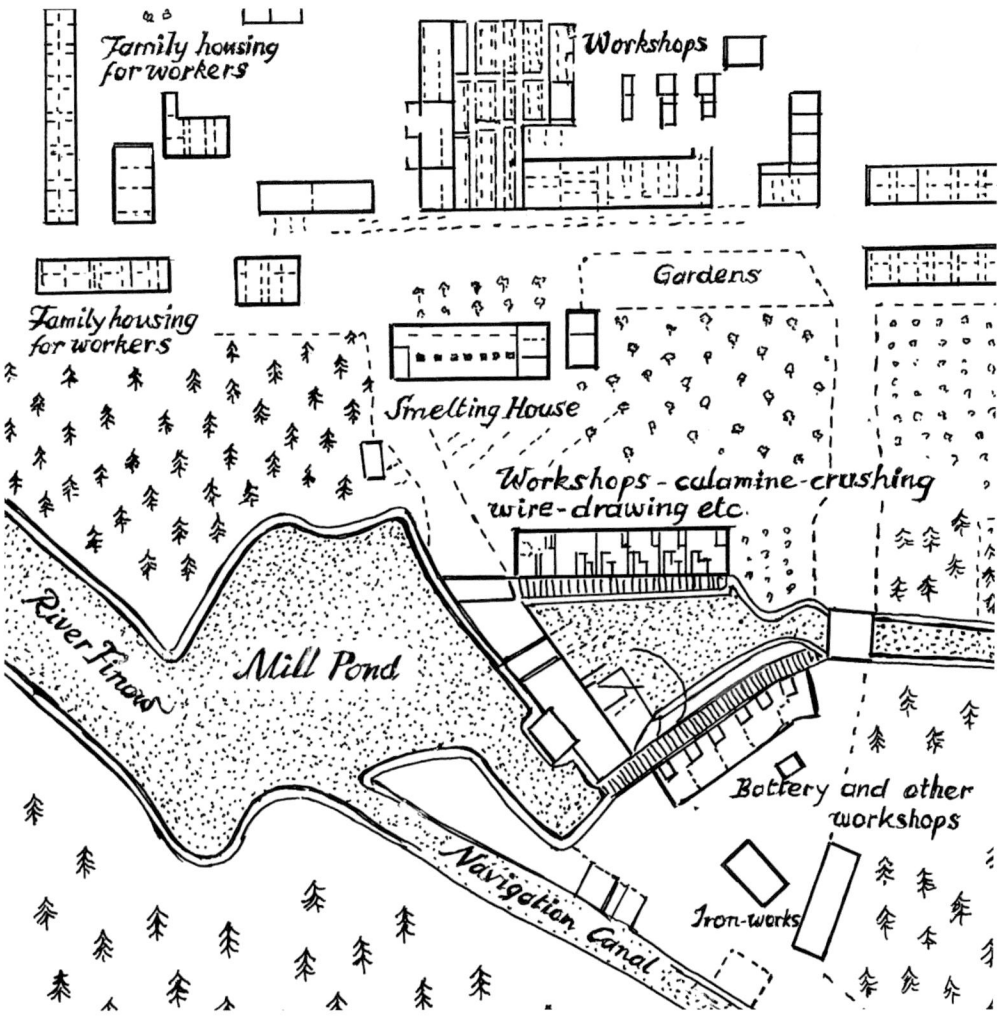

Figure 75. Plan of Hegermühle brass works, Finow-Eberswalde, Brandenburg 1784

monks' fish-pond at Ilsenburg, Harz,[188] but in the 18th century, chemical and noise pollution were reported more frequently. By 1736, Thomas Harwood of Tern House was well aware that 'a fforge Hammer going in the Night at so small a distance is a very great Newsance'. Harwood had earlier been warned of 'Great Damages that had been done in severall places by smelting works, the Smoak and Stench whereof had in many places kill'd Beasts, and destroyed Trees in the Neighbourhood'. Harwood had been asked whether noise from hammers at the proposed brass-works might disturb his family at night, and whether the smoke might set back his gardens or fruit trees. Harwood thought not. He had persuaded himself, wrongly as it later turned out, that these

---

[188] Priesner 1998: 23

BRASS FROM THE PAST

Figure 76. Surviving Hegermühle buildings: *above,* remains of the 1739 furnace-house walls *below,* 1724 workers' housing.

things would not bother him.[189] His greed for profit had blinded him to the reality of pollution.

Pollution was a problem in mainland Europe too, though less is heard about it, though the European brass industry flourished at both long-established and new sites. In 1701, the Hohenzollern cause was strengthened when Friedrich-Wilhelm of Brandenburg (reigned 1713-1740) merged his dukedom with Prussia, to become Friedrich-Wilhelm I, 'King in Prussia'. For his Hegermühle brass-works at Finow-Eberswalde, north of Berlin, he hired about 20 brass-makers, battery-workers, finishers and wire-drawers from Holstein, Silesia and the Harz.[190] From 1702, Friedrich Müller from Halle leased the mill but, after 1709, when Friedrich-Wilhelm I of Prussia forbade the export of brass, domestic demand from Prussia alone was too low. Friedrich Müller therefore went bankrupt and the state took over the brass works. From 1714, three Huguenots named Aureillon, Didelot and Lejeune ran the mill, working 12 furnaces continuously to produce different grades of brass, cast every 12 hours into thick plates for battery work, or thinner sheets to slit for wire-drawing. Two terraces of workers' cottages survive from 1724, and the early works was defended by two towers and a palisade.[191]

In 1730 two merchants named Splittgerber and Daum ran the Hegermühle. They recruited skilled Harz brass workers and added two furnaces, a battery hammer, wire works and numerous workshops. They exported to Russia, Poland, Spain and the Ottoman Empire, and supplied brass to Prussian armaments factories at Potsdam for fifty-five years. In 1736, Friedrich-Wilhelm I, hoping to keep his good Hegermühle

---

[189] Shropshire Archive: MS, 112/12/20/28 (1732)
[190] Kreisarchiv Barnim, Brandenburg: R 3842, typescript, J.A. Beling, 1769
[191] Fischbach 1786: 80

brass from Habsburg grasp, again forbade the export of Brandenburg brass without royal permission. Prussia's main internal brass markets were in Pomerania, Magdeburg and Halberstatt.[192] In 1736, to fulfil domestic demand, the Hegermühle produced cauldrons, coach and window hinges and fittings, assorted sizes of brass pins, door knobs, locks, door-plates, nails, tea and coffee jugs, hooks, chains, clocks, taps and stop-cocks.[193]

In 1739, fire destroyed major Hegermühle workshops, but they were rapidly rebuilt. From 1740, brass was stamped with the eagle emblem of Friedrich I 'the Great' (reigned 1740-1786), and was again exported, but only to markets outside the rival Habsburg Empire. This stimulated a strong brass trade to Marseilles, Livorno, Lisbon, Smyrna, St Petersburg, Archangel and Moscow. Hohenzollern Prussia needed brass for its war against Habsburg Austria, so in 1746 Friedrich I opened a second 32-kilometer-long Finow Canal, leading from Stettin (further west) past the brass works to Berlin, constructed to facilitate east-west transport.[194] The factory became renowned for its production of parts for Prussian armaments.[195]

Through strategic marriages, the Prussian Hohenzollerns acquired distant, smaller, satellite territories. The Duchy of Julich, including Stolberg, fell under Hohenzollern rule in 1701 and now supplied Prussia with three-quarters of its brass. The Stolberg brass industry therefore saw its heyday in the early 18th century, using about 6000 tons of zinc ore for brass-making in 1726.[196] In 1734 a fashionable Stolberg spa visitor, Count Carl-Ludwig von Pöllnitz, visited the brass works and described crucibles – 'filled with fresh plates of raw copper and even with scrap pieces of household brass, and onto these are thrown a certain amount of calamine, melted together with the metal ... and as the material becomes molten, the workers immediately attempt to remove and throw out the overflowing scum, using ladles. While this work is in progress, the copper part of the mixture is refined and combines with the metal that already contains calamine. Once it has reached this refined state, it is poured into a crucible or from one pot to another. The brass-making is considered complete when this process has been performed nine times.'

He also comments ... 'The copper that was red before melting, as it is when it arrives from the mines in Sweden and elsewhere, takes on a glistening colour like gold after alloying with the calamine. It is the most beautiful metal.'[197]

---

[192] Fischbach 1786: 81
[193] Kreisarchiv Barnim, Brandenburg: MSS, A11 HistAE 2853, folio 4; A11 HistAE 4976/63 folios.2v, 3, 3v, 7
[194] Aurich 1906: 15
[195] Menzel, Christian, 2007: *Das Messingwerk und das alte Hüttenamt*, www-docs.tu-cottbus.de, 8
[196] Roderburg 1927: 15
[197] von Pöllnitz, 1737, 1998 edition: 162-165, translation, author

The writer, an eighteenth-century fashion-seeking fop and man-about-town, had sought out this popular spa so as to be seen at the forefront of current trends.

From the late 18th century onwards, fashion leaders flaunted brass pocket 'toys' like penknives, chatelaines, corkscrews, snuff boxes, miniature mirrors and spectacle cases – to be discarded when the fashion passed. In 1712, English artificers were said to have 'acquired a singular Reputation Abroad, particularly in France, where our English Watches, Clocks, Locks, Buckles, Buttons and all sorts of English Brass Toys, are in great Esteem'.[198] The English word 'toy' then meant an inexpensive knick-knack - a trifle more than a plaything.[199] In 1725, Daniel Defoe, mentioned that the best 'Watches, Clocks, Jacks and Locks – and especially Toys and gay Things' were made in England and were exported to 'Holland, France, Italy, Venice and to all parts of Germany, Poland and Muscovy'.[200] An eighteenth-century brass advertising token for 'Dutch & English toys'[201] shows a London shop interior, where a child gazes at shelves full of 'toy' metal birds and animals. Stolberg, the fashionable spa, remained a leading European brass-producer until the mid-18th century, after which Scandinavia and England seized more of its former market share.[202]

Like Stolberg, calamine-rich Iserlohn, had (for lack of an heir-apparent to the Duchy of Julich) disintegrated during the 1609 War of Julich Succession, and had fallen under Hohenzollern rule. Iserlohn lay 20 kilometres south of Dortmund in the Ruhr area (now Westphalia), formerly in the county of Mark within the combined Duchies of Julich-Cleves-Berg. From 1701, the Prussian king, Friedrich Wilhelm I, ordered Iserlohn to supply calamine for his Hegermühle brass works to make brass equipment for the War of Spanish Succession (1701-14). After this war, Iserlohn sent calamine to the Stolberg area in return for semi-prepared or sheet brass.[203] In 1712 Iserlohn and its tightly-packed half-timbered housing burned down, Fire was a serious hazard where buildings were built of timber, so brass-fitted fire-fighting appliances were in demand (a surviving 'Fire Squirt' or syringe from a London church, dated c.1750, has cast brass nozzles, barrel and side handles).[204] Four years after the Iserlohn fire, its brass-founders were casting buckles, a craft introduced to Iserlohn from England in about 1715.[205] By 1722, 22 Iserlohn founders, using brass from the Stolberg area, produced not only buckles, but clasps, household equipment, bridle, harness and

---

[198] British Library: MS, 8245 A 18, folio 8 (1712)
[199] Noel-Hume 1969: 313
[200] Defoe 1725, 1983 edition: 290-291
[201] British Museum: CM 1870-5-7.4511
[202] Roderburg 1927: 15
[203] Bettge 1987: 262
[204] Gilbert 1969: 6
[205] Schulte 1938: 648

Trade and technology

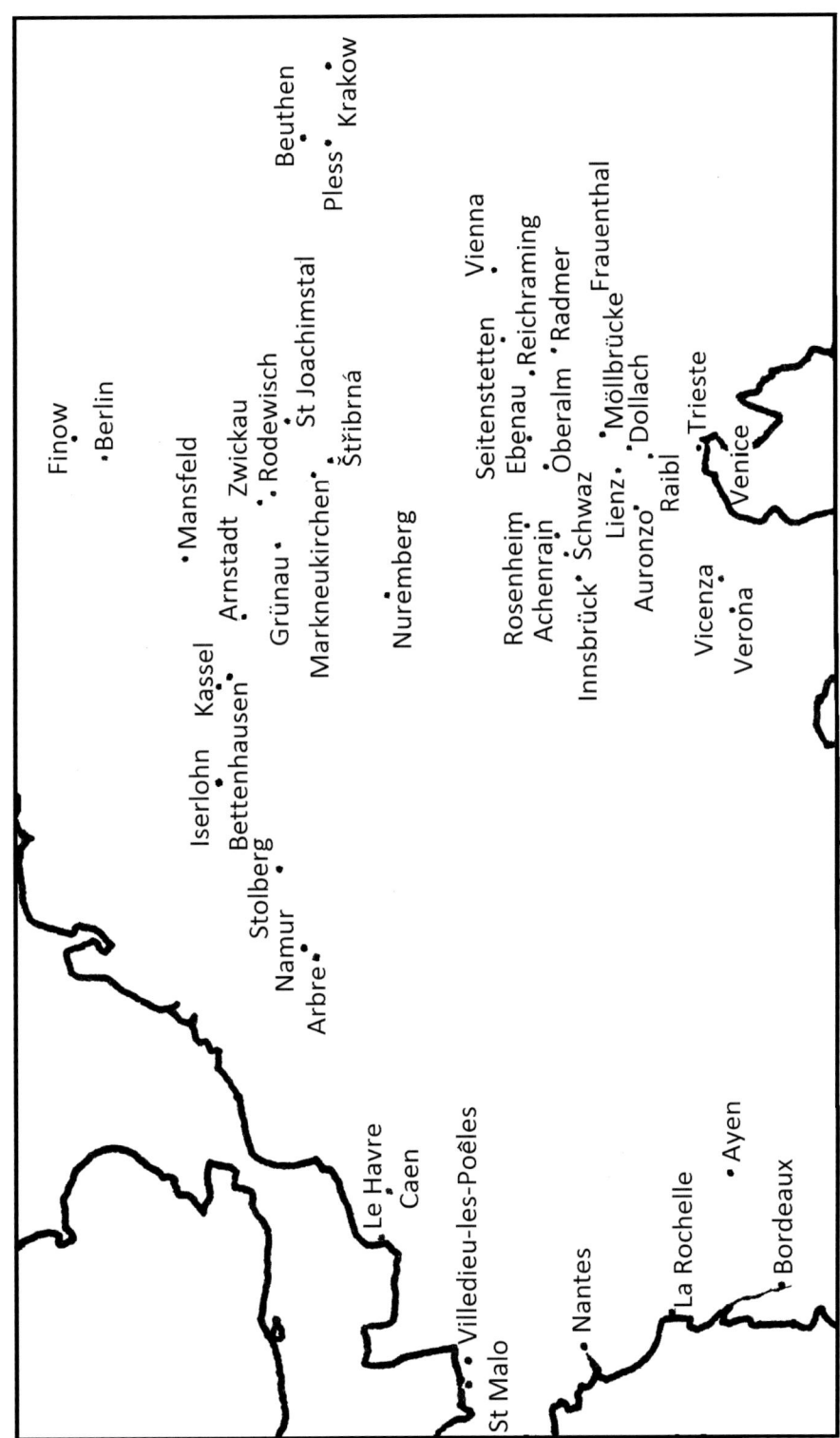

Figure 77. Europe, map showing some sites mentioned in the text

carriage trappings, bugles, military accoutrements, trunk-locks and hinges for the Prussian army, and, from 1720, brass pins.[206]

In 1736, as a first step towards calamine-rich Iserlohn producing its own brass, a consortium of local merchants erected calamine-calcining works up the steep, wooded Grüne valley,[207] where over six and a half tons of crushed calamine ($ZnCO_3$ ore) was roasted daily in sixteen calcining ovens. Three stepped calamine-washing pools were built on a hillside stream on the western slopes of the *Brandkopf* outside the old town walls. Brass foundries and a battery hammer were built downstream. By 1738 fine brass wire was produced at Iserlohn, for making pins, needles and hair pins[208] together with hooks, chains, tacks and hand-embossed tobacco boxes for export to Portugal, Spain and elsewhere. Their competitors were Swedish and Dutch craftsmen, who also made brass pocket or table-top tobacco-boxes and lids.[209] In 1749, a newly formed Iserlohn authority started a brass works on a stream in Hemer village, two kilometres east of Iserlohn.[210]

The mainland European brass works described so far were under Hohenzollern control, but many others were in Habsburg territory. The Namur brass mill fell to the Habsburgs in 1713. From 1695, hand brass-beating at Namur had been replaced by water-powered battery hammers. They were of various kinds – rounded for shaping up basins, and flat-headed for thinning plates to cut into strips for wire-making – and they weighed from nine to 23 kilos. Some had a tip 10 cm wide for thinning strips and others had curved, beak-like tips, weighing from nine to 14 kilos, for beating out concave bowl or cauldron shapes. The cauldron linings were afterwards hand-hammered and polished, to remove battery hammer marks.[211] A mill with five brass battery hammers and a wire works was located three kilometres up the River Burnot, a tributary of the Meuse, at Arbre, 18 kilometres south of Namur.[212]

By 1764, the Namur furnace-house had ventilated cementation brass furnaces below ground level, fuelled with wood and coal,[213] each furnace consuming 150 kilos of coal every 12 hours. The furnace-charge consisted of sixteen kilos each of scrap brass and Swedish copper, 27 kilos of well-crushed calamine, and nine to 11 kilos of powdered, sieved charcoal, dampened to prevent the brass from burning. This produced a brass plate weighing 38-39 kilos. A ladder led below ground through a furnace-hole, for servicing the furnace and ducts.[214]

---

[206] Hildenbrand 1983: 37, 50
[207] Bettge 1987: 255
[208] Hildenbrand 1983: 20, 33, 50-51
[209] Noel-Hume 1969: 311
[210] Klostermann 1996: 31-32
[211] Diderot, and d'Alembert 1758, volume 9: 219-220
[212] den Ouden, 1987: xxv-xxviii
[213] Duhamel du Monceau 1764: 60
[214] Diderot and d'Alembert 1758, vol. 9: 218; volume 6, 1, plate 7

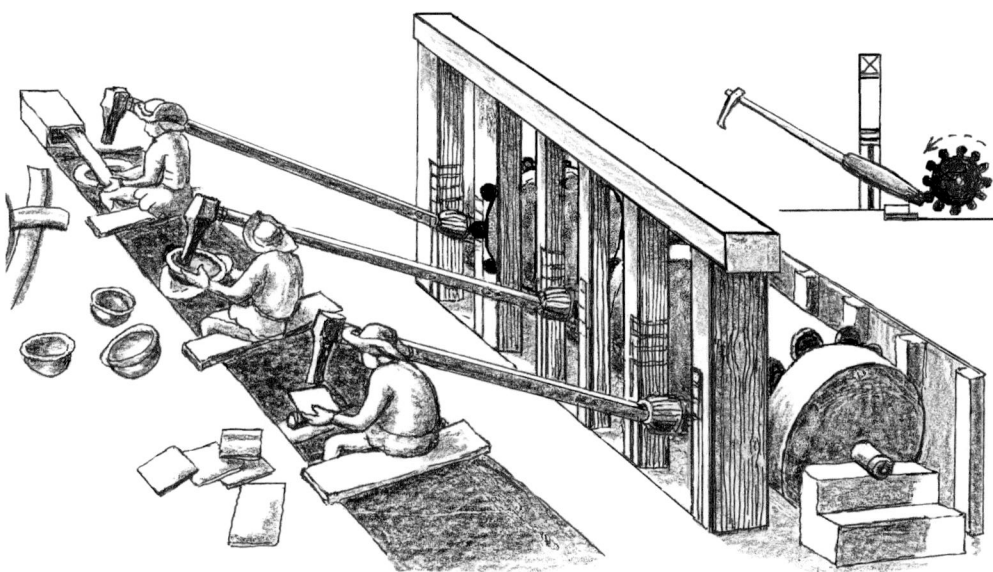

Figure 78. Battery hammers, late 18 century: *back*, forging strips for wire-drawing, *centre*, hollowing out cauldrons, *near*, flattening plate brass for shaping into drinking vessels. *Encyclopaedia* 21, 330

At Bettenhausen brass works, near Kassel, Hesse, also in Habsburg territory, the ducted furnace ventilation holes are still visible. Brass was becoming very profitable, so in 1716 Count Carl von Hesse charged a higher lease (800 taler) for his brass-works. During the Seven Years War (1756-63) factories closer to Kassel went up in flames, but Bettenhausen brass-works was spared, being just outside the city. It produced flat sheets to be worked into armour, sabre hilts, sheaths, wire, semi-prepared cauldrons and pots. Its battery works and coin mint used rolling-mills (like the one introduced at Bjurfors in Sweden in 1671).[215] The upper and lower roller-axles of these efficient machines were rotated at different speeds by separate water wheels, producing fine sheet brass.

By 1711, further south, Nuremberg was under Habsburg influence. Hammer brass-mill, Laufamholz, on the outskirts of the city could supply the outstanding scientific instrument- and clock-makers of Nuremberg. In the early 18th century, gentlemen amateurs wished to be seen to use brass scientific instruments, including sectors, parallel rules, armillary spheres, microscopes and telescopes. Hammer mill was expanding to meet this demand, and was acquiring timber for new buildings.[216] In 1713, Magnus Volkamer and merchant Karl Forster held Hammer Mill and battery works. After Volkamer died, his brother-in-law, Johann Forster, took control, marrying Volkamer's daughter in 1729. He was succeeded by his Forster son and grandson, so,

---

[215] Knoke, Horst, 1990: die Mühlen in Bettenhausen, www.agathof.de/6-Mühlen_in_Bettenhausen-1.cfm, 4-6
[216] Stadtarchiv Nürnberg: MS, 455/16, Hammer, Laufamholz

as the Volkamers died out, the Forsters took over, aided by substantial bequests. As the family prospered, they built workers' apartments and accommodation for the sick and disabled.

Further south, the mineral-rich Habsburg territory of Silesia (Śląsk) lay in today's south-west Poland. In 1709, Count Jakub von Flemming, an influential courtier of Augustus (1670-1733, Elector of Saxony and newly-crowned Polish king), built a brass-works at Jakobswalde (Kotlarnia) in Silesia. It was powered by a tributary of the River Klodnice flowing through the Slawentzitz (Slawiecice) forests west of Beuthen (Bytom, north-west of Krakow). Flemming added a (tax-exempted) sheet-brass forge, wire works and mirror factory, and recruited skilled workers from Habsburg regions like Saxony, Thuringen and Steiermark. Timber came from the estate forests, calamine from the prolific Szarlej mines north-east of Beuthen and copper from the Banat of Temeswar (in today's Romania).[217]

In 1714, Count Magnus von Hoym, owner of rich calamine mines near Beuthen, bought the Slawentzitz estate, including Jakobswalde brass-mill.[218] In 1717, Flemming, who still ran this brass works, installed a battery hammer and sold brass to Poland and Sweden. A subsequent tenant, Johann von Janisch, improved the mill's profit, and in 1741 built his own brass-mill.[219] The rich resources of Silesia tempted Frederick II of Prussia to try to wrest it from Maria Theresa of Austria, its Habsburg heiress. In 1742, during the ensuing War of Austrian Succession (1740-48), Silesia, including Jakobswalde brass works, fell to Hohenzollern Prussia. The disputed Krnov (Jägerndorf) district (now in the Czech Republic), was the only part of Upper Silesia to be retained by Habsburg Austria after 1748.

In neighbouring eighteenth-century Saxony, zinc-oxide furnace accretions and zinc metal were still collected as a by-product of lead- and silver-smelting, a practice extending east to the Carpathian Mountains. In 1737, Henckel (1678-1744), a Freiburg mines specialist, wrote that at Chemnitz, Saxony, 'the heat from the combustion chamber smelted out from the lead a metallic form of powdered furnace zinc' that could be alloyed with copper 'to give it the colour' (i.e. to make brass).[220] In 1759, Caspar Neumann (1683-1737), chemist at the Prussian court, described purpose-built reservoirs set into the lead-furnace front wall above a gutter carrying out-flowing molten lead. A large flat stone enclosed the inner side of each reservoir, leaving a few chinks for zinc vapour to enter. Another clay-sealed outside stone was constantly doused with cold water to cool and condense the fumes, causing more metallic zinc to form. After the 20-hour lead smelting, a workman struck off the outer stone, letting the mercury-like zinc flow out into moulds consisting of saucer-shaped depressions

---

[217] https://pl.wikipedia.org/wiki/Kotlarnia; location, 50° 16' 38 N, 18° 22' 13 E
[218] Fechner 1903/2013: 362
[219] Perlick, Alfons, 1972, in *Deutsche Biographie* volume 9, deutsche-biographie.de/sfz54009.html: 670
[220] Henckel 1737: 310

in sand, to form hemispherical ingots.[221] The sublimed zinc metal produced was useful in brass-making but less pure than distilled zinc, still the better product for making higher-zinc brass.

Two brass-works in Saxony, heavily involved in the Hohenzollern-Habsburg conflicts were the Rodewisch and Ellefeld mills in the Auertal (Auer valley), southern Saxony, uncomfortably close to the border between warring factions. By 1704, brass yields at Ellefeld had increased to about 13 tons annually. Timber for charcoal and fuel ran short, so in 1708 a brass-maker from Holstein spent a year introducing coal-fired reverberatory furnaces to a new furnace-house at Rodewisch. The Rodewisch foundry had numerous casting-stones for large plates of brass, producing ten plate-castings per furnace every week – about 1000 plates a year from two furnaces. Good caking (coke-producing) coal was needed to avoid contamination, so experiments were made with coal from the nearby Zwickau mines.[222]

Around 1740, the owners of Rodewisch brass works expanded by converting a mill on a strong mountain stream at Jägersgrün (12 km south of Rodewisch) into a subsidiary battery works. One water wheel on this stream could power at least three battery hammers to beat out brass sheet for wire-drawing,[223] providing the thousands of pins required by lace-makers for the flourishing Erzgebirge (Kruzné Hory) cottage lace industry. The brass-wire makers cut, etched, polished and sharpened good, rust-proof pins, producing, in 1754-5, about eight and a half tons of fine lace-makers' pin-wire – three times the quantity produced 50 years earlier. North of Saxony, one Marcus Fulda took over the Grünau brass works in 1715, but shortages of Thuringen Forest copper affected local mills at Grünau, Hockerode and Arnstadt, whose brass furnaces produced only 30-40 tons of brass annually.[224]

Further south-west, in the Tyrol, the sole heir to the depleted Achenrain brass mill, Carl Oswald Aschauer, was syphoning off its profits to pay his private debts. His brass market meanwhile shrank, due to an epidemic among French customers, and his rents and debt interest soared. In 1707, state authorities inspected Achenrain mill, made a map survey,[225] and appointed an administrator (for a map of Austrian sites see Chapter 4). As valuable stock piled up, workers shifted away, unofficially, to a newly founded brass-works at Rosenheim.[226] In 1725, the bankrupt Carl Oswald Aschauer surrendered to administrators in Schwaz, dying a few years later. In 1740 the Tyrolean public treasury shouldered part of the Achenrain brass works debts.[227] A technical

---

[221] Neumann 1759, translation by Lewis: 122-123
[222] Gericke 2008: 85-86
[223] Gericke 2008: 91
[224] Hachenberg and Ullwer 2013: 57-58
[225] Vienna Hofkammer Archiv: Film 152, Bundle 79, folios 16, 22, 23 (1706-1725)
[226] Priesner 1997: 54
[227] Urbanner 1972: 49

manager lived on site to maintain control, and a business manager gradually restored the mill's finances by negotiating with the Bolzano market and exploiting new sales opportunities offered by brass tableware fashions.[228] Before 1725, the factory sold semi-prepared brass wares, but new Viennese markets spurred it to produce a wider range of finished objects, in competition with Frauental works, 40 kilometres southwest of Graz. By 1748, the Achenrain debts and business capital were repaid, and demand for brass was rising.[229]

Frauental brass works, near Deutschlandsberg in Steiermark (Styria, eastern Austria), was erected in 1714 by Earl Ferdinand von Zehentner and financial partners, in a well-timbered area with convenient water transport via the River Lassnitz. However, the mill lacked capital and produced poor-quality brass. Within a few years, two partners went bankrupt, so from 1752, the state managed the mill, improving the buildings and erecting a new wire workshop and battery hammers.[230] Lack of capital thwarted Reichraming brass works, also in Steiermark, but help was at hand.

In 1705, Franz Rieser gave up Reichraming brass works, and decline set in, so, from 1725, workers started to disperse and in 1730 the works was sold. The great change in its fortunes came in 1739, when a merchant, Franz Lang, made a gift of Reichraming brass-works and Radmer copper mines to the great Benedictine abbey of Seitenstetten, where his son was a monk.[231] Inventories dated 1731, 1743 and 1771 detail the activities carried out in each workshop, warehouse, office and furnace-house. Refined and unrefined copper for Reichraming was delivered in half-ton casks, mainly from Radmer, 50 kilometres further south, and from Kalwang, slightly further, and later from Dognascau (now in Romania). Zinc ore, both zinc carbonate and zinc silicate, arrived in barrels from the Tyrol and Kärnten, to be sorted into ore types described as fine, red, paired, mixed, crusty or white, before being crushed in the calamine mill. Charcoal was burned near the mill, using timber from the richly forested mountains towering above the valley. Earthen crucibles of fireproof clay from Passau, on the River Donau (Danube), were fired, then dried well to make them last.

At Reichraming, copper, calamine and charcoal were carefully weighed before being layered into the crucible, the proportions being varied according to the type of brass planned. Brass to be refined for forging was re-alloyed several times. Non-refined brass was used for casting, and certain mineral salts were added to produce specific brass types. Impurities thrown up during alloying were skimmed off, using two skimming ladles. The furnaces were lined with a mixture of sand and crucible clay. Tongs for gripping and lifting the hot crucibles were suspended from the roof by chains, and

---
[228] Köfler 1962: 375-376; Urbanner 1972: 49
[229] Urbanner 1972: 49
[230] Ucik 2006: 181-182
[231] Brunnthaler 2000: 114-115

Figure 79. Reichraming brass mill, 1763: 1 furnace house; 2 calamine mill; 3 wire-drawing; 4 brass battery; 5 smelting; 6 charcoal; 7 stables; 8 manager's mansion; 9 workers' dormitories; 10 channel supplying water to mill wheels, Seitenstetten Abbey, by kind permission

men wearing shields and guards used hand-held tongs to pour the material from the crucible into moulds.

Cast brass ingots were forged into shape by three battery-men sitting at anvils each served by trip-hammers. They beat the ingot lengthways, annealed it, and then beat it crosswise to create sheet metal. The brass was often annealed on a hot stone rather than over an open charcoal grate, but it was sometimes cold-forged. Two trolleys carried hammered bowls and kettles to the annealing oven. After casting, specialist brass-scrapers in a separate workshop used knives to pare off uneven spots, the brass shavings being carefully recycled. The next stage was etching or cauterising in a cauldron by treating the surface with acid.[232] This pickling process cleaned the metal, leaving it sparkling. Wire was drawn using a hand-powered winch. Water cleansed the wire of acid, before it was polished and coiled into rings. Sheet metal was lightly rolled and packed in wooden crates to be sent to Steyr, where cutlers both held together and decorated their knife hafts with brass rosettes, inserted by a worker using an iron punch tool (*Augeneisen*).[233]

From 1741-43, Abbot Paul Vitsch of Seitenstetten Abbey (1729-1747), a man of local Steyr stock with knowledge of mining, managed the copper mine and brass works.

---

[232] initially potassium bi-tartrate, and later ligneous acid
[233] Seitenstetten Abbey: MSS, Inventaria von Reichraming und Radmer 1731-1876, Box 53/O, folios 2-9

The abbey took full possession from 1745, including almost 29,000 gulders of debt, and brought stability to the venture for the next century and a half. Timber came from thousands of square kilometres of forest land held in trust from the Steyr estates. The next abbot, Dominik Gussmann (1747-1777), and his competent works-manager brought Radmer mines and Reichraming brass works into profit.[234]

By 1708, further south in the duchy of East Tyrol (now south Tyrol) Lienz brass works, had a calamine-crushing mill, scraping and wire-drawing workshops and a manager's house with a clerk's office, living room, bedrooms, kitchen, cellar and granary. Sadly, the mill burnt down that year, but the Tyrolean treasury placed the Schwaz mining authority in charge, and the rebuilt Lienz brass works returned to profit, producing brass ingots, plate and wire. Remote Lienz brass mill sourced its ores from alpine areas little known further north. It used two types of zinc ore, calamine ($ZnCO_3$) and zinc-rich needle-like or kidney-shaped hemimorphite,[235] mined in the Jauken (in Carinthia, 20 kilometres east along the Drau valley), Raibl (50 kilometres south, in Carinthia), and Auronzo (north Veneto, Italy, 80 kilometres across the mountains). Copper ores for Lienz came initially from local mines, but by 1769 local supplies failed to match demand, so copper was brought from Klausen (Kitzbühl, Tyrol). Charcoal came from Leisacher Gries (across the mountains to the east), and from nearby Pölland, but it grew harder to acquire the 11,400 cubic metres of timber needed annually. Lienz exported its brass products through Trieste, to the Levant and southern Italy.[236]

The Habsburg family of Holy Roman Emperors, based in Vienna, were interested in brass production and the profits to be gained from it, so, as wealthy leaders of fashion, their palaces gleamed with brass objects and gilt brass. Society as a whole underwent significant changes during the early 18th century, and there is nothing like fashion for reflecting social change. The communal religious dimensions of the European medieval and early baroque era were now superseded by a compartmentalised social order, with emphasis on the individual.[237] Households observed regulated time schedules (requiring brass clocks), eighteenth-century life became measured and managed by the time of day, and people entertained according to the rules of polite society. Spatial rules were introduced (requiring survey instruments), and middle-class living-spaces became divided into rooms with different functions, often dictated by gender and class. Musical attainment was encouraged, so chamber instruments became popular for music-making. In 1697, ninety kilos of brass virginal wire arrived in Britain from Germany.[238] Brass wire gave a sweeter, warmer tone, so virginals and spinets were strung with brass strings only, but when harpsichords (often with over

---

[234] Brunnthaler 2000: 115, 117, 120
[235] ($Zn_4Si_2O_7(OH)_2 \cdot H_2O$)
[236] Heinricher 2006: 9
[237] Leone 1988: 235-261
[238] Houghton 1693, 1728 edition, volume 2: 184

60 strings) became popular in the 18th century, the lower strings remained brass, but the shorter strings from the C above middle C might be steel.

The extravagant trends of the Habsburg court favoured showy fashion products, including buttons, buckles, fittings and sword-hilts, designed to attract and impress. The popular imagination had been seized by individual self-awareness and the desire to express themselves through their possessions. Brass artefacts were circulated to a wider public by networks of peddlars, who promoted new goods and sought out new markets, so the latest brass items became popular because they were more widely accessible. Since the early 18th century, a person's social standing was enhanced by the practice of astronomy, time-keeping, measurement, planning, and scientific pursuits. Gentlemen amateurs needed to be seen to use brass scientific instruments, armillary spheres, microscopes and telescopes, and brass mathematical and survey instruments, including dividers,[239] protractors, sectors and parallel rules.[240]

The Habsburg rulers at their Viennese court led or embraced these aspects of fashion and social change. They competed not only with the Prussian Hohenzollerns, but with the brass trade of independent states such as Salzburg, whose Oberalm and Ebenau brass-works competed with newly-built rival Habsburg brass-works at Frauenthal and Rosenheim, and with Achenrain and Reichraming, their traditional rivals, all four within the Holy Roman Empire. Many brass-works in (today's) Austria and Germany were restricted by local laws prohibiting buying, selling, recruiting or working outside their own duchy or state. The laws were imposed to stop the drain of expertise and valuable resources and to ward off competition. In 1723, the Vienna court cabinet placed heavy import duties on Salzburg brass entering the Holy Roman Empire, and a 1737 Habsburg edict (the *Kremsbrücker Aufschlag*) blocked Salzburg from exporting brass tax-free through Habsburg-dominated Tyrol to its traditional Italian markets. It was a common European strategy to protect home manufacture by imposing duties on goods imported across state borders, but Salzburg could not sell enough brass within its own borders to sustain production, so its trade gradually declined.[241]

Rosenheim, the first Bavarian brass works, was founded in 1717, after Johann Heffter obtained a 'privilege' to operate a brass mill near existing copper battery works. Its two major drawbacks were that it owned no forests for charcoal, nor strong water power, copper, calamine or crucible clay. Coal fuel could be used, but the use of coal fuel for brass-making, long used in China and more recently at Stolberg and in Britain, was very expensive for Rosenheim since coal had to be transported 30 kilometres from an inefficient mine at Miesbach. The existing water power (which already drove other mills) was strengthened by new weirs, so Bavaria managed to make plate and

---

[239] Fox and Barton, 1986, excavated finds, VI W9 and VI P9
[240] Museum of the History of Science, Oxford: 48965, *c*.1730
[241] Priesner, 1997: 59-60

BRASS FROM THE PAST

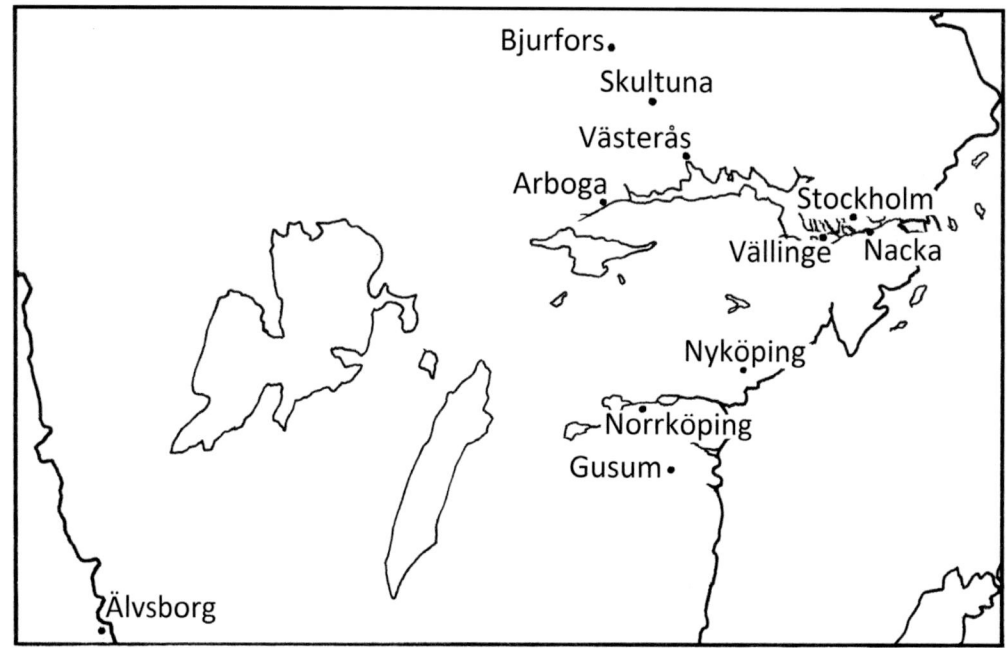

Figure 80. Sweden, map showing some sites mentioned in the text

sheet brass, wire and ingots. In 1718, the Rosenheim factory employed 36 workers, producing about 40 tons of brass, but little is heard of it later in the 18th century.[242]

In 1740, Stříbrná (Silberbach) brass factory in nearby Bohemia, ran very short of local copper, despite being near the Kraslice copper works. From 1735-55, the Counts Roslitz, new Habsburg owners of the Kraslice district, opened new copper mines to revive the industry, but they proved unprofitable to work. In 1742 Stříbrná brass mill still appears on a map, but production was low.[243]

Northern areas, including Holstein and Sweden, had now begun to suffer export problems due to wars in Europe. In 1729, in Holstein, outside Lübeck, Trems brass-mills made sheet brass, wire and cauldrons, operating two furnaces, two battery hammers and annealing, etching and polishing workshops (though water shortage sometimes slowed production).[244] Throughout the 17th century, Sweden had been a very strong brass-making nation, but it now lacked surplus copper, which had been used up funding its wars. After a catastrophic cave-in at the great Falun mine in 1687, Swedish output had never fully recovered, so, although Sweden still exported copper, less was left for its home industries.[245] The copper shortage caused heavy setbacks to

---

[242] Priesner 1997: 71, 81-83, 298
[243] Jaroslav Zaplatal personal communication
[244] Warncke 1949: 213, 215
[245] Hekscher 1954: 85-86

its brass works, some of which virtually ceased production. By 1710, even the annual output of the mighty Norrköping brass works had sunk from 168 tons of brass to 84, with only 15 of its 21 furnaces working.[246]

When the Swedish government took over the dilapidated Skultuna site and its debts from Abel Reenstierna, new tenants were hard to attract. So, in 1715, when King Karl XII needed money for his Great Northern War with Russia (1700-21), the Crown auctioned Skultuna brass mill and estate to Dutch brothers, Claes and Petter Matheson.[247] Skultuna began to produce items for the growing Swedish domestic market, including brass mortars, brass bowls, wider at the top and curving to a narrower base, lidded brass milk-cans; flasks with narrow inward-curved necks; bulbous jars, drinking-bowls and smaller rounded cast bowls.[248] Independent Swedish founders and braziers also worked brass. In 1720, an immigrant German brass worker in Malmö, southern Sweden, produced sword belts, buttons, riding tackle, coach furnishings, saddle and sledge fittings, coffin-furniture and platters.[249] By 1740, Skultuna obtained granite casting slabs from St Malo, fitted into an iron frame and smeared first with clay to allay the heat, and then with cow dung to help release the set brass from the casting-slab mould. France supplied fireproof clay for making crucibles in summer, when they could be dried and stored for winter.[250] In 1765, Petter Matheson's son-in-law and former servant, Carl Browall, became manager and started rebuilding Skultuna mill.[251]

Gusum brass mill, south of Norrköping, was leased in 1707 to an energetic Dutchman, Petter de Gomsez, who acquired labour and equipment from the recently closed Vällinge brass mill, and added workshops at Gusum for producing pins, buttons, buckles and other small brass fashion items. From October 1710, wartime shortages halved the supplies of best copper, which lowered output, but Gusum mill consumed about 156 tons of best refined copper monthly for alloying brass. Large reserve stores of casting-stones, crucible-clay, English, Polish, Hungarian and Flemish calamine, tallow, wood, coal and charcoal, were kept in warehouses at the mill and at nearby Valdemarsvik harbour.[252]

During the 1720s, two widows of former Gusum mill-owners, sisters-in-law Sofia Spalding and Maria Westerberg, ran the affairs of Gusum mill amicably together with considerable business acumen. In 1731, they constructed new factory premises, a manor house with accommodation for both families and a small local church. The collaborative spirit, however, faded with the deaths of these steadfast and enterprising

---

[246] Forsberg 1953: 57
[247] Forsgren 2010: 19-20
[248] Erixon 1969: 298
[249] Erixon 1978: 19
[250] Gadd 2007: 7
[251] Forsgren 2010: 20
[252] Forsberg 1953: 56

women, leaving the estate open to friction and family disputes. In a 1736 agreement, the Spalding family inherited eight furnaces, a foundry and 12 wire-drawing benches, while the Westerbergs gained four brass furnaces, a foundry and ten wire-drawing benches. However, wire-drawing ceased and the manor, farms and buildings were divided by a high boundary fence. Though the output of the separate mills continued for some years, they had to diversify to survive.

In 1742, for example, Maria Westerberg's son Eric revived brass pin manufacture at Gusum, Gusum soon become the nation's largest pin producer, employing 200 people. The scale of the Gusum pin output upset Stockholm pin-sellers, who implored officials to ban the sale of Gusum pins, but Eric Westerberg continued trading them. To produce pins, the mill alloyed brass and cast plates, which were sent to Middle Mill, rolled into thin sheets and slit into strips. The strips travelled two kilometers by barge to be drawn to coarse wire in a workshop, before returning to another workshop to be drawn to fine wire, and then transported to a pin workshop to be cut into short lengths. Finally, the pins travelled to a tip-sharpening and burnishing workshop. Researcher Karin Forsberg calculated that the journey from start to finish comprised 11 kilometers. Bundles of pins and special pin-heading blocks for twisting pin-heads from brass wire were delivered to the homes of local old women and children who added the pin-heads and packed the finished pins in paper bundles. Gustaf Spalding, owner of the rival Gusum mill, also tried pin-making, but failed.[253]

At Bjurfors, Christina Cronström had managed the brass works after her husband died in 1680, but her own death in 1708 occurred at a time of harsh economic climate. Karl XII's overseas conflicts (Great Northern War 1700-21) were ruining the nation's finances, and in 1709 Bjurfors brass works fell into a 'deplorable state'. Brass prices had fallen, whilst copper, needed for coinage, rose in value. National poverty reduced domestic demand, and wartime shipping hazards made export difficult. Isak Funck, Christina Cronström's son-in-law and former chamberlain to King Stanislaus of Poland, eventually purchased Bjurfors brass works and initiated chandelier production.[254] In 1734, the Swedish botanist, Linnaeus, sketched the water-powered rolling-mill at Bjurfors.[255]

Nacka brass mill, near Stockholm, was leased from 1688 by Abel Reenstierna, who, already deeply in debt, used calamine, casting stones and copper as collateral for a mortgage on the mill. Export was impossible in wartime, so Nacka's new products became buckles and shiny brass buttons – each military jacket needed a dozen. Reenstierna mechanised button-making by borrowing labour-saving equipment from the Mint (possibly a rolling mill to stamp out brass discs), operated by cheap casual labour, which replaced skilled journeymen. This practice aroused complaints from the

---

[253] Forgren 2010: 102-103
[254] Forsgren 2010: 126
[255] Sahlin 1934: 110-111

now-destitute expert brass-workers, the factory fell into in disarray, and the buttons and buckles sold below agreed guild prices. Reenstierna was so heavily in debt that his Norrköping and Nacka mills both became forfeit to the state.

In May 1704, three state officials arrived to claim the contents of Nacka mill, breaking a window to rush in and seal off the office, workshops and warehouse. Reenstierna was jailed, spat upon by his many creditors. The forerunner of the State Bank (*Rijksbank*) tried to sort out the damaged workshops and casting-stones; coal spoiled by fire, and barrels leaking calamine. The bank restored the workshops to button and buckle production, but refused all back pay to the workers until the goods were sold. Most men moved elsewhere, but a few stayed on in return for food and shelter at the mill, for which they paid. Many consumed more than they could afford, and fell into debt, but they had to stay, because any new employer would be legally forced to shoulder their debts. By 1707, three kilns were operable at Nacka, but the wire-works were often idle and made poor quality wire. In 1710 production ceased when several Nacka staff died of plague, then rife in Sweden. In spring 1712, the bank failed to sell the mill as a going concern and by 1718 Nacka's brass-making days were over.[256]

Norrköping brass mill too, was badly affected by the Great Northern War, sparked off by Russia's determination to reduce Sweden's considerable land possessions in Europe. Like Nacka, Norrköping brass works, owned by the bankrupt Abel Reenstierna, was managed from 1704 by the State bank. On 30 July, 1719, the Russians set fire to Norrköping, then the second largest city in Sweden, reducing it to ashes. The brass works was destroyed and the Russians plundered over 16½ tons of highly-prized brass.[257] In 1721, the brass works was granted eight years freedom from taxes, to help it recover from the Russian attack and a subsequent severe flood.[258] Water-flow to this mill was often depleted by summer droughts, and was sometimes brought to a standstill by winter ice. In 1740, although brass output was slight and of poor quality, a Stockholm merchant, Jacob Forsberg, bought and revived the mill, injecting new capital and re-scheduling the sale and supply of raw materials. Norrköping once again became the foremost Swedish brass mill, and stamped its rings of brass wire with a lion and shield showing its improved quality. Forsberg had the Holmen Tower rebuilt, following destruction by the Russians, but in 1754, aged 45, he suddenly died.

In 1705, the fully operational brass-mill at Nyköping, slightly further east, had been bought by Dr Matthius Iser, a Stockholm property-owner and chaplain to King Karl XII.[259] He at once provided badly needed workers' housing in Nyköping, but Karl XII's Great Northern War caused shortages of labour and materials until by 1710 copper shortages were acute and the mill staff was halved. In 1713, Iser tried to carry out

---

[256] Forsgren 2010: 77-80
[257] Molbech 1820: 66-9
[258] Riksarkivet, Stockholm: typescript Register, folio 226, Norrköping, Resolution B,
[259] Riksarkivet, Stockholm: typescript Register, folio 661, Nyköping, (30 June 1705); folio 1616

necessary mill repairs but in 1714, after the death of King Karl XII, the Russians seized the chance to harry the eastern Swedish coast and storm Nyköping, burning it down and destroying most of the brass mill. The Russian ravages forced many brass-workers from Norrköping and Nyköping to seek work elsewhere. By 1727, two Stockholm merchants and major ship-owners, Paul Heublein and Claës Wittmack, had bought the ruined Nyköping mill, but only rubble and fire debris remained, and the skilled brass workers had gone. In 1732, however, the Cabinet recommended rebuilding Nyköping brass mill, which started production in 1734 with nine furnaces. It was soon employing 40 workers to cast chandeliers, operate battery-hammers to make cauldrons and other wares, and draw brass wire, its main product.[260]

War or antagonism between different parts of Europe during the first half of the 18th century held up brass production in many areas and this was a bleak period for most. However, one positive outcome was the movement of skilled brass workers. Because of hostilities, skilled craftsmen and furnace-men moved across borders as refugees, diffusing their expertise and spreading awareness of new technology. The situation often forced brass-mill owners to seek out new business strategies, spread their sources of supply and discover different markets. The later part of the century would see increasing global competition, not only between brass works but between trading nations. Trade expansion was already diffusing the knowledge and influence of previously alien cultures and fashions, which itself suggested new uses for brass.

[260] Forsgren 2010: 60-61

Trade and technology

Byzantine or Coptic low-zinc, leaded brass bowl, probably made in Egypt in the 5th-6th century, one of several high-status burial goods in a solo, pagan Anglo-Saxon burial dating to the 6th-7th century. This elite burial was recently found by chance near Snettisham, eastern England. photo © Ian Cartwright, School of Archaeology, University of Oxford

Chapter 8

The turning tide

*c.*1750-1800

The oceans, seas and rivers of the later 18th century were filled with shipping under sail, carrying cargoes from one end of the world to another, though storms, pirates and enemy privateers ensured that not all was plain sailing. This maritime world was itself a major consumer of brass. The huge loss of shipping from navigational error, for example, started the race to invent an accurate device for finding longitude at sea, which led, in 1762 and 1764, to trials of John Harrison's ingenious brass chronometer H4. Watchmakers John Arnold and Thomas Earnshaw meanwhile modified his ideas to develop smaller, simpler and more practical sea-going brass chronometers, and by 1796 Earnshaw's instruments proved reliable enough to be reproduced for regular use at sea.[1] Brass wire sieves and large brass containers were ordered for safer use with naval gunpowder because brass was non-sparking.

Figure 81. British naval brass equipment: *left,* late eighteenth-century sextant; *centre,* barrel-spigot (tap) with bucket-hook, 1780-1815; *right,* careening block, c.1809, with central reinforcing sheave (plate) and brass coak lining the rope-groove

British Admiralty records reveal the wealth of brass used in the days of sail. Apart from navigational instruments, orders included brass rowlocks (oar-sockets),[2] locks and hinges for ship furnishings.[3] In 1739, the *Success* fire-ship, was ordered to ensure that the 'fastenings of the boxes in store are brass'[4] – many examples of naval rust-resistant brass box-corners and fastenings survive. Due to increased travel, eighteenth-century trunks and boxes, vital for safe storage, had ornamental brass hinges and locks, some with pierced brass lock-clasps.[5] In July 1768, an officer of the *Ferret* cutter, from Plymouth, requested brass cocks for pumps to counteract the stench from bilge-

---

[1] British Museum: 1958,0921.16, Earnshaw marine chronometer 245, 1780-1800,
[2] The National Archive: MSS, ADM 106/942/25 (1741); ADM 106/1162/61 (1768); ADM 354/145/167 (1752)
[3] The National Archive: MSS, ADM 106/1043/121 (1747); ADM 106/1145/129 (1766); ADM 106/1128/177 (1763);
[4] The National Archive: MS, ADM 106/910/58 (1739)
[5] Musée des Beaux Arts, Bordeaux: *Nature Morte à la Vielle*, oil painting, Roland de la Porte, 1724-93

water, and the Navy constantly requested brass wire, coaks (inner linings for pulleys, against rope friction), pump-chambers, stop cocks and spigots (taps).[6] A surviving late eighteenth-century brass barrel-spigot includes a bucket hook.[7] Some were tempted by the value of brass – one unfortunate shipyard bricklayer hanged himself after being locked up for stealing a seven-pound (3.3 kilogram) brass weight.[8]

From 1767 to 1800, Thomas Ripley, a craftsman and chandler in Wapping, London,[9] is one of many who sold brass octants, sextants and quadrants to the British Navy,[10] and, on a much larger scale, a new 2.44-metre-high wall-quadrant was made in 1759 for Greenwich astronomer James Bradley, its more durable brass frame designed to replace an over-strained iron one.[11] Late eighteenth-century popular telescopes included Gregorian reflecting telescopes with lenses aided by curved mirrors to help focus light from the stars; and equatorially mounted telescopes moving parallel to the plane of the equator. Other late eighteenth-century instruments with brass parts included apparatus to demonstrate physics or 'Mechanical Powers'; a transit instrument to be poked out of a window for measuring the positions of the stars and planets, and a portable electric generator for treating distressed patients.[12]

The expanded world-wide sailing-ship trade to the American colonies, the Guinea Coast and the Far East, increased the demand for non-magnetic, rust-free, brass ship's fittings, but their use was not always merely utilitarian. In 1769, the commodore's private cabins on board his ship the *Montague* at Plymouth were given extra brass fittings befitting his superior rank.[13] A few years earlier Captain Alexander Hamilton's eye was caught by the ostentatious five-and-a-half-ton brass cannon, displayed on the quayside at Canton (Guangzhou).[14] Sporting guns had decorative brass escutcheons; sailors polished their ship's brass fittings, and the words 'top brass' still have a strong ring to them. Brass, though a comparatively cheap alloy, embodied some of those impressive, durable and showy qualities calculated to impress in the naval or military arena.

In 1741, Nehemiah Champion of the Bristol brass works proposed brass sheathing for ships, and on 3 April of that year, Deptford and Woolwich naval officers were ordered to conduct trials with sheet brass for sheathing wooden ships, because the copper

---

[6] The National Archive: MSS, ADM 106/1168/71 (1768); ADM 106/942/40 (1741); ADM 106/946/148 (1741); ADM 106/919/42 (1740) ; ADM 106/1136/138 (1764); ADM 106/1158/238 (1767); ADM 106/1165/19 (1768); ADM 106/1168/71 (1768)
[7] National Maritime Museum, Greenwich: ZBA 0439 (1780-1815)
[8] The National Archive: MS, ADM 106/947/155
[9] National Maritime Museum library, Greenwich: HNL/7/201 (1)
[10] Dunn 2006: 24-25
[11] National Maritime Museum, Greenwich: AST0971, quadrant, James Bird, 1759
[12] Museum of the History of Science, Oxford: 14164 (*c*.1780), 12666 (*c*.1790) and 70367 (1770)
[13] The National Archive: MS, ADM 106/1182/163 (1769)
[14] Hamilton 1727: 296

content in brass has properties which destroy microbes and algae. Although the officers opined that 'salt water will canker the brass and the nail-heads, and the edges cannot be closed perfectly', trials proceeded. Admiral Vernon approved sheathing on ships on October 30, 1741 by, sending an order from Cuba that unsheathed ships should be sent home forthwith.[15] A young visitor to Warmley brass factory, Bristol, wrote the following undated account, entitled 'The Progress of Mr Champion's Brass works'. After describing various processes, he wrote:

…we next saw whole pieces of Brass brought out of the Furnace in Waggons; Brass rolled out extremely thin and intended to cover the bottoms of Ships, in order to prevent their being worm eaten or growing rotten.[16]

Warmley brass works made both brass and copper sheathing for ships, and, although brass ingredients were cheaper than copper, if fuel and alloying time were included, the cost was similar. However, it gradually became apparent that pure copper was softer and therefore easier to work, rivet and seal onto the curved hulls of ships. The British frigate, HMS Alarm, was sheathed in copper in 1761,[17] and the Welsh copper magnate, Thomas Williams, made sure that merchant ships were sheathed in his abundant product. So, from then on, brass sheathing rapidly gave way to copper.[18]

The great eighteenth-century sailing ships were, however, equipped with brass sheaves or 'shivers'[19] for pulley-blocks, including careening blocks.[20] Large circular pulley-blocks made of lignum vitae had a deep, brass reinforcing groove, or sheave, let into the rim for the rope to run through, and coaks - brass plates - were screwed onto both flat sides to hold and support the axis-pin on which the pulley-block turned. Extra strong careening blocks were used when keeling ships over for hull repair or cleaning, and righting them afterwards, for example when repairing the hull of Captain James Cook's *Endeavour* on his far-flung voyage to Australia in 1770.

China, too, remained a distant destination for sea voyages from Europe. Under the Manchu Qing dynasty, China was probably the wealthiest nation in the world, with more than thirteen million square kilometres of territory and a huge population. During the Chinese Qianlong period 1736-1795), the Baoyun Pavilion, built at the Summer Palace, Peking (Beijing), included leaded brass (23-27% zinc) in its construction.[21] China and Japan saw the profits to be gained from exporting to the west. Among

---

[15] The National Archive: MSS, ADM 106/948/6 (1741); ADM 106/947/229 (1741)
[16] Badminton Muniments: MS, Fm K 1/4/1, undated, 1749-1767, by courtesy of his Grace the Duke of Beaufort
[17] www.revolvy.com/main/index/.php?s.=HMS%20Alarm20(1758)
[18] *The Gentleman's Magazine* 31, 1761: 533
[19] National Maritime Museum, Greenwich: AEL 0453 (c.1809)
[20] The National Archive: MSS ADM 106/491/354 and ADM 106/384 (1696); ADM 106/943/72 (1741); ADM 106/1038/51 (1746)
[21] Zhou Wenli 2016: 12

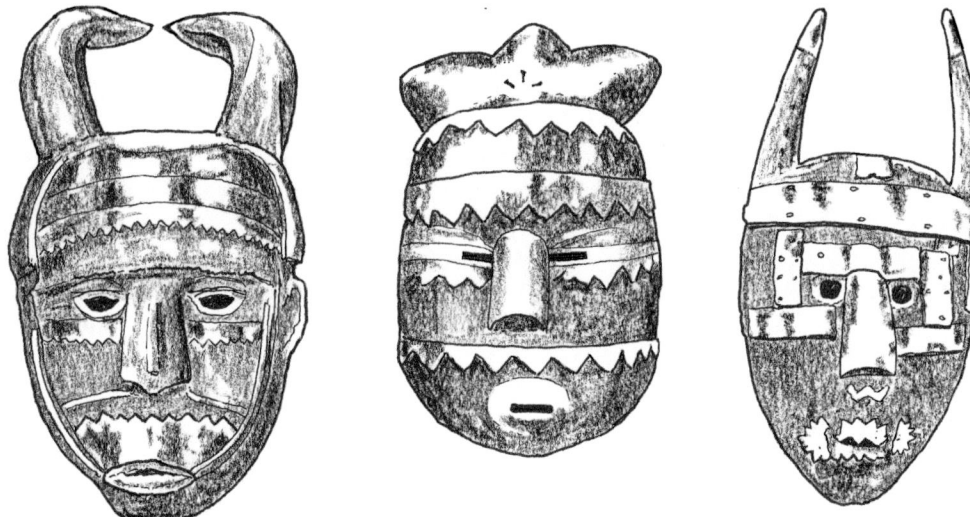

Figure 82. Ancestral masks from Temne, Sierra Leone, with applied brass strips

Chinese exports around 1780-1800 were items like glass cameo snuff bottles with brass-mounted stoppers.[22] Japan produced swords, held to be the soul of the Samurai, some embellished with brass parts, including Edo-period *tsuba*, or sword-guards, depicting plants or entwined figures.[23] In Burma (Myanmar), the necessary materials were carried to brass-working sites by pack animals. The male ox leading a merchant pack convoy through mountain passes would carry the large bell on its back for the herd to follow. A great brass Shan ox-bell survives, hung in a wooden framework to harness it to the ox. Pastured cattle wore small cast-brass bells, usually trapezoidal or semi-circular, with the clapper held on a wire. Small brass objects have been found in pagoda relic chambers dating to the 1760s, and brass was worked in the Tampawadi quarter, Mandalay, from 1783.[24]

Along the Guinea Coast of Africa, brass was still used for purposes not envisaged by its exporters, notably by the Oba of Benin to express power, and by Ashanti peoples to commune with powerful spirits. Further west, particularly at male initiation rites, men of the Temne people of northern Sierra Leone wore expressive light wooden masks[25] embellished with sheet-brass strips and geometric shapes, to represent ancestral spirits. Brass was a metal traditionally held to represent chieftaincy, and its decorative use was evidently integral to communion with the spirits of the ancestors.[26] With new, shining brass parts, the masks must have been spectacular.

---

[22] Birmingham Museum and Art Gallery: 1885M1849 and 1927M180
[23] Birmingham Museum and Art Gallery: 1930M952 (1740-90)
[24] Fraser-Liu 2010: 125-128, 134-135
[25] British Museum: BM1952 AF 7.15; BM 1953 AF 25.1; BM 1954+23.3502; BM 1954+23.3503
[26] Hart 1987: 69, 73-74

India, meanwhile, was undergoing internal power struggles. During the second quarter of the 18th century, the Mughal Empire gradually declined under pressure from the Hindu Maharatha. The weakened Mughal leaders tolerated the Hindu religion, encouraging its artistic creativity. Surviving eighteenth-century Hindu brass figures include images such as 'Durga slaying the buffalo demon' from Madhya Pradesh and a shiva linga (sacred phallus) shrine from Moharashtra, cast in three sections. The brass craftsmen of Rajasthan produced a brass image of a Rajput girl, Dipalakshmi and a seated figure of Tirthankara, dated 1752, both made using distilled zinc.[27] Holy Ganges water from Varanasi (Benares), revered by Hindu worshippers and pilgrims, was carried in elegant brass water vessels (*chambu*).[28]

In India, in 1756, the European Seven Years War sparked off parallel hostilities between the French and English in India (the third Carnatic War). Local Nawabs (Muslim princely supporters of the emperor) still administered large parts of Tamil Nadu, Karnataka and Andhra Pradesh, but by 1760, the Hindu Maharatha Empire had spread over much of the Indian subcontinent. By the later 18th century, however, the British East India Company controlled Bengal and Karnataka, where they installed a Muslim ruler. Although such hostilities may have thwarted creative Indian brass-making, Islamic influence introduced new brass scientific instruments to India,[29] including a celestial globe made by Persian astronomer Nasir al Din al Tiusi in 1790-91, with a pierced hole to locate the earth, and relief engraving of the stars in their constellations.[30] Further east, in Borneo, the Dayak people of Kalimantan had a centuries-old tradition of using brass to produce ear-lobe weights, strings for their lutes (plucked with an anteater's claw), brass and silver bracelets and large sets of brass gongs.[31] They also made brass containers for betel and other valued commodities, sometimes with traditional designs of monkeys, flowers and giants, highlighted by a background of black tree-resin[32], though these products have been hard to date.

As in India, Anglo-French conflict was active in northern America, where the French, backed by Native Americans, fought the English colonists. England, meanwhile, protected its own brass industry by refusing its colonists in America the right to alloy their own metal,[33] so from 1750-70 more than a fifth of British copper and brass was exported to the American colonies. Before 1792, however, American brass founders, especially in Philadelphia, were casting industrial and domestic cauldrons, pots and pans, including large vessels for distilling rum, brewing and sugar refining.[34]

---

[27] Biswas 1993: 325
[28] Ashmolean Museum, University of Oxford: EA X.280; EA 2003.63 (1700-1800); EA X-2 108 (1750-1800)
[29] Serrao, Judith, 2007: www.mangalorean.com/browsearticles.php?arttype.features&articleid=
[30] Durham University Oriental Museum: DUROM 1994.1
[31] Thiessen 2016, 28, 32
[32] Museum Nasional, Jakarta, 25269
[33] Iron Act, 1750
[34] Hyde 1998: 7

Figure 83. Decorated bowl or container from Borneo, thought to be Dayak

Caspar Wistar, a prosperous Philadelphia brass button maker from 1725-52, hampered by England's restrictive rules against metal production, cast buttons from scrap brass, finishing them in his own foundry. In 1750 one of Wistar's former apprentices set up a brass button factory in New York City[35]. To gild brass buttons, five grains of gold per button was added to a mercury mixture, brushed onto the buttons, and roasted in a furnace.[36] At James Geddy's Foundry in colonial Williamsburg, Virginia, his sons David and William, advertised brass harness-buckles, hinges, surveyors' compasses and 'curious' fenders and fire-dogs, cast in sand moulds.[37] Brass items excavated at colonial Williamsburg, Virginia, included coach hinge parts, harness-leather ornament and a saddle terret (c.1750-90). By the 1760s the Geddy family workshop employed 12 to 15 craftsmen, engaged in seven

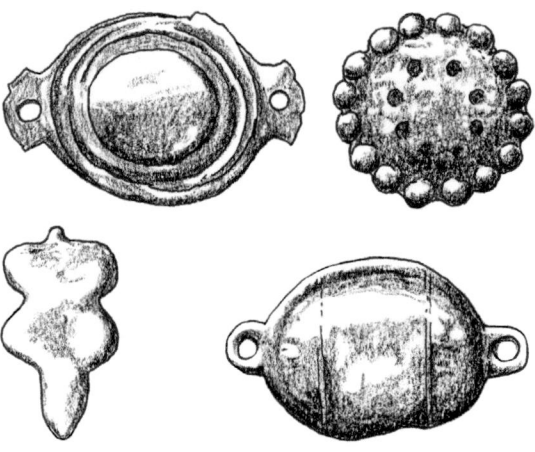

Figure 84. Eighteenth-century brass harness bosses from Colonial Williamsburg, Virginia

---

[35] Whisker 1924: 195
[36] www.thebuttonmonger.com/content/A_history_of_buttonsv1.pdf
[37] Dan Berg and Hassel 1992: 9-12

or eight different crafts, mainly brass casting. American craftsmen were in short supply, so it paid to diversify to compete with imported goods.

By 1851, Bristol, as a brass exporter, was rapidly being overtaken by the port of Liverpool. Indignant English brass workers meanwhile complained that the North American colonies were not only 'setting up manufactories among themselves' but smuggling in quantities of Dutch, German and French manufactured goods 'by the connivance of our custom-house officers'.[38] It became fashionable, and cheaper, to silver solid brass artefacts, including candlesticks, rather than using solid silver – one survivor being a cast and pierced brass soap-holder, *c*.1750, with traces of silvering.[39] In the 1790s, brass surfaces might be tinned, to prevent harmful reactions with some foods.

By the end of the century, guns and gunpowder were so popular that the second amendment to the United States Constitution (1791) declared the right to keep and bear arms. Guns and pistols included moving or decorative parts made of brass, and brass blunderbuss barrels were current in the later 18th century.[40] Most European rifles, fowling-pieces and pistols carried brass cast and engraved decoration, and even country gunsmiths achieved elaborate ornament, including trigger-guards[41] and brass escutcheons on flintlock sporting guns.[42] The Canadian fur companies traded spare brass trigger guards, butt plates, side plates and ramrod holders embellished with rococo designs, some of which were excavated from the King's Stores, Quebec (destroyed in 1760),[43] but these are rare survivors.

Relatively few early brass artefacts survive. Firstly, most brass artefacts lack heirloom appeal, so are discarded after the owner's death, because brass was neither as precious as silver or gold, nor so readily engraved. Silver surfaces were easily engraved with refined designs – but cementation brass was not so well adapted for displaying the finely detailed pictures, inscriptions and crests suited to heirlooms. A second reason for the lack of surviving brass was due (as already discussed) to scrap brass always being in demand to add to brass-making-crucibles to lower the melting point and thus speed up the alloying process. Any out-of-fashion brass objects, factory clippings, waste material or workshop-floor dust were sought for melting down in the next batch of brass, which is why utilitarian survivors are rare.

Brass recycling is sometimes documented in passing. In 1697, the writer and apothecary John Houghton mentions 'old plate brass' being added to crucibles, and

---

[38] British Library: MS, T.1144 K.15
[39] Victoria and Albert Museum: M.141-1939
[40] German Historical Museum, Berlin: permanent displays
[41] Colonial Williamsburg Foundation, Virginia: trigger-guards, 282; 22 8; 22 927
[42] Noel-Hume 1969: 217
[43] Marsette 2003: 34

implies that most brass artefacts were re-melted, by observing that 'the greatest consumer of copper is [brass] pins, because they seldom return to the melting pot'.[44] In July 1651, British naval ordnance officers were ordered to deliver any 'brass and unserviceable brass ordnance in their custody' to the Tower of London foundry for melting down to make thirty-six guns for new frigates,[45] though at that period the word brass might have been used loosely, meaning brass or bronze. In 1765, a London brass-candlestick customer wrote to Birmingham manufacturer, Matthew Boulton, with the words 'Waste material returned' worth £4 13s. 6d,[46] and in 1787, brass-maker Thomas Williams discussed with Matthew Boulton the economics of returned scrap brass.[47] American foundries, prohibited from alloying brass, melted down many scrapped imported brass goods for re-casting. The 1761-3 daybook of harness-maker Alexander Craig, of Williamsburg, Virginia, shows him selling scrap brass to the nearby Geddy Foundry to re-cast into objects.[48] Most foundry workers who made small everyday items like thimbles worked only with scrap brass.

Figure 85. The restored Geddy Foundry, Colonial Williamsburg, Virginia

Out-of-fashion objects made of brass alone were easily melted down. For this reason, brass survives more often in artefacts made of other materials besides brass – for example in fine furniture with brass handles, hinges or inlays; or wooden-cased clocks with brass faces. Surviving examples include oval furniture knobs, which grew popular from about 1750, to be succeeded later in the eighteenth century by flattened round knobs. From 1750-75, drawing rooms contained armchairs, light tables and sofas with brass casters. Small brass tacks were used to secure upholstery, and brass decorated the surfaces of mid-century high-status furniture such as mahogany wine-coolers banded with brass,[49] or inlay on tables of 1740-50.

In 1750-51, however, two important customers, France and Spain, banned the import of British manufactured metal goods in favour of their own, which curtailed the trade

---

[44] Houghton 1693, 1728 edition: 190
[45] Calendar of State Papers Domestic 1651: 283
[46] Birmingham Archive: Boulton MS, 3782/1/13, (1765)
[47] Birmingham Archive: Boulton MS, 3782/12/73/72 (1787)
[48] Dan Berg and Hassell 1992: 25
[49] National Trust, Erdigg: ERD F.99-6

of British brass craftsmen, particularly in Birmingham.[50] Brass goods needed to be durable and weatherproof for harness, and coach fittings.[51] In the later 18th century, coach and sedan-chair leather, often exposed to the weather, was secured with large rust-proof circular-headed brass tacks. In 1769 John Pickering of Birmingham advertised decorative chasing in gold, silver and brass on 'Coffin Furniture, and ornaments for Coaches, Chariots, Sedans and other Carriages'.[52] Decorative coach and coffin fittings were stamped out of thin sheet brass.

The brass plates supporting Chippendale furniture handles (1750-75), were still in bat's wing style, but with more upward-turned wings, some stamped in fretwork-like Chinoiserie-style lattice patterns.[53] From about 1750 to 1800, mahogany chests of drawers, rather than having one backing plate to a whole handle, had individual handle-posts each with a circular collar or, after 1765, rococo-style rosette discs.[54] During the 1785-1800 Hepplewhite-period loop handles followed the outline of their thin oval brass backing-plates – the handle-ends turning sharply inwards – and brass door furniture followed a similar pattern. In the 1760s, furniture keyhole plates were small and oval, moulded rococo, rectangular or plain brass[55]. Around 1760 long-case clocks might have plain faces,[56] reflecting a more classical taste, or ornamental corner spandrels.[57] Certain clocks had complex functions – Matthew Boulton, for example, received a request in 1766 for a 'pair of Figure plate Dyes' ... 'one Dye for the Lunar and one ditto for the Kallendar'.[58] Later eighteenth-century brass picture frames and umbrella parts carried die-and-stamp decoration, also popular for coffin furniture.[59] Changing fashion in furniture and utensils can only have boosted brass production.

Domestic brass artefacts excavated from mid-eighteenth-century domestic and brass foundry contexts in Colonial Williamsburg, Virginia, include thimbles, spherical bells, tacks, buttons, buckles, hinges, taps, curtain rings, keys, andiron finials and sets of weights,[60] showing that the English settlers had a simple but comfortable life-style. In coastal states closer to New York, where Dutch influence and trade remained strong, surviving eighteenth-century domestic objects include higher-status brass bird-cages, candle sconces, snuffers, sugar sifters, chandeliers, warming pans, fire-grates, fenders, fire-tongs and barbers' basins.[61]

[50] British Library: MS, T 1144 K (7) (1751)
[51] Colonial Williamsburg Foundation, Virginia: 5187 ANH; 5185 ANH, carriage door hinges,
[52] Patent Specification 920, Pickering 1769
[53] Colonial Williamsburg Foundation, Virginia: CWF 0194.195A, drawer-handle plate,
[54] National Trust, Chastleton: CHS F/64, 1784,
[55] Colonial Williamsburg Foundation, Virginia: CWF 2718 13.1, keyhole plate
[56] Birmingham Museum and Art Gallery: 1953.5.345.2
[57] Museum of the History of Science, Oxford: 39570
[58] Birmingham Archive: Boulton MS, 3782/1/15
[59] Hamilton 1926, 1967 edition: 268
[60] Colonial Williamsburg Foundation: Geddy ES.1539.19B; Lightfoot 0017 908GA; 20 A27; 3-C
[61] Schiffer 1978: 38-162

Numerous European Atlantic ports, including Amsterdam, Rotterdam, Bristol, Liverpool, Cadiz and Lisbon exported to both America and the Guinea Coast of Africa.[62] In 1754, Baptist Mills brass works, Bristol, sold 80 to 90 tons of brass pans annually to the Guinea Coast. The diameters of the pans varied from 30 cm up to 1.22 metres for the largest, with 50 or 60 sizes in between.[63] The merchant account books of Nehemiah Champion, of Baptist Mills brass-works, confirm his vigorous trade to the Guinea Coast and the West Indies during the 1760s,[64] his ships returning from the Caribbean laden with sugar, coffee and ginger, even while Bristol trade was being rapidly overtaken by Liverpool.

By 1754, the 48 Baptist Mills brass furnaces took calamine from the Mendip Hills in Somerset, Flintshire or Wales, and they were perpetually fired up, some for years. Cracks and holes in pans and kettles were soldered with neat, nearly invisible patches, before being smoothed on a lathe.[65] Baptist Mills produced a wide range of brasses – an analysed sample of workshop floor debris,[66] laid down over time, contained minute fragments of varied brass (15 to 35 per cent zinc). Analysed samples from one part of the site[67] contained 0.23 per cent arsenic, possibly due to a Cornish copper ore source, and 11 analysed samples[68] were low-zinc brass for casting or gilding purposes, ideal for the Birmingham button trade. Similar results were obtained for later eighteenth-century figurines made in Birmingham by Matthew Boulton, a regular buyer of Baptist Mills brass.[69] Metallic zinc, added to molten brass in precise amounts, allowed consistent production of specialised brasses, and a gilding layer applied to the surface was thought to enhance the finished golden look. In France in the 1760s-70s, high-status chandeliers, and furnishings of bright brass (18-20% zinc, with 1% each of lead and tin for casting) were gilded with a gold-mercury amalgam (ormolu), then burnished to a high sheen. Boulton used a warmer-toned 10% brass as a substrate for gilding. Hammered-down high-zinc brass leaf – 'Dutch metal' – was sometimes applied to non-metallic surfaces[70].

Baptist Mills was distilling its own metallic zinc by June 1767, the year before Boulton opened his Birmingham factory.[71] Nehemiah Champion had imported distilled metallic zinc from 1731, when his son William was experimenting with zinc distillation in the Baptist Mills laboratory. Metallic zinc was often included in brass for musical instrument strings, to improve yield-stress so that tightening the string

---

[62] Tibbles, Anthony 2000: www.liverpoolmuseums.org.uk/ism/resources/slave.trade
[63] Berg and Berg 2001: 144-145
[64] Bristol Archive: Champion MS, 12455 (4)
[65] Berg and Berg 2001: 145
[66] Morton, 2009, 274, Chris Salter analysis
[67] 1990 excavation, Ian Beckey and Mike Baker, Conduit Place, Bristol, NGR ST 6003 7442
[68] Morton 2009: 274, Chris Salter analysis
[69] Goodison 1999: 101-168
[70] Peter Northover, personal communication
[71] Gloucester Archive: Badminton MS, D421 B 1, June 1767

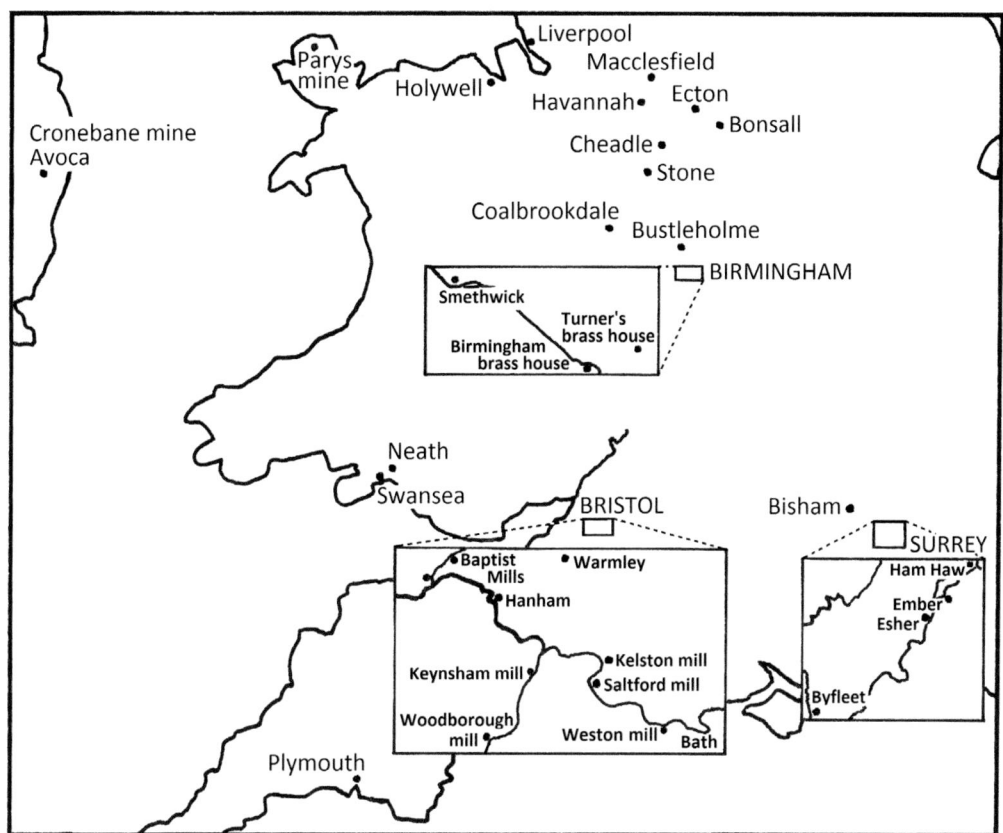

Figure 86. England and Wales, map showing some sites mentioned in the text

did not reduce its elasticity (plastically deform it).[72] Two fragments of excavated fresh cold-drawn brass wire excavated from Baptist Mills were high in zinc and might therefore have been intended for instrument strings.[73] By the 1760s-70s, alongside its American and West African trade, Baptist Mills produced specialist brasses for Birmingham craftsmen. Having no port, Birmingham could not ship out heavy items, so it produced many small, light-weight fashionable brass goods involving maximum craftsmanship.

This was the European age of ostentation, with accessories like brass buttons, buckles and snuff boxes in high demand, and French brass craftsmen produced very tempting, well-designed examples.[74] Matthew Boulton therefore encouraged drawing and design skills in his apprentices. In 1762-4, his factory pay-roll included a mould-turner, stamper, buckle-filer, tong-forger, pattern-makers, die-sinkers, candlestick-polishers, burnishers, hook filers, metal rollers and a woman 'paid for cutting 50 button

[72] Peter Northover, personal communication
[73] Peter Northover analysis for the author
[74] Styles 1994: 543

cards'. His Boulton and Fothergill factory purchases included 'skin for rubbing buttons on, a pair of bellows and three lathes'. Metallic zinc came from Bristol, fireproof Stourbridge bricks lined the furnace, and 288 leather knee-chapes were ordered to protect workers' knees from furnace heat. By 1770, Birmingham's 44 buckle-makers and 83 button-makers worked with varied grades of brass.[75]

In October 1762, a fashion-conscious Londoner suggested to Matthew Boulton that 'your platina [high-zinc brass] Metal would make Beautiful Candlesticks ... and if they were perforated or filigreed it would still add to their Beauty'. In 1766, Nathaniel Jeffrys, a London customer, ordered 'French plated candlesticks', two pairs of pillar candlesticks and 'Bed chamber candlesticks – 12 good and sound'.[76] Other eighteenth-century lighting items included portable candlesticks with ring handles, conical snuffers, brass wall sconces and wick-trimmers (a pair of scissors with one blade attached to a box holding a pad).[77] During the 1750s, besides trying to follow the fluid styles fashionable in silver, brass-founders produced candlesticks with plain stems, sometimes with twisted knops. They made both tripod and single-stem candlesticks and, by the 1770s, candlesticks with petal, neo-classical square or sometimes round bases and, from about 1780, oval bases.

Figure 87. Candlesticks: *right*, late eighteenth century square-based candlestick: *left*, Georgian petal-based type candlesticks, *c.* 1780 and *c.* 1760

The energies and inventiveness of small local founders, braziers, wire-drawers and pin-makers drove forward the industry in utilitarian and rural products, as opposed to high fashion. An eighteenth-century brass spherical (rumbler) bell excavated at Williamsburg, Virginia,[78] was made in a village workshop in Aldbourne, Wiltshire, England, and similar bells (*c.*1740), sometimes containing a pea, have been excavated in England. Eighteenth-century brass sheep bells were often numbered, for example from 1 to 12. Sheep bells both helped shepherds to locate the flock and the flock to follow the leading sheep, and some iron sheep bells were brass-coated to sweeten the tone.[79] Purer brass has a sweeter tone, but the more lead it includes, the duller the

---

[75] Birmingham Archive: Boulton MS, 3782/1/34
[76] Birmingham Archive: Boulton MS, 3782/1/1/13 and 14
[77] Colonial Williamsburg Foundation, Virginia: CWF A 20, candle snuffers,
[78] Colonial Williamsburg Foundation, Virginia: 3-C and ER 1346.198
[79] John Wilson, farm tools collector, personal communication

sound. Brass features in a list of mill machinery at Mill Lane, Maidstone, dating to 1752, which mentions 'Brasses', 'Brass Steps', and 'other Tackling'.[80] Clay pipes were produced from brass moulds, and brass vessels were used in chemical laboratories. Even a few engraved brass dog collars survive from the mid-18th century,[81] and '24 Doggs Collar' appear in one Bristol brass-founder's inventory.[82]

By the 1760s, Birmingham's button industry, with a strong export trade to Europe and America, had few rivals and earned huge profits.[83] Elaborate Birmingham brass buttons, some incorporating shell and glass, were intended specifically for the French male fashion market.[84] Whereas contemporary cloth buttons could be made at home, brass buttons had to be produced in a highly organised factory. From 1760, gilt, silver and pinchbeck buttons (10-13 per cent zinc) became popular, using literally tons of low-zinc brass. Two surviving flat Birmingham buttons of about 1770 have gilt-brass open-work borders.[85] British military officers had to buy their own brass and gilt-brass military buttons,[86] brass shoulder-belt plates and helmet badges (lower ranks wore horn buttons). The orator, Joshua Steel, wrote to the manufacturer Matthew Boulton, on 29 October 1762, 'I think your opinion of your Platina metal [a pure, fairly high-zinc brass][87] is too modest. My Buckles are extremely admired … Tradesmen in the City asked me … if my Buckles were not Gold, so that if you could propose a platina Hilt, I fancy the Officers would like it very well'.[88]

German chemist, Johann Christian Wiegleb (1732-1800), wrote in 1789 that to produce the specialist high-zinc gilding brass known as Princes Metal, the copper in the crucible was first covered with a thick layer of powdered charcoal and fused by heating, then metallic zinc was directly mixed into molten brass, stirred, and the brass quickly poured out. Forge-tolerant hard solders were likewise prepared by adding one eighth, one-sixth or even a half part of melted metallic zinc to melted brass. Hard solder for silver was made by melting together equal parts of silver and cementation brass, then adding a one-fifteenth part of melted metallic zinc.[89]

In 1767, at Warmley zinc-distillation plant, a chemically-minded bishop, Dr Watson, observed William Champion's 'circular kind of oven', containing six retorts, each about 1.22 m high. 'Into the bottom of each pot was inserted an iron tube, which passed through the furnace floor into a vessel of water'. The zinc that condensed in

---

[80] Goodsall 1957: 128-129
[81] Blaise Castle Museum, Bristol: TA 177
[82] Bristol Archive: MS, FCI 1733-54/8-11 (July 23 1733)
[83] White 1977: 68
[84] Sinead Byrne, curator, personal communication
[85] Birmingham Museum and Art Gallery: 1953FIS 1770-1800
[86] Birmingham Museum and Art Gallery: 29053, 296153 and Gaunt button collection
[87] Ure 1839: 165
[88] Birmingham Archive: Boulton MS, 3782/1/1 (1762)
[89] Wiegleb 1789: 438

small particles in the water was re-melted, cast into metallic zinc ingots and sent to Birmingham.[90] This suggests that Champion had followed Angerstein's 1754 advice to modify his zinc collection process by shortening the condensing tubes and leading them into water. Excavated Warmley distillation residues reveal metallic zinc droplets surrounded by zinc oxide and zinc sulphide.[91] By the end of the century, the industrial distillation process spread further into Europe.

At Warmley, William Champion used steam engines to raise a head of water by pumping recycled water round a circuit. In 1749, he installed a Newcomen-style steam engine, which derived its pumping-energy from the atmospheric pressure present in a partial vacuum (condensing hot steam by suddenly injecting cold water into it).[92] Four steam pumps raised water 3.66 metres from a deep lake and forced it through underground passages to an upper triangular reservoir.[93] From there the water flowed back down to drive water wheels that powered battery works, slitting mills and (from 1761) the wire-drawing machines in the wire mill. The young visitor to Warmley, quoted earlier in this chapter, wrote that

pieces of Brass and Copper' are 'hit by battering hammers into Dishes, Pans or any form whatsoever. The Pump which conveys Water to work all these Indians [engines] with, brings up 120 hogsheads [roughly 3,000 litres] in one Minute, and this Pump is set into motion by a fire Indian' [steam engine]. 'Brass plate is divided by a large pair of Shearers into very narrow pieces which being again put into the Fire and rendered ductile is drawn thro' a small hole into Wire of any size with which wire pins are made; others are cut into Rods which being sent to Guinea are exchanged for Negro's. The Wire ... is cut into proper lengths for different sized Pins, then pointed at each end by a Stone, the Heads are then put on by little Children of five year old, and then boiled an hour and a half in Lead, which makes them white'.[94]

In 1754, Warmley had twelve cementation brass furnaces (lined with 23cm-thick refractory brick),[95] fifteen copper furnaces, battery-, rolling- and slitting-mills and four zinc distilleries. Warmley cauldrons and pans all went to the Guinea coast. In May 1767, the naturalist Sir Joseph Banks (1743-1820) noted the many water wheels at Warmley powered by one small brook, with the help of two of the largest steam engines in England. The biggest one, with a cylinder diameter of 1.88m, raised almost four thousand litres of water to an upper reservoir with each stroke, at about nine strokes a minute.[96]

---

[90] Watson 1786: vol. 4, 39
[91] Dungworth and White 2007: 79-81
[92] www.animatedengines.com/newcomen.html
[93] Warmley, English Heritage Report, 1998, South Gloucestershire Council area advice Note 30, 1-8
[94] Badminton Muniments: MS, Fm K 1/4/1, undated, 1749-67, by kind permission of his Grace the Duke of Beaufort
[95] South Gloucestershire Sites and Monuments Record: 12715, Robin Stiles, typescript, 1986
[96] Banks 1767, 1898 edition: 22

Warmley was a 'vertical industry' with all processes carried out on site, including coal-mining, ore preparation, zinc distillation, copper smelting and brass making, casting, battery, wire-drawing and pin-making. One of its two satellite mills, Kelston battery and rolling mill[97] was established in 1764 on the River Avon to operate coal-fired reverberatory annealing furnaces.[98] Outer fuel cavities enclosed an inner clay-smeared chamber, insulating the brass artefacts inside from harmful sulphurous coal fumes. The artefacts were stacked on a trolley with revolving top, propelled into the annealing oven on metal rails.[99] Bitton, the other Warmley satellite, was a brass battery mill, near a coal source on the small River Boyd.[100]

The 1768 inventory of a wealthy investor in the Warmley Brass Company, merchant Thomas Goldney of Bristol, includes 'two brass hearths, brass shovel and tongs', 'two brass candle branches over the chimney'; 'two large leather trunks ornamented with brass nails'; 'one brass figure of a Lyon cast over the chimney'; '1 brass fire grate, brass fronted, and brass fenders', and '3 brass pans'.[101] A surviving pestle, mortar and skimmer date to around 1750.[102] Another English inventory, dated 1761, includes a 'Brass skimmer; ladle and slide; pair of brass scales and four weights; brass mortar with iron pestle; brass tinder box' and '7 brass flat candlesticks'.[103]

In July 1761, the over-ambitious William Champion proposed 17 new furnaces at Warmley. On meeting with Champion, the Baptist Mills partners, incensed by his aggressive behaviour 'could not tell how to contain themselves without being in a violent passion and behaving ungentlemanlike'.[104] A Warmley shareholder angrily called the Baptist Mills company 'grand oppressors, who have ... played us off and bamboozled us all by turns', and suggested charging Baptist Mills more for coal. By 1765, William had expanded too fast for his cash flow. He tried to defeat competition by forming a cartel with his suppliers, aiming to collect all the copper ore currently smelted in Cornwall, Liverpool and London and smelt it at Warmley instead, in 80 copper furnaces.[105]

Unsurprisingly, William Champion's grandiose expansion scheme attracted no investors, so failure was imminent. In 1767 his factory stocks included brass 'Kettles, Pans, Guinea Kettles, Guinea Neptunes, Guinea Rods, Guinea Manillas', wire, ingots, plate brass, pins and zinc metal. The two main shareholders pushed for bankruptcy, to recoup what they could. But William Champion contrived to withdraw capital behind

---

[97] National Grid Reference ST 6946 66805
[98] Patent Specification 867, 1767
[99] Swedenborg 1734: 354
[100] National Grid Reference ST 6812 6977
[101] Bristol University Library: Goldney MS, DM 1398/ A
[102] Blaise Castle Museum, Bristol: TB 3243 and T 5018
[103] John Rylands Library, Stafford: Stone MS, X 1/7 (1761)
[104] Gloucester Archive: Badminton MSS, D 421 B 1, July 7, July 21, Sept 3, 1761; May 18, 1761
[105] Gloucester Archive: Badminton MSS, D 421 B 1, n.d., 1760s, 25 March 1765; D 27100 QP 13/1-2

their backs.[106] In 1769 the Baptist Mills brass company took over Warmley, with its coal supply and more modern plant.[107] By 1779, most equipment was sent to Bitton, leaving behind only five or six zinc distillation furnaces.[108] In 1781, Baptist Mills transferred its copper smelting to Warmley,[109] where brass making ceased, though some zinc-distillation continued until the late 1840s,[110] Warmley was probably the metallic-zinc source for Abel and Levi Porter in Connecticut, who, by 1802, imported zinc from England to produce brass by direct mixing. Together with the three Grilley brothers, brass button makers, they built what may have been the first American brass making mill, in the Naugatuck Valley, and soon added a brass rolling mill.[111]

James Emerson, William Champion's manager, remained at Warmley, but in 1779 he set up his own brass mill and zinc distillery at Hanham[112] on the snaking River Avon south-east of Bristol, with three calamine houses, a crucible pottery, charcoal-house, ore-grinding mill, four brass furnaces, a forge and pump-house. He distilled both zinc carbonate (calamine) and zinc sulphide (blende) ores in four distilleries, using vertical, downward-distillation retorts and produced brass by directly mixing zinc metal and zinc oxide with copper. An English chemist, Peter Wolff, wrote a description of Emerson's process to the Italian geologist Giovanni Arduino (1714-1795), inspiring Arduino, in January 1781, to draw an impression of a distillation furnace for calamine ($ZnCO_3$) and blende (ZnS),[113] which suggests that he was considering zinc-distillation and brass-making in the Vicenza or Verona areas of north Italy, where he had studied volcanic alpine minerals.

Emerson's furnace charge for a nine-crucible furnace was about 24.5 kg granulated copper, 12.25 kg granulated metallic zinc, 4.5 kg calcined calamine and 3.5 kg ground charcoal. The metallic zinc was kept separate while a handful of the above ingredients were mixed and placed in the crucible, followed by 1.4 kg of granulated metallic zinc. The crucible was filled with layers of the mixture, with zinc metal sandwiched between. This produced, on average, 37 kg of 'pure fine brass' but calamine was often omitted and only zinc metal used.[114] In 1786 Emerson's described his distilled zinc as white and bright,[115] and he used it to produce high-quality non-magnetic golden-coloured brass, particularly suitable for ships' compasses. Birmingham brass-makers

---

[106] Gloucester Archive: Badminton MSS, D 421 B 1, Oct. 20 and Nov. 26,1767; April 26, 1768
[107] *Farley's Bristol Journal*, March 11, 1769
[108] Rudder 1779: 663
[109] Birmingham Archive: Boulton MS, 3782/12/26/16
[110] Hereford Archaeology report, 1995, section 5, Outline history of Warmley site
[111] Anon, 1998, The history of brass making in the Naugatuck Valley, in *Innovations*, March 1988, online newsletter of the Copper Development Association, www.copper.org/publications/newsletters/innovations/1998/03/naugatuck.html
[112] National Grid Reference ST 635 718
[113] Vaccari 2007: 160; figure 6, quoting Biblioteca Civica, Verona: MS, G.Arduini, b.670, IV.b.9,
[114] *Patent Specification* 1207, 1738
[115] Watson 1786: 45

were now out to defeat Bristol. The sharply-declining main Bristol Company undercut Emerson, whose success threatened them. In 1803, therefore, further undercut by cheaper imports of metallic zinc now produced in Upper Silesia, Emerson's zinc distillation works went bankrupt.[116]

The Upper Silesian breakthrough was the distillation of zinc in horizontal (rather than vertical) retorts, using reverberatory furnaces. The technology had already been used for other substances, for example, in 1656 the German chemist, Johann Glauber, had described the distillation of saltpetre in horizontal retorts.[117] In 1798, Johann Christian Ruhberg, a works inspector, in partnership with Johann Friedrich Böttger (of Meissen porcelain fame), introduced the process. Together, in 1799, they built Europe's first horizontal-retort zinc-distillation plant, which proved more continuous and efficient than Champion's downward-distillation process. Ruhberg's works (Wessola glass works) located at Pless (today's Pszczyna) about 65 kilometres west of Krakow in Upper Silesia (see map, chapter 7), became an efficient, flourishing business.[118] Two other shorter-lived zinc-distillation plants, built in 1799 and 1801 by Bergrath Dillinger at Döllach in the Austrian Möll valley, used Champion's vertical downward-distillation technology,[119] and in 1805 Abbot Daniel Dony would introduce a modified horizontal-retort process in Belgium.[120] Innovations were prompted by the high eighteenth-century demand for brass to satisfy fashion trends and trade to both the Guinea Coast and America.

After 1763, when an Anglo-French Treaty awarded North America to Britain; the port of Liverpool gained a stronger American brass market than Bristol. As the English slave-trade transferred from Bristol to Liverpool, brass-making flourished at three northern brass-mills – Cheadle, Macclesfield, and Holywell. By 1771, Liverpool had more than doubled its exports to Europe, sending 335 tons of brass to West Africa and exporting large quantities of brass to North America and the Caribbean. In 1776 Liverpool exported 226 tons of brass to the West Indies. This expansion due in part to the rise of Birmingham brass-making and partly to the entry of Thomas Williams, the copper mining magnate, into brass-making and the slave trade.[121] In the 1770s, 100 slave-trading ships a year left Liverpool annually for the Guinea Coast but in 1788 only 37 Bristol ships sailed there.[122]

The first English brass works to respond to new Liverpool trade opportunities was the Brookhouses mill at Cheadle, Staffordshire. Thomas Patten, owner of a copper-

---

[116] Eveleigh 1995: 19
[117] Glauber 1656: 62
[118] Jacobs 1889: 349
[119] Hollunder 1888: 372
[120] Ingalls 1903: 6, 306
[121] Harris 2003: 10, 12
[122] Barratt 1789: 189

smelting works at Bank Quay, Warrington, founded the brass mill at Cheadle in 1755 to use his copper by making ingot brass for midlands brass-founders. Cheadle obtained calamine from both Staffordshire and Derbyshire and his crucibles held two parts calamine to one of Derbyshire copper. Crucibles were made from refractory (heat-resistant) Stourbridge clay mixed with local clay and powdered ceramic waste from the Staffordshire potteries. On July 22, 1761, Robsahm, a Swedish diarist, sat sipping a midday beer outside the Wheatsheaf Inn at Cheadle, and learned from an immigrant brass-worker that the current expert brass-maker at Cheadle was the son of a man named Keys, recruited from the Netherlands, one of several skilled workers called Keys now employed there.[123] Initially, Patten's eight Cheadle furnaces produced 125 tons of brass a year, but by shortly after 1800, his 36 furnaces alloyed 600 tons of brass annually, for wire, sheets or three-kilo ingots.[124]

In 1766 Thomas Patten started a new brass battery and wire works at Holywell, Flintshire, North Wales, to produce goods for West Africa – 'black' manillas for currency, 'bright' manillas for arm and leg bracelets, brass neptunes for salt evaporation, and brass pans for gold-panning and 'for various other purposes'.[125] In 1780, Patten bought an additional Cheadle site at Bedbrook Farm,[126] by a tramway from Woodhead coal source. The Cheadle works used granulated copper, a practise becoming adopted throughout Europe (by 1762, for example, Upper Hungarian copper was granulated).[127] By the 1770s, Cheadle added distilled zinc to its brass, to produce the popular warm, golden-coloured pinchbeck (10-13 per cent zinc) and brighter, high-zinc brass for gilding. Brasses which contain more than 20 per cent zinc, have a cooler, yellower appearance.[128]

By 1830, the Cheadle works had abandoned the cementation process entirely and produced various grades of high-zinc sheet brass for making 'pale' wire for pins, containing about 33 per cent zinc, and 'best' Cheadle brass wire, made from 50 per cent zinc and 50 per cent copper, for the webs of paper-making machines, and for strings for pianoforte and other stringed instruments.[129] From the 1770s, pianos, previously rare, were more widely played in boudoirs and drawing-rooms. Berlin forte-piano makers acquired high-zinc brass by the early 19th century, and became celebrated for keyboard instrument-strings. Brass was used for decorative hinges,[130] handles and lid-hooks for harpsichords,[131] and, around 1770-85, for brass piqué decoration

---

[123] Kunglige Biblioteket, Stockholm: MS, M 260, folios 102-103, Robsahm
[124] Aitken 1866: 237
[125] Pennant 1796: 175, 180, 211
[126] National Grid Reference 0145 4330
[127] van Musschenbroek 1762: 537
[128] Craddock 1978: 8, 11-12
[129] Aitken 1866: 237, 239
[130] Oxford University Bate Collection of Musical Instruments: x 982
[131] Law, Dave. 2007: website: traditional-brassware.co.uk

for the French hurdy-gurdy.[132] The French preferred 27%-zinc cementation brass for chandeliers, and other ornamental items but English chandeliers were normally made of shiny 25-30%-zinc brass including metallic zinc. Matthew Boulton gilded his ornamental brass figures with an amalgam of powdered gold and mercury applied onto a solution of mercury nitrate, then heated until the mercury vaporised, leaving the gold.[133]

Other brass mills sprang up in response to Liverpool's takeover of the main trans-Atlantic and Guinea-Coast slave trade from Bristol. In August 1758, Charles Roe, a successful silk-mill owner at Macclesfield, Cheshire, gained permission to enclose 1335 square metres of waste land near Macclesfield Common coal-workings. On this plot, Roe erected a brass works[134] powered by a stream (since blocked by canal-building) descending to the River Bollin below.[135] A windmill crushed calamine from Flintshire and Shropshire, delivered by river and canal as far as Northwich (Cheshire).[136] Roe had to cast around for copper sources. Initially, his men mined copper at nearby Alderley Edge, and then, for ten years, from Ecton mine, Staffordshire.

In 1769, a journalist descending into Ecton copper mine described the 'rattling of wagons, noise of workmen boring beneath your feet' and 'explosions in blasting'. As he climbed down 150 metres 'by ladders, lobs and cross pieces of timber let into the rock', the 'constant blasting of the rocks, ten times louder than the loudest thunder, seems to roll and shake the whole body of the mountain'. Down below, he met the 'glimmering light of candles and nasty suffocating smell of sulphur and gunpowder'. About 60 stoutly-built miners, a 'merry and jovial lot', worked six-hour shifts, naked to the waist.[137] For nine years from 1758, Charles Roe obtained 110 tons of copper annually from Coniston in the Lake District,[138] and in 1763 he took a lease to speculate for copper at Parys Mountain, Anglesey (North Wales).[139] By 1767, local Macclesfield coal ran short[140] and Alderley Edge mines flooded, so Roe transferred his Alderley Edge miners to some great copper seams just discovered at Parys Mine.

Charles Roe had a brother-in-law who owned a plantation in the West Indies,[141] who needed a supply of copper and brass 'neptunes' (flat, broad-rimmed, Guinea Coast salt-pans) for merchants to buy slaves to work his plantation. The brothers-in-law operated at either end of the now-flourishing Liverpool slave trade. In 1768, Roe

[132] Victoria and Albert Museum, London: 95-1870
[133] Peter Northover, personal communication
[134] National Grid reference SJ 922 729 – SJ 924 731
[135] Cheshire and Chester Archive: MSS, DCH FF/13 and LBM 1/3, f. 7
[136] Chalenor 1953: 144
[137] *Gentleman's Magazine* 1769: 60
[138] Chalenor 1953: 141
[139] University College of North Wales, Bangor: MS, Mona 3544
[140] Smith, Mark, 2007: miningmemorabilia.co.uk/AIMC.htm *National Mining Memorabilia Newsletter* 12, 1998
[141] Bentley-Smith 2005: 331

started a copper-smelting works by the River Mersey, sending his refined copper to Macclesfield by canal via Northwich. His cargoes of calamine, copper, windmill-cogs and windmill-sails appear in local canal records, and from 1769 28-kilogram brass neptunes were transported back down the canal from Macclesfield to Liverpool.[142] From his warehouse in Sparling Street, Liverpool, Charles Roe exported many copper and brass pans and manillas for the Guinea Coast. Late eighteenth-century manillas (arm or leg bracelets) and a necklace and bracelet of coiled brass wire survive in Liverpool.[143]

In 1785, Thomas Williams forced Charles Roe out of his lease of the rich Parys-Mountain copper mine, so Roe bought Cronebane copper mines near Avoca, County Wicklow, Ireland, transferring his Anglesey miners over there. Roe and the Cheadle Brass Company shared a calamine-calcining (slow-roasting) plant at Holywell, Flintshire. After Charles Roe's death in October 1787, his son William Roe took over, and in 1788 a rich copper seam was struck at Cronebane.[144] Roe's Liverpool copper-smelting works were sold in 1792, to be replaced by new copper works at Neath Abbey, near Swansea, conveniently close to prolific South-Wales coal fuel sources.

In 1795, Macclesfield brass works brick-kilns and potteries produced white furnace-bricks and deep copper-smelting crucibles, and the calamine-crushing windmill stood close to long low buildings where calamine was 'repeatedly washed in running water' from three large reservoirs fed by springs from the nearby hills.[145] Tall buildings around a large courtyard were used for granulating calamine, refining copper, alloying brass and making 'pan bottoms, brass wire', and quantities of brass nails. In 1801, a factory sale notice mentioned twenty-nine workers' houses with gardens; two thousand crucibles; no less than 78 Cornish granite casting-stones, and 30 tons of prepared calamine.[146] The works covered almost five hectares,[147] 36 times the area acquired in 1758.

Charles Roe established two satellite brass works, the first in 1766 at Bosley, Staffordshire,[148] on the River Dane by a road from the Staffordshire coal-fields. The 7.82-metre fall of water at Bosley's upper mill turned five waterwheels,[149] powering four brass battery mills and a rolling mill. The 4.57-metre lower weir turned one waterwheel, powering copper rolling mills and hammers.[150] Charles Roe's second

---

[142] Cheshire and Chester Archive: MS, LNW 11/2 (1771-1772)
[143] Liverpool Maritime Museum: 2006, liverpoolmuseums.org.uk/maritime/slavery
[144] Smith, Mark, 2007: miningmemorabilia.co.uk/AIMC.htm, *National Mining Memorabilia Newsletter* 12, 1998
[145] Aiken, 1795: 438,
[146] Auction sale notice, 1801, in private hands
[147] Chester and Cheshire Archive: MS, QDE 2/10, 1804
[148] Bonson 1995: 17; National Grid Reference SJ 9141 6460
[149] National Grid Reference SJ 9145 6465
[150] Auction sale notice, printed by Bailey of Macclesfield, in private hands

satellite mill site lay downstream at Havannah,[151] four kilometres north-east of Congleton, Cheshire, where, in 1762, he set up dams, weirs and mills. Before 1770 he had created a broad four-metre-deep weir at Havannah,[152] and brass and copper mills[153], planning the site as a model industrial hamlet, with seven workers' cottages and gardens. When sold in 1801 the mills had five large water wheels to power brass wire-drawing and copper-sheet rolling.[154]

At Holywell, Flintshire, North Wales, Charles Roe of Macclesfield was active alongside Thomas Patten of Cheadle brass works, John Champion from Baptist Mills, and Thomas Williams (1737-1802) of the Parys Mine Company, Anglesey. They all harnessed the force of the Holywell Brook that gushes forth from a legendary pool named St Winefride's Well to descend Greenfield Valley. The water flowed strongly at 18,000 litres a minute at a constant year-round temperature of 7.2°C, descending rapidly (70 metres in 2.4 kilometres),[155] before flattening out across a floodplain to flow into the Dee estuary.

The Holywell site, known as Greenfield or, in Welsh, Maes Glas[156] had direct maritime access to the busy port of Liverpool, which by 1752 already sent 58 slave-trading ships a year to Africa and traded directly to America.[157] From about 1728, Holywell landowner and Cheadle Brass Company shareholder, Thomas Barker, supplied Cheadle with calamine from veins in the limestone hills on his Holywell estate.[158] In 1740-4 Thomas Patten of the Cheadle Brass Company leased two plots of land, including part of the Holywell Brook, Forge Pool, a dam, two waterwheels, a charcoal house and several workers' houses.[159]

In 1758, John, eldest son of Nehemiah Champion of Baptist Mills, Bristol, set up a distillation plant on a former lead works site east of the Holywell Brook, experimenting with zinc sulphide ore to make zinc oxide for brass.[160] Zinc sulphide (ZnS, 'blende' or 'black jack') was, as mentioned earlier, too sulphurous for direct use in the cementation process. John Champion washed the zinc sulphide ore, broke it into hazelnut-sized pieces, and immediately roasted it with charcoal in domed reverberatory furnaces to remove sulphur and produce zinc oxide. He patented a brass-making process[161] using

---

[151] National Grid Reference SJ 8693 6464
[152] Murgatroyd 2003: 169
[153] Staffordshire Archive: MS, estate plan D (W) 1909/A/8/1-10; leases 1909/E/6/8 and 1909/E/21/3;
[154] *Manchester Mercury,* auction sale notice, October 20, 1801
[155] Davies and Williams 1986: 8
[156] National Grid Reference SJ 1907 7692; Scheduled Ancient Monument F 160
[157] Parkinson 1952: 88
[158] Berg and Berg 200: 321
[159] Flintshire Archive: D/MT/233 (1744)
[160] Pennant 1796: 203
[161] Patent Specification 726, 1758

Figure 88. Greenfield Mills, Holywell, Flintshire, north-east Wales

one part of roasted blende (ZnS), one part calcined calamine (ZnCO$_3$) and two parts charcoal, to produce ingots containing 30% zinc.[162]

In 1765, Thomas Patten took on the lease of John Champion's plant, mill pool and weir, where he installed four copper and brass battery mills, each water-wheel driving six battery hammers. He produced his first brass there in August 1766. By 1767, Patten transferred his copper-smelting works to Holywell from Warrington. Thomas Williams, however, empowered by the rich Anglesey copper resources of his Parys Mine Company, could sell copper cheaper than Patten, and forced him out of his Greenfield brass-making enterprise. In 1786 Thomas Williams bought Thomas Patten's Holywell brass works and re-named it the Greenfield Copper and Brass Company. The only remains of Patten's brass mill are a 7.5m-diameter wheel-pit and a large rectangular floor of dressed sandstone blocks.[163]

At his Greenfield mill, Thomas Williams added new foundries and furnaces, channelling water from the six water-wheels back into the Holywell Brook through a network of culverts and covered gullies. He connected battery and annealing pits into the culvert system.[164] In the 1780s, he built 35 spacious workers' cottages, including a terrace called 'Battery Row', overlooking his large mill pool.[165] In 1796, he employed about 50 brass workers who produced brass ingots, plate-brass and wire, using six

[162] Mosselman 1825: 489
[163] Davies and Williams 1986: 28-29
[164] Morgan, D., 1979, Excavations at Greenfield Mills, Holywell, Archaeology in Clwyd 2, 10-11
[165] Davies and Williams 1986: 28

melting-houses with 24 furnaces. Copper granules came from Williams' own copper refineries at St Helens (Lancashire) and Swansea (South Wales); calamine came from Thomas Pennant's local mines, and timber from his woods. The brass produced was used to 'finish goods for Africa, America and most other markets', shipped out from the company's large Liverpool warehouses.[166] Williams was a determined and well-capitalised businessman, never coy about broadcasting that his Holywell brass works supplied the slave trade.

Greenfield Mills, unfortunately, were not as green as their name. Thomas Patten's 1743 lease had forbidden the processing of metal that was 'sulphureous so as to be poisonous or harmful to man or beast…'.[167] However, Samuel Johnson, who visited the Greenfield site in 1774, found the atmosphere unbearable, commenting 'I was less weary, and had better breath, as I walked further'.[168] In 1796, this polluting works, now run by Thomas Williams, was 'a great cluster of vast square chimneys, the discharge of the tremendous volumes of thick black smoke rising from the making of brass'. Workers had to wash their hands before eating, and needed 'careful ablution of the body at least once a week, to rub off the copper dust, which adheres to their bodies, and occasions violent eruptions of a green colour'.[169] At Thomas Williams' copper mine on Anglesey 'not a blade of any sort was live where the Smoke reaches'; and the burning of the Ore has 'destroyed every thing of the vegetable kind within its reach and such is the stench of it as well as its tendency to suffocation that no mortal being can think of living near such works but those who are employed in them.'[170] In the early industrial period, it was hard to foresee the effects of pollutants and mechanical noise. City corporations evidently tackled the problem robustly, but private landowners, for whom the brass-works represented high rent income, clearly faced a dilemma still familiar today – financial profit with pollution or pollution-free with little profit? Thomas Williams had few qualms about pollution.

In 1788, the ambitious Thomas Williams leased Temple Mills on the River Thames at Bisham in Berkshire, where one William Ockendon had manufactured brass and copper since 1759. A map dated 1777 captions Bisham 'neare this place are the Temple Mills noted for making brass kettles and pots'.[171] Thomas Williams used Temple Mills to roll copper sheets for sheathing naval ships and making bolts to secure them, and to hammer copper pans and bottoms for rum distilleries. Although he worked mainly copper at Temple Mills, a brass wire-drawing mill also operated there until the 1790s.[172]

---

[166] Pennant 1796: 206-7
[167] Flintshire Record Office: Holywell MS, D/MT/233
[168] Johnson 1774: 75
[169] Pennant 1796: 209
[170] University College of North Wales, Bangor: MS Mona 3544, folio 17
[171] Bowen and Kitchen, 1777: Berkshire map
[172] Lyson 1806: 199

Figure 89. Turner's Brass House, Birmingham, 1753

Birmingham in the British Midlands, a major customer for Bristol brass, also became Bristol's main brass-making rival. In 1740, the Turner family of Birmingham erected a brass-works at the south end of Coleshill Street, then east of Birmingham[173] (on the site of today's Matthew Boulton College, Aston University),[174] to supply the slave trade and the city's multiplying brass workshops. Turners' Brass House used copper from Wales, and, by the later 18th century it obtained English calamine from Warkworth and Bonsall in Derbyshire. At that time Derbyshire supplied much of Europe, due to restricted trade with the Altenberg (Vieille Montagne) calamine source during the French Revolutionary wars (1792-1802) and Napoleonic wars (1803-15). Brown, yellow and white Derbyshire calamine ($ZnCO_3$) occurred in narrow seams near the surface, often mixed with lead ore. Another company bought and cleaned it, crushed it to the size of nuts and roasted 500-600 kilograms at a time in a reverberatory furnace. The large pieces were separated out and the small ones sieved, dried, ground to a powder and sold.[175]

Turners' Brass House in Birmingham exported manillas in 'immense quantities' to settlements on the Calabar and the Bonny Rivers (Niger delta).[176] In 1754, the factory produced only casting-brass, using Welsh copper. In 1725, only about 400 tons of copper ore had been mined in Britain, but in 1760 output reached 2000 tons, and

---

[173] Bradford, Samuel, 1750; and Hanson, Thomas, 1778 and 1781, plans of Birmingham
[174] National Grid reference SP 0757 8722;
[175] Chaptal 1807: 235-236
[176] Aitken 1866: 273-274

in 1780 peaked at 5000 tons thanks to Thomas Williams' very productive Anglesey mines in North Wales. By 1800, Anglesey copper had flooded the market, so copper prices and output dropped.[177] Turner's Brass House obtained abundant coal from just north-west of Birmingham, and its nine coal-fired furnaces each held nine 35cm-high crucibles charged with 18.6 kg copper, 22.7 kg calamine (with charcoal) and about half a kilogram of scrap brass, which yielded 34 kilograms of brass. One firing lasted ten hours, using just over 150 kilograms of coal per furnace, and yielding about 300 tons of casting-brass annually. At this date, despite the huge effort involved in producing brass, copper was more valuable, selling for 12 pence a pound (£100 a ton), while brass sold for only 10 pence a pound (£83 a ton).[178] For wire-drawing or forging, Birmingham craftsmen bought in higher-quality brass from Bristol, Macclesfield or Cheadle, but local founders cast Turners' brass into everyday objects such as fenders, hinges, door knobs, coat hooks and knockers. The British army and militia turned to Birmingham craftsmen for brass sword-hilts,[179] harness decoration, gun-parts, spurs, saddle terrets and bridle bosses.[180]

From 1769 Birmingham industries gained new vigour when a new canal was opened to the Staffordshire coal-fields.[181] This advantage prompted the Bristol and Cheadle companies to cut prices in an attempt to undermine Turner's Brass House, but the Birmingham founders and forgers vehemently supported their local factory. Finally, however, Turners' Brass House had to amalgamate with the Bristol and Cheadle consortium, rivals for the same Midlands markets, which incensed many Birmingham investors, angry at the price-fixing ways of the Bristol company. Turners' Brass House continued until 1810, when it closed.

Twelve kilometres north-west of Birmingham, Bustleholme Mill,[182] Bescot, Wednesbury, stood on a mill canal off the River Tame. In 1732, a West Bromwich iron dealer bequeathed his two mills (used for iron rods and seed-oil) to his son, Jesson Lowe, who sold oil-cake to a wealthy local family.[183] From 1737-49, Joseph Wood, a prominent local brazier, sold brass kitchen and table ware to the same family.[184] By 1754 he had leased Bustleholme Mill from Lowe and established a brass-works with two furnaces, acquiring copper from Bristol and calamine from Derbyshire.[185]

---

[177] Hornsby 1989: 98
[178] Berg and Berg 2001: 38-39
[179] National Trust, Erddig: ERD.MIL 4 i-lxviii
[180] Colonial Williamsburg Foundation: spurs, 0252 29-a-a; 3-10-16A 1; 3-11-18C; 2.4; bosses, 2.18; 0786; 0185 ANH; 2-4; 2.18 (1761); 0786 ANH; 0185 ANH
[181] Hornsby 1989: 98
[182] National Grid Reference SP 0216 9477
[183] Birmingham Archive: Bustleholme MS, 3145/277/16
[184] Birmingham Archive: Bustleholme MSS, 3145/266/14; 3145/55; 3145/267/20; 3145/269/14; 3145/271/24; 3145/273/7; 3145/278/36
[185] Liverpool University Library: Angerstein MS 7.1 (22), folio 30

In 2007, bluish-purple marks left by hot zinc were still visible on a dressed stone at Bustleholme Mill site.

Late in the 18th century a new Birmingham brass works was planned. Matthew Boulton's advisor, Peter Cappa, campaigned to locate it near the prolific South-Wales coalfields, but Birmingham businessmen refused to sponsor a brass works outside their own city. Public meetings were held, committees appointed and subscription lists opened.[186] By 1781, money was raised for a central Birmingham Brass House, with an imposing neo-classical façade facing Broad Street.[187] It was built next to the new Birmingham-to-Wolverhampton canal, providing easy access to coal, copper and calamine sources previously beyond reach.[188] Birmingham enthusiastically adopted James Watt's steam engine, invented in 1781 'to power mills and other machinery'.[189] Once production began, the price-fixing Bristol and Cheadle cartel had to drop its brass prices by £28 a ton.[190] The Birmingham Brass House continued into the next century but was largely demolished by 1866.[191] A separate, short-lived brass works and foundry started at Smethwick[192] in 1790, with its own wharf on the Birmingham Canal.[193]

Steam power had removed dependence on flowing water power. By 1771, the Trent and Mersey Canal was open through to Stone, nine kilometres north of Stafford. In 1794, local solicitor George Vernon, and brass-maker Richard Keys, laid out Stone brass works on two-thirds of a hectare of land,[194] with 119 metres of canal frontage.[195] The brass works, a complete walled model factory, was equipped with a steam engine. The calamine-crushing mill, however, was driven by water from a stream,[196] canalised into an elevated water-trough to pour into the buckets of an overshot water wheel (much more efficient than an undershot wheel, because its power is increased by gravity). Richard Keys and two younger Keys relatives working at Stone were descended from the refugee family recruited 50 years earlier to Cheadle brass-works.[197] In 1798, the brass works used copper from Swansea, and calamine from Derbyshire and Flintshire. Workers fetched coal by canal from beyond Stoke-on-Trent, propelling their barges through the dark, narrow Harecastle tunnel (then the longest in Britain) by pushing

---

[186] Birmingham Archive: Boulton MSS, 3782/12/25/73, folio 91; 3782/12/26/33, folio 69
[187] National Grid Reference SP 0605 8666
[188] Smith, Piggott, 1828: map of Birmingham
[189] British Patent number 1306
[190] Hutton 1835: 188
[191] Birmingham Sites and Monuments Record, Report number 20517, 06/04/2006
[192] National Grid Reference SP 0204 8890
[193] Warwickshire Museum Field Services report, 2000
[194] National Grid Reference SJ 9223 3260
[195] Staffordshire Archive: MS, D 1406/3, undated typescript
[196] Ordnance Survey 1.2500, Staffordshire sheet 30nw2, 1875-1886
[197] Staffordshire Archive: Vernon MS, D (W) 1826/41 and D (W) 1826/57

against the tunnel sides with their feet.[198] In 1803, George Vernon sold the brass works,[199] which finally closed in 1825.

The brass works produced about 300 tons of brass and 200 tons of brass wire coils annually. Both the wire and copper mills were worked by a 36-horse-power steam engine, and the 1825 Stone sale notice and site plan[200] show that this factory, on its restricted site, had a copper refinery, calamine crushing mill, calcining house and store, charcoal house and mill, large counting houses, workman's cottage, forge and three-stall stable for the barge horses. The manager's brick house had 'two Parlours at the front, Kitchen, Brewhouse, Cellar, comfortable Bed Rooms, and Attics'.[201] A later photograph shows discarded retorts ranged against a fence, their large size implying direct mixing of metallic zinc with copper.[202] Stone brass works succeeded because the highly experienced Keys family knew their processes and markets well enough to plan and equip every aspect of the brass-making process within a limited space.

The British brass industry, with its strong transatlantic trade, finally flourished in the 18th century, being less afflicted by conflict and wars than its counterparts in mainland Europe. In fact, British brass-works profited from supplying brass gun parts, buttons, sword-hilts, ship's equipment and horse accoutrements to armies fighting on fronts overseas, whilst avoiding the threat of destruction themselves. Sweden, likewise, suffered less than mainland Europe, but was still recovering from economic problems and Russian incursions earlier in the century.

From 1756, Bjurfors brass-mill in northern central Sweden was leased by Gustaf Kierman (1702-1766) who had, in his youth, carried out the love-triangle murder of a pin-maker, his employer. At Bjurfors brass mill, Kierman constructed a dam, channel and forge in the same grey stone as the existing furnace houses and wire-drawing workshops. In 1760 the mill operated smoothly and was soundly built. Evidence for substantial stone mill buildings, including a wheel-pit, can still be discovered in the Bjurfors forest and behind the manor, with its tranquil reflective lake. Kierman became a dynamic Stockholm merchant, entrepreneur, politician and director of the Swedish Levant and East India Companies. He persuaded the National Bank to print surplus banknotes, to facilitate borrowing against property, for quick profits. Kierman plunged into this market, producing raging inflation, with dire consequences for his employees and creditors. As punishment, Kierman was fined 60 barrels of gold, lost his high office (and his brass works) and was sentenced to life imprisonment in the notorious Marstrand Castle, where he died. The great Swedish botanist, Linnaeus, had recorded Kierman's earlier murder crime in a notebook, to illustrate his moral Theory

---

[198] Holmes, 2005: 11
[199] Staffordshire Archive: Vernon MS, D (W) 1826/41
[200] *Staffordshire Advertiser*, sale notice, 1 October 1825
[201] William Salt Library, Stafford: plan SC D/1/14, 1825
[202] Staffordshire Archive: photograph C/P/65/6/1/83/23/2

of Just Retribution. Linnaeus would, as the writer Nils Forsgren suggests, have seen Kierman's fate as just retribution indeed.[203]

With Kierman gone, his son-in-law, Johan Didrik Duwall took on Bjurfors mill, visited in 1768 by Crown Prince Gustav III. The prince noted in a letter to his mother that inscriptions on Bjurfors brass snuff boxes reflected growing discontent over the country's current miserable poverty. Sniffing snuff, which derived from American exported tobacco, was popular among both sexes. Skultuna and Bjurfors mills had manufactured brass snuff-boxes from earlier in the 18th century and introduced new decorative punching and stamping techniques after 1760, to make Sweden's boxes more competitive.[204]

Bjurfors mill was subsequently leased by three merchant foundry-owners, who produced brass smoothing-irons, chandeliers, and candlesticks. The mill's expert mould-maker, Carl Norman, invented a brass 'calendar-box'. Swedish merchant businesses commonly advanced money to brass-mills before buying their export products. By leasing out official ownership to merchant-houses who could afford to invest in mill improvements, brass-works shareholders avoided responsibility for supervision and debt. At Bjurfors and Skultuna, both manufacture and buildings improved in the hands of Stockholm merchant houses. A Lübeck merchant, Joachim Wahrendorff, (1726-1803), leaseholder at Bjurfors from 1776, built a manor, foundries and kilns and repaired the workshops. A neat row of workers' timber-built accommodation probably dates from this period. He experimented with cheap zinc ore from Nasafjall, Lapland, rather than from Upper Hungary (Slovakia), but it produced impure, brittle brass.

Bjurfors mill prospered until the French blockade of trade to and from England during the French Revolutionary Wars (1792-1802), which paralysed Sweden's vital export trade and forced up copper prices. This resulted in a dramatic downturn at Bjurfors by the turn of the century. Impoverished Bjurfors workers requested support for old and destitute widows and the children of large families, while the wire-drawers and chandelier-founders cried out for clothing allowances, and the furnace-men called for beer to drink during the long hours spent over heat.[205]

Skultuna, twenty kilometres further south across lakes and forests, remained a major Swedish brass works, producing brass sheet, wire, chandeliers, candlesticks, snuffers, cauldrons, wash-tubs, milk cans, graters, plate candlesticks, needles, tankards, tea- and coffee-pots, censers, plate warmers, warming pans, snuff-boxes, chamber pots, tobacco boxes, bells, saddle fittings, bits, sleigh-bells, whip-shafts and harness fittings. Cutting brass metal with shears was strenuous and labour-intensive, but Browall

---

[203] Forsgren 2010: 126-129
[204] Forsgren 2010: 130-131
[205] Forsgren 2010: 132-136

Matheson, the manager at Skultuna, introduced his friend Sven Rinman's invention to revolutionise the process. In 1770, Sven Rinman, a gifted engineer, invented a mechanical slitter, which proved a major technical advance for the entire Swedish brass industry. This water-powered cutting-machine slit the cast brass plate into nineteen strips, reducing slitting time from six day's work by five men to half a day's work by two – a striking Swedish economic innovation.

Brass works like Skultuna not only worked their own brass into artefacts, but sold it on to Swedish brass craftsmen, who cast brass objects in workshops nation-wide. By about 1780, eleven brass foundries operated in Stockholm alone, and there were many more out in the country. The most skilled brass founders cast items like candelabra and chandeliers from multiple moulds. Swedish braziers, smiths and founders made brass tools, sponge-boxes, savings-boxes, combs, buckles, frames, hooks, needle-cases, time-pieces, wind instruments, drums and mortars.[206] Mortars mattered to households, because food, due to storage problems, often became stale and had to be flavoured with crushed aromatic herbs.

At Skultuna, after Browall Mattheson died, his quiet, educated son took over, but in 1780 the estate went up in flames, burning the mill, workshops and homesteads to ashes. The rebuilt Skultuna mill was among the first to be insured under a 1782 provincial fire insurance scheme. Brass came into full production again, but the naval forays of the French Revolutionary Wars (1792-1802) prevented exports. To avert disaster, Skultuna once more focused on the home market, dramatically increasing its output of domestic goods, especially candlesticks.[207] To make ordering easier for its customers, its diverse range of candlesticks was listed in printed promotional catalogues, by size and type. Brass household wares were now mass-produced in factories more cheaply than individual artisans could make them. A Skultuna pin-factory was built, and the brass works craftsmen started to die-punch decorative relief patterns onto brass sheet,[208] until the mill became the largest Swedish producer of brass-wares.

In 1790, however, a severe fire again largely destroyed the Skultuna factory, then owned by Carl Jacob Adlerwald, who rebuilt it and later replaced the cementation process by direct mixing of zinc metal with copper.[209] A watercolour dated 1795, shows the Skultuna rolling-mill with a water-wheel and pyramidal-shaped furnace chimney.[210] Large eighteenth-century charcoal barns survive today, painted with red iron oxide residues from the Falun mine. Travellers passing though Arboga in west central Sweden (south-west of Skultuna), recorded brass battery mills in operation

---

[206] Erixon 1969: 300
[207] Forsgren 2010: 20-21
[208] Riksarkivet, Stockholm: MS Register 13, folio 77, 14 May 1774
[209] Lundgren, Helga, April 1991: Skultuna bruk och skogen. www.hembygd.sew/index.asp?lev=15124
[210] Skultuna Bruks Museum: J. G. Herstedt 1795, watercolour of Skultuna rolling mill and furnace-house

there in both 1765[211] and 1808,[212] which indicates that brass was still forged at Arboga, even if no longer alloyed there.

From 1761 to 1790, Nyköping brass mill, south of Stockholm, was managed by Carl Indebetou, who raised production to a high level by introducing Sven Rinman's revolutionary slitting tool. From 1798 onwards, Nyköping brass mill cast up to 777 candlesticks a year, as well as coffee-pots, snuff boxes, fenders and spittoons, and in 1799, the new Swedish Post Office ordered 149 brass post-horns from the mill.[213]

From 1752, further south at Gusum brass works, Eric Westerberg manufactured 150 different designs of table knives and forks, including models with brass-decorated handles, However, by 1767 state manufacturing subsidies had ceased, and cheaper foreign tableware came onto the market, so knife production at the divided Westerberg and Spalding factories at Gusum petered out. The Gusum mills remained in divided ownership until the end of the century, but in 1812 the debilitating competition between the two families at last came to an end, and the two mills united,[214] allowing the Gusum brass works to continue until 1988.

North of Gusum, the Norrköping brass works on the powerful Motala river were, by 1750, very extensive, Rebuilt after the marauding Russians had moved on from the Swedish coasts, it was now entered through the imposing Holmens Tower. In 1756, however, as economic conditions worsened, Norrköping mill was sold to a French Huguenot, Jean Henri Lefébure (1708-1767). Through inheritance, marriage, and thrift he had become immensely rich, owning several iron mills and a shipping company. An active politician, on the board of the national bank, the *Riksbank*, he influenced national fiscal policy and motivated the ingenious Sven Rinman to revolutionise brass making at Norrköping. The Swedish Board of Trade sent Rinman to observe European thimble-making machinery, then a closely guarded secret.[215] Rinman bribed his way into factories to learn more, and in 1758-9 he set up a foundry with machinery for producing about 500 thimbles a day (including open-ended thimbles for shoemakers and saddlers), enough for Swedish needs.[216]

In the later 18th century, Norrköping brass-works had sixteen furnaces, calamine works, foundries, a rolling-mill, battery mill, wire-drawing and annealing workshops, and housing for 100 employees. The independent craftsmen (founders and smiths) of Norrköping protested that factory craftsmen used cost-price brass to make wares to hawk unlawfully around the local area. The mill, they claimed, not only priced

---

[211] Fenning and Collyer, 1765, volume 2: 94
[212] Pinkerton, 1810, volume 6: 567
[213] Forsgren 2010: 62
[214] Bondeson 1965: 34, 85
[215] Forsgren 2010: 91
[216] Helmfrid 1954: 123

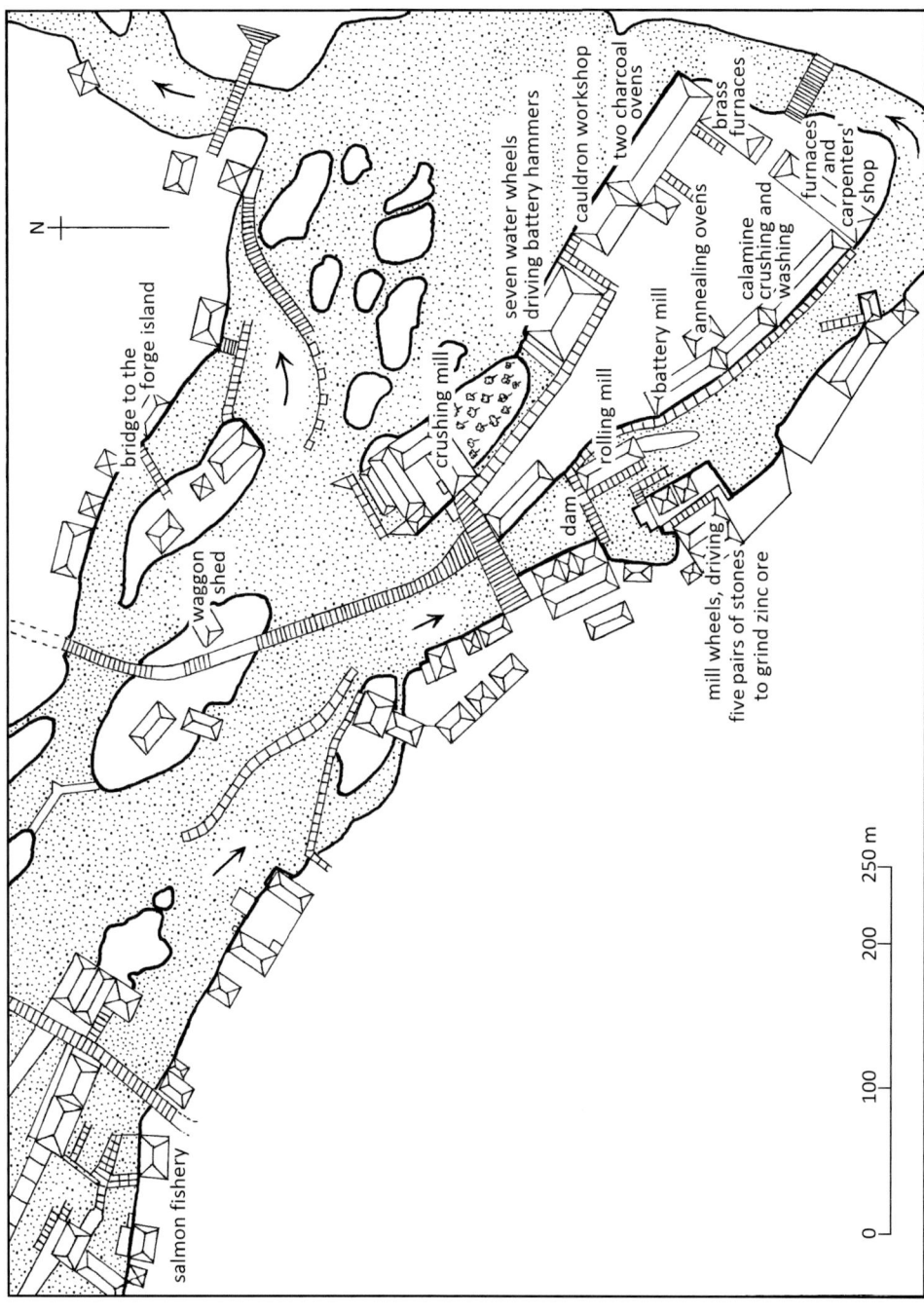

Figure 90. Plan of Norrköping brass works, based on a plan by Pontelius, 1751

its brass higher to independent craftsmen, but even poached their skilled apprentices and journeymen. From 1754 onwards, Norrköping foundries cast brass clocks, watch-cases, snuff-boxes, ornamental flat-irons, lamps, light arms, candlesticks, scissors, harness and carriage furniture. The items were sold in Norrköping shops or through Lefébure's Stockholm merchant business.

Jean Henry Lefébure died in 1767, having artificially controlled the state exchange rate without authority. His son, Jean Lefébure, inherited his father's many debts but nevertheless managed to rescue much of Norrköping mill's assets. In 1772, government officials finally took the side of the independent craftsmen.

In 1772, the Norrköping thimble plant produced over 50,000 thimbles a year, but progress was halted by a major fire in 1784, and a flood in 1788 which left water

Figure 91. Brass keyhole plate cast at Norrköping c.1780

so deep that a 'boat with six oars' could be rowed round the brass mill. The American Revolutionary War (1775-83) and French Revolution (1789-99) had an even more devastating effect on the Swedish brass mills, still over-dependent on exports. Due to hostilities, neither shipping nor overseas markets were available, and Norrköping brass mill could not survive on the sale of thimbles alone.[217] In 1778 Jean Lefébure sold the mill to merchant Elias Pasch, who, from 1781-1790, through all the disasters, successfully continued to sell Norrköping brass to France. By about 1780 Pasch added ornate brass drawer- and keyhole-plates to his output. From 1791-1795 brass output rose to 2327 tons (13,688 *skeppspund*, the largest annual amount recorded for any pre-1800 Swedish brass works), the bulk being shipped to Rouen,[218] but by the end of the century only 177 tons of Norrköping brass were manufactured annually, declining until by 1812 the works was almost idle.[219]

France's major factory at Villedieu-les Poêles, Normandy, used only scrap or ready-made brass, collected and sorted into type, ready to melt down for casting. The main eighteenth-century French slave-trading (and hence brass-trading) port was Nantes,

---

[217] Forsgren 2010: 91-94
[218] Klingström 2009: 157-158
[219] Molbech 1820: 69

on the navigable River Loire, followed by Bordeaux, La Rochelle and Le Havre.[220] A French calamine source existed in 1764 at Ayen, west of the Massif Central.[221] French furniture-makers are famed for their *boulle* brass and turtle-shell or wood marquetry, for example on console tables of *c.*1755-60.[222] An elaborate armoire (1770) by Charles Boulle himself is decorated with brass scrolls, foliage and cascades of flowers,[223] and others copied his technique. Many surfaces, such as gilded furniture-mounts, were overlaid with 'false gilding' – thin plates or finely beaten leaves of brass glued on and then burnished.[224]

Near Namur (then under the Austrian Habsburgs) brass was revived in 1723 by Henry de Bivort, with one or two mills on the Arbre stream (now in French-speaking Belgium), where Jean-Francois Tresoigne had operated two mills since 1700. In 1726, the widows of Bivort and Tresoigne took over management of the brass mills, employing 226 workers by 1738. In 1765 the crucibles of each Namur furnace were charged with 16 kilos of old brass, 16 of copper, and 27 of well-crushed calamine, to which were added 9 to 11 kilos of powdered, damp charcoal, to produce 38 to 39 kilos of brass,[225] giving an estimated zinc content (by weight) of 23%. In 1770 Bivort's children built three more mills and a wire-drawing workshop by the Arbre.

Central European brass works meanwhile suffered disruptive political upheaval and conflict at different places and times. The brass works at Trems near Lübeck, in Holstein, was badly affected by the short Danish-Russian War of 1762, instigated by the insane Peter III of Russia, born Duke of Holstein-Gottorp. He started this conflict during the Seven Years War, to conquer Schleswig for his native Holstein, and it afflicted Trems brass-works, already weakened by competition from the Hohenzollern-owned Hegermühle brass-works (at Finow-Eberswalde, north of Berlin). Nearby Lübeck still had a calamine mill in 1786[226] but Trems brass-works finally foundered when its exports were cut off by the French Napoleonic Wars (1803-15).[227]

Silesia (Czech-Polish border area), on the other hand, suffered during the War of Austrian Succession (1740-48), when the Hohenzollerns fought a localised First Silesian War (1740-42) to gain zinc-rich Upper Silesia, part of the Bohemian kingdom inherited in 1526 by the Austrian Habsburgs. Hohenzollern King Frederick II of Prussia, meanwhile, had his eye on mineral-rich Silesia as a vital zinc ore source for his main brass-mill, the Hegermühle, north of Berlin, and in 1742 he managed to annexe Upper Silesia through the Treaty of Breslau. The new Finow Canal, constructed by Frederick

---

[220] Tibbles, Anthony, 2000: www.liverpoolmuseums.org.uk/ism/resources/slave.trade
[221] Duhamel 1764: 68
[222] Victoria and Albert Museum: W.44.1,2-1947 and W.23&B-1975
[223] Musée du Louvre, Paris: OA5441
[224] Wiegleb, 1789, vol. 2: 578
[225] Diderot and d'Alembert 1758, volume 9: 218
[226] Gunther Meyer, personal communication; Lübeck Archive, map AHL Bauhof 281, 1766
[227] Warncke 1949: 215-18

II from 1743-1746, passed alongside his Hegermühle brass works. Over 32 km long and with 20 locks, the canal improved communications with Stettin and Berlin.[228]

In 1756 the Austrian Habsburgs lost patience and tried to wrest back zinc-rich Silesia, a conflict which triggered the Seven Years' War (1756-63) in which the Prussian Hohenzollerns managed to range Britain and Hanover on their side, while the Habsburgs were backed by France, Russia, Sweden and Saxony. The Hegermühle was pressed into producing brass for Prussia's war against Austria, so in 1756 a large new rectangular, twelve-furnace-house was built. In 2008 the limestone walls of this building were still standing to upper ground floor level, with furnace remains inside.[229] Several workers' cottages also survive. The Brandenburg army used brass produced at the Hegermühle for embellishing its elegant rococo pistols and rifles dating to about 1730-80, which were assembled at Potsdam.[230] In 1779, the mill employed 64 brass workers,[231] and by 1786 it produced brass wire, pans and kettles, tea and coffee jugs, and sheet brass in various thicknesses for use in making taps, stop-cocks, military drums, French horns, hinges, locks and door furniture.[232] Hegermühle brass wire became celebrated for strings for musical instruments, including harps and harpsichords, which suggests that the factory was adding zinc metal to its brass to ensure hardness and strength. In 1786, the brass works was taken over by the Royal Berlin Mining and Smelting Company who transferred some of the work to Potsdam. By 1839, rolling mills and other machinery operated at the Hegermühle,[233] which remained operational until 1945.

After Prussia had seized Silesia in 1742, the Jakobswalde (Kotlarnia) brass-works expanded under the Hohenzollerns, gradually gaining a calamine-crushing mill, four new brass furnaces, a wire works, a further battery hammer and a spoon works, but within 20 years Silesia was once again a battleground between the Hohenzollern Prussians (who won) and the Austrian Habsburgs. After 1782, the new proprietor of Jakobswalde brass-works, the Prussian duke Friedrich Ludwig of Hohenlohen-Ingelfingen erected a works office to oversee all his brass and iron businesses in the Jakobswalde area.[234]

At Iserlohn, the Prussian satellite possession in the Ruhr district, Johann Caspar Lecke established a brass consortium in 1749 to fund calamine mining, brass making and the production of marketable brass articles. In 1751, the Iserlohn town authority leased land to the consortium at Grüne bei Iserlohn, a deep, wooded valley near the town,

---

[228] Aurich, H., 1906. *Die Industrie am Finow Kanal*, booklet, Eberswalde, 15
[229] Herr Roland Gabsch personal communication
[230] National Trust, Dyrham Park: DHM W 1363; DHM W 57/246, /248; DHM W 59/306
[231] Woiwode 1996: 24
[232] Fischbach 1786: 80-83
[233] Ure 1839: 169
[234] Piersig, 2011: http://de wikipedia.org/wiki/Jakobswalde, 29-31

where a brass-mill and calamine-crushing works were set up the following year. A calamine-washing unit was built on a hillside stream on the western slopes of the Brandkopf, a hill rising steeply from the valley. Crushed calamine was roasted with charcoal in circular calcining mounds. When built up, each mound weighed over six and a half tons, and calamine was roasted for up to three days to steam out impurities like sulphur. The heat was then raised further to burn off arsenic, cooled and roasted again. For brass-making, eight crucibles were charged with a weight ratio of 40 copper to 60 calamine, and 25 charcoal. From sheet brass produced in the Grüne valley, the town's craftsmen drew fine pin wire, and cast buckles and clasps. Friedrich Wilhelm II of Prussia protected Iserlohn brass production by forbidding it any import of brass goods or export of scrap brass. From 1755, rolling mills were installed,[235] and an eighteenth-century plan shows the increasingly convoluted extent of the calamine mine tunnels beneath Iserlohn.[236] In 1756, a brass battery works with six trip-hammers was built at Hemer, three kilometres east of Iserlohn, to produce hollow vessels like pans and cauldrons, and finer brass sheet suitable for snuff boxes.[237] Iserlohn, as a Prussian territory, once again supplied Prussia with brass military harness fittings.

From 1760, close to Iserlohn, at Hemer and Sundwig, Johann Lobbecke, with two brothers from Utrecht, produced brass thimbles and from 1790 they made brass pins. By 1798, merchants were exporting brass worth 7000 talers annually from the factory at Iserlohn, sending it to Spain, Brabant and France.[238] Iserlohn craftsmen produced huge quantities of articles for fashion and for the slave trade, besides domestic cauldrons, candlesticks, chandeliers, and hinges, locks and knobs for furniture.[239]

Brass snuff or tobacco tins produced in Iserlohn were individually hand-engraved or hand-embossed until 1754, when a mechanical stamping technique was introduced to speed up production. The designs embraced biblical, hunting and social themes, town views, shipping and battles. One Iserlohn brass tobacco box, depicting the Battle of Breslau (1757), bore an inscription praising Friedrich the Great of Prussia (1712-1786), himself an enthusiast for these boxes.[240] The last Iserlohn stamped tobacco box dates from 1777.[241] The Dutch were major tobacco importers, and Dutch pocket or table-top tobacco-boxes often had brass lids.[242] Besides tobacco, sugar was a major eighteenth-century import from America, which also gave rise to new brass equipment, including

---

[235] Klostermann 1996: 35-38
[236] Klostermann, Rolf, 16 October 2013: Historische Einordnung', unpublished presentation at Iserlohn Museum, 10
[237] Hildenbrand 1983: 51
[238] Schulte 1938: 151
[239] Hildenbrand 1983: 38
[240] Victoria and Albert Museum: IS 289-1897
[241] Iserlohn.de/kultur/messing
[242] Noel-Hume 1969: 311

crystallising vessels for making sugar-candy, sugar tongs,[243] and brass sugar-cutters for slicing loaf sugar.[244]

Like Iserlohn, Stolberg was now in a Prussian satellite territory, but its already wealthy brass-makers did not adopt all the new inventions used elsewhere. In 1773, Stolberg imported Norwegian copper by sea from Røraas mines, through the port of Trondheim, as well as copper from Chile, Peru,[245] Upper Hungary, Sweden, the Levant and Cornwall. Stolberg's calamine still came from the Altenberg (Vieille Montagne) and some from Cornelimünster a few kilometres distant.[246] Stolberg remained a leading European brass-producing centre until the third quarter of the 18th century, after which England and Scandinavia took over much of its former European brass market.[247] The Prussian brass works were in northern and eastern Europe, whereas the Habsburgs' brass works were mainly in central southern Europe.

During the Seven-Years War (1756-63), the brass works of southern Saxony were affected by the constant movement of Prussian troops back and forth over the Saxony-Bohemia border (the ridge of the neighbouring Erzgebirge, Krušné Hory or Ore Mountains). The Prussians held and occupied Saxony throughout the war, causing havoc in the foothill settlements, including Rodewisch.[248] By 1759 the Prussian demand for military brass spurred the Rodewisch brass works to peak production, using refined copper from Mansfeld. Around 1762, however, Mansfeld copper deliveries were disrupted by hostilities, so brass production declined, reaching a standstill for several months. Under Prussian rule, the Saxon currency lost value, but workers, who faced severe hardship, managed to negotiate 'holiday pay' during the shut-down. They even held a spontaneous strike, resulting in a preferential pay-scale for wire-drawers over other workers. In 1764, after the Seven-Years War, the Mansfeld copper supply was gradually re-instated.[249]

In Saxony, many eighteenth-century brass musical-instrument makers were centred in Markneukirchen. They were descendants of seventeenth-century religious refugees from Krasliče, Bohemia, who had fled the Habsburg persecution of Bohemian Protestants. Markneukirchen was only 15 kilometres from Štřibrná brass works in Bohemia, and its nearby Krasliče copper refinery.[250] In 1764, Štřibrná brass works was alloying cementation brass, and operating battery hammers and wire-drawing equipment.[251] It obtained calamine from Nuremberg. Poland and England,

[243] Schulte 1938: 89
[244] Museum of English Rural Life, Reading: 51/543.6.3
[245] Schleicher 1974: 36
[246] Chaptal 1807: 286
[247] Roderburg 1927: 15
[248] Gericke 2008: 127
[249] Gericke 2008: 132
[250] Hachenberg 1998: 119-120
[251] Galon 1764: 55

and received casting stones from further east along the Krušné Hory (Erzgebirge) mountain chain. In its later years it obtained copper from both Mansfeld (Germany) and Loket (Bohemia).[252] Although Stříbrná was so close to Markneukirchen, across the border, a Saxon 'privilege' system forced the instrument-makers to use brass from Saxony, which meant from Rodewisch 45 kilometres away. The musical-instrument makers preferred Stříbrná brass, because Rodewisch brass (evidently higher in lead and other impurities) cracked during hammering and annealing, and weighed more. Stříbrná brass yielded five or six extra sheets of brass per hundredweight, enough for four extra pairs of musical horns.

The instrument makers therefore protested vociferously at having to buy Saxon brass, or else pay exorbitant taxes for nearby Stříbrná brass from their ancestral Bohemia. The Markneukirchen instrument-makers' outcry was supported by colleagues in Dresden[253] (also in Saxony). In 1786, Stříbrná still produced brass wire,[254] but by 1795 when the instrument-makers at last won the right to purchase Stříbrná brass, it was too late, because the local Eibenberg copper mine, which supplied Krasliče, was almost spent, and by 1792 the Stříbrná brass works virtually, if not totally, ceased production. A new wire-drawing workshop, erected in 1797, failed to meet the rising demand throughout Saxony for fine brass wire for pins for the Krušné Hory lace-makers and also for textile workers, a need felt as far as Leipzig, 110 kilometres further north.[255]

Brass musical instruments proliferated in later eighteenth-century Europe. The 1612 Nuremberg bass trombone was described earlier and, by about 1750, orchestral horns (as in Markneukirchen) were made in Nuremberg, popularised by Handel, who wrote scores for them. In the later 18th century, Nuremburg trumpet-maker Georg Heinrich Rodenbostel worked in London, sharing a workshop with horn-maker Friedrich Hofmeister. Mozart also wrote scores for basset horns, which have brass key-work and bells.[256] Few hunting-horns appeared in England before the late 18th century, but German hunting-horns were of the brass half-moon bugle type.[257] Brass, being a harder, stronger alloy than bronze, gave musical instruments better acoustic resonance, and brass sheet metal is easier to work into instruments than bronze.

The Nuremberg brass works, outside the city at Hammer, Laufamholz, was acquired in 1772 by Nuremberg merchant Johann Volkamer who built himself a large timber house there.[258] In 1793, his son-in-law Georg Christoph Forster took over the factory, supported by generous Volkamer donations and bequests.[259] A view, dated about

---

[252] Jaroslav Zaplatal, personal communication
[253] Hachenberg 1998: 124
[254] Treixler 1929: 122
[255] Gericke 2008: 149
[256] Oxford University Bate Collection of Musical Instruments: x 605, x 72; x 606, x 400, x 486-7, x 489
[257] National Trust: CHS/M/112
[258] Stadtarchiv Nurnberg: MS, 455/1, March 1772, Hammer, Laufamholz
[259] Wittek 1984: 124

1800,[260] shows the mill in very rural surroundings, but by 1820 Hammer, Laufamholz, had 140 inhabitants and 37 dwelling-houses, and was the largest industrial site in the Nuremberg area.[261] Originating from the early 17th century, Hammer mill workers had a social relief programme including free housing for ill or disabled workers, widows and orphans.[262] New rolling mills would be installed from 1804, when a *Lahngoldschmiedhaus* (a foil forge) stood by the river, for rolling or beating out fine brass foil, which could only have been achieved with high-zinc brass. By 1834 even finer sheets of brass foil were traded from Laufamholz to Calcutta (Kolkata) and Bombay (Mumbai) to cover temple and mosque domes.[263]

West of Nuremberg, in the independent state of Salzburg, competition from Habsburg competitors was growing too strong for Oberalm brass-works, which closed in 1805. The sister factory at Ebenau held out until 1844, well after Salzburg fell into Habsburg hands in 1816. Production also continued at the Habsburg brass works at Rosenheim, Upper Bavaria, which by 1803 was hammering out large brass beer pans for brewing.[264]

In 1764, at Achenrain, near Brixlegg in the Austrian Tyrol – another Habsburg territory – local calamine and copper were alloyed for brass to be cast into buckles and bells at Schwaz, further up the Inn valley.[265] Achenrain's two reverberatory furnaces each held 12 crucibles (more than the usual nine) and a new cementation process started every 12 hours. Wood fuel was fed from below onto an iron grill above an ash-collecting tray, and the flame rose through a square hole to be deflected and circulated around the crucibles. The oval outer furnace was almost 3 metres (9 *pieds*) long, and the inner hearth, containing crucibles, was 1.5 metres (5 *pieds*) long. Five draught holes in the furnace fanned the flames. It was raised a metre above ground and surmounted by a metre-high vault. Brick-lined reverberatory furnaces like this were said to last for 60 years without repair. The first brass cementation process took place in the larger Achenrain furnace, while a smaller furnace refined two crucibles of the original brass by a further cementation process. Achenrain produced about 85-122 tons (1,750 *quintaux*) of good brass annually.[266] Molten brass was poured onto a metal slab (not granite as elsewhere), coated with a clay slip. The cast sheet was cut with shears into 31 narrow brass rods for wire-drawing, each weighing about two kilos (*quatre livres ½*).[267]

---

[260] Annart, F. A. *c.*1800, copper engraving, 'Prospect des unweit Lauffenholz gelegnen guteingerichteten Schmelz und Hamerwerkes 1½ Stunde von Nürnberg entfernt',
[261] Hussennether 2010: 5
[262] Wittek 1984: 122-124
[263] nuernberginfos.de/muehlen-nuernberg/industriegut-hammer.html (2011)
[264] Priesner 1997: 83
[265] Galon, 1764: 68
[266] 1 quintal (centner) may be calculated assuming 112 pounds to a centner and 17.86 centners to a ton – at that period in Austria. In reality, the weight varied by area, date and context. Galon wrote that, in this area, 1 *quintal* = 140 *livres*
[267] Galon 1764: 94

Deeper into Habsburg territory, the alpine brass works at Lienz, southern Tyrol, employed in its heyday (1760-90) over 80 factory workers, charcoal-burners and woodmen. In 1775 it had a calamine mill, crushing, polishing, etching and scraping workshops, office, battery works, furnaces, foundry, crucible pottery, stabling, workers' cottages, apartments and bath-house. In 1786 over 600 tons of brass was exported from Lienz to Italy, Switzerland, France, Germany and Turkey, necessitating the purchase of more forests for timber.[268] In 1787, Lienz brass works was buying copper from many parts of Austria, and produced almost 109 tons of brass, of which about 100 tons were exported.[269] In 1797 many Lienz brass-workers were called up for military service in the Napoleonic Wars and in 1798 a great fire for a second time reduced the brass-works to dust and ashes.[270] Before the wars, Lienz lay on the trade route from Trieste to Germany but after the wars trade was re-routed elsewhere, so the brass works remained closed.

Around 1700, the Möllbrücke brass mill, sixty kilometres east of Lienz on the River Möll, a tributary of the Drau in Carinthia, was bought by Baron Sternbach, a local business proprietor. The forested, mountainous location of Möllbrücke mill meant that it suffered from physical disasters more than human conflict. In 1756 it burned to the ground, and brass production declined after rebuilding, so in 1762 the state agreed to purchase it on behalf of the current Habsburg empress, Maria Theresa. She and her family entourage visited the works in 1765, just after the quality of Möllbrücke brass had been judged too poor to sell in Venice. In 1769 the empress decreed the disposal of all state brass works and factories, including Möllbrücke, but it was largely swept away by floodwater in 1771, leaving too little for the state to auction, so in 1775 it was officially closed down.[271]

By 1760, Reichraming brass works (230 km further north), owned by the great Seitenstetten Abbey, now made enough profit for the abbey to purchase art treasures. Two oil paintings dated 1763 show, respectively, the façade of the brass-works buildings and a cut-away view of the interiors, demonstrating the working procedures – furnace, foundry-work, battery-work and wire-drawing, followed by etching, cleansing and preparation for dispatch.[272] By the later 18th century, cutlery styles were plainer, so demand for Reichraming brass rosettes for knife-handle sheaths reduced. Copper supplies from nearby Radmer were declining in quality, so the Abbot of Seitenstetten leased Kalwang copper mines and refinery from another monastery. Furnaces in the forested Reichraming mountain valley posed a fire hazard, so one brass fire engine and three wooden ones stood ready. In 1782, when religious freedom was granted,

---

[268] Kofler 1962: 379
[269] Heinricher 2006: 9-10
[270] Heinricher 2006: 10.
[271] Ucik 2006: 177
[272] Unknown artist, two oil paintings of Reichraming brass works, 1763 (150 cm x 50 cm), Seitenstetten Abbey

a school was built for workers' children. By 1828, Reichraming mixed distilled zinc directly with copper to produce brass.²⁷³ Further south, (40 kilometres south of Graz, near today's Austria-Slovenia border) lay the rival Frauenthal brass works, a stone-built three-storey mill on the River Lassnitz. In 1791 it had a rolling mill, possibly Austria's first.²⁷⁴ Many European brass mills were by now either in decline for reasons of disrupted trade, natural disaster or war, or still in production having converted to using distilled zinc.

Brass trade across the Atlantic would soon be under threat too. In the 1760s, Britain attempted to regain money lost in conflict by passing a Stamp Act requiring Americans to pay taxes directly to England, simultaneously favouring British interests by imposing or withdrawing duties on certain imports. Over the next decade, these laws provoked American indignation and political unrest to the point of rebellion, so militia units formed ready for the revolution that broke out in 1775.²⁷⁵ France and Spain sided with America, while Britain enlisted four thousand German auxiliaries, including a contingent from Hesse, wearing tall mitre caps with large brass front plates,²⁷⁶ probably made at Bettenhausen. In April 1775, before the Revolutionary War, American hero Paul Revere famously rode from Boston to Concord with news of the approaching English. Already an expert goldsmith and engraver,²⁷⁷ Paul Revere supplied brass trimmings for the American Light Infantry,²⁷⁸ and in 1797 brass fittings, cogs and sheaves for the frigate *Constitution*.²⁷⁹ American Independence, declared in 1776, was just one historic reason for European brass trading faltering by the early 19th century. In 1783, with the Treaty of Paris, European brass makers lost their profitable American market, so the tide was turning. Brass furnishings excavated from Colonial Williamsburg, Virginia, include drawer-handle plates of 1785-1800 bearing American eagles.²⁸⁰

A second reason was the rising anti slave-trade movement of the later 18th century, which culminated in the 1807 Abolition of the Slave Trade Act, and the end of Caribbean slave trafficking in 1811 – largely halting brass trade to the Guinea Coast. Thirdly, the French Revolutionary Wars introduced social change, which further restricted markets by altering personal attitudes to fashion, finery and adornment. In the west, brass was soon to be made in fewer, more mechanised, factories, with zinc distilled in horizontal retorts, but it would also meet competition from newer alloys.

---

[273] Brunnthaler 2000: 119-121
[274] Allgemeine Encyclopädie für Kaufleute und Fabrikanten, 1838, Leipzig: 579
[275] www.bl.uk/onlinegallery/features/americanrevolution/timeline.html
[276] Keagle 2009: 36-37 (average zinc content, 28% wt)
[277] Goss, 1891: 544
[278] Massachusetts Historical Society: Revere Family Papers, Wastebook 2, Boston 1783-1797
[279] Massachusetts Historical Society, special collections: MSS, Revere, 280c 1797
[280] Noel-Hume 1969: 230-231

Evidence for brass varies over time, relying, before written accounts start to emerge, on relatively rare high-status archaeological finds. From the first millennium AD, impressive figurines, statues and temple, mosque or church equipment are gradually joined by secular items such as ewers, cauldrons and basins. During succeeding centuries, brass is increasingly used for personal adornment, as buckles, buttons, bangles or sword hilts, or accompaniments to high living, such as pandans, hookahs, astrolabes, clocks or horse trappings. By the 17th century, metallic zinc is more widely known and used in the West, and there is more evidence for the influence of industrial rivalries, the slave trade and war on brass production. By the 18th century, more is known about individual users and makers of brass, and their quest for wealth through improved technology and industrial-scale production.

The centres of production shifted with the centuries, moving from the prehistoric Middle East towards the Mediterranean and the Roman provinces, then reaching a high point in the Himalayan, Persian and Indian regions. In the medieval period, the focus was in north-western Europe and Byzantium, but by the 17th century China, India, Sweden and the central southern European states were active in making and working brass, whereas in the 18th century, Britain and Stolberg rose to peak production.

Brass, this glistening golden-coloured man-made alloy, took time, trouble and money to produce, and the resulting pollution and forest devastation were all too obvious. Clearly, the similarity in appearance between brass and gold had much to do with its popularity from prehistory onwards – it was so often used to influence or impress others. Though brass often appeared in sacred contexts, it gradually became popular for everyday utilitarian purposes. Indeed, in the days before stainless steel and plastics, brass had practical advantages, being attractive, hard-wearing, malleable, ductile, corrosion-resistant, non-magnetic, spark-free and slightly germicidal – a combination of qualities un-paralleled in one alloy before the 19th century. Brass was therefore used in many ways, and was traded widely even, sadly, in exchange for slaves. The brass-makers and their managers ranged from monks to emperors, and brass drove them to emotions ranging from devotion to greed. Brass production was deeply influenced by war, economics, politics, religion, fashion, weather extremes and availability of raw materials, and the initial input to any brass industry invariably came from expert outsiders, often refugees from conflict or persecution. This under-reported alloy has certainly had an unusual and volatile history, not least because, rather than being dug from the ground, people had to make it, and then strive to invent better ways of making it. It is this essential human input that makes its history so rich and varied.

Twisted-stem brass candlestick cast at Skultuna brass works c.1700, photo © author, courtesy of Skultuna Bruks Museum, Sweden.

Glossary

alloy	a deliberately-made metal compound containing at least one metallic element, usually created to achieve a desired property or set of properties
anneal	re-heat a piece of metal object to alter its microstructure. In the context of this work, it generally means to restore the ductility of cold-worked metal through recrystallisation but it could also mean, for example, the homogenisation of a piece of metal
battery	use of trip hammers driven by water power, or later by steam, to beat out metal into flat or hollow shapes
brass	an alloy of copper and zinc
bronze	an alloy of copper and tin
calamine	zinc carbonate, Zn CO3, ore
calcining	roasting ore slowly in the presence of carbon (charcoal), often for days, to rid it of unwanted elements such as sulphur
casting	producing blanks or shaped metal objects by pouring molten metal into a prepared mould, in which it cools and solidifies in a form dictated by the mould
cementation	here, a method of brass-making, during which the zinc vapour formed during the reduction of its ore is diffused at high temperature into solid copper, to form the alloy
crucible	a refractory vessel, for melting metals and alloys
distil	remove unwanted impurities from a metal such as zinc by heating it in a retort until the metal turns to vapour, which rises to the top of the retort, is collected, and solidifies as pure metal
distillation *per ascensum*	zinc vapour produced by distillation is collected as it cools to a liquid in saucer-like pockets at the top of the retort, to be collected from there, or run off down a tube into a collection-vessel
distillation *per descensum*	zinc vapour produced by distillation is collected as it cools to a liquid at the top of the retort, and descends through a tube inserted vertically through the retort. The tube leads out through the base of the retort, allowing the molten zinc to flow out, but only after the bung has been charred away by furnace heat
foundry	a workshop or building where metal is melted and then cast in moulds to produce a variety of artefacts
gunmetal	a variable mixed alloy containing about 88% copper, 8-10% tin and 2-4% zinc. Leaded gunmetal might contain up to 7% lead

latten, lattin	derived probably from *alaton* (Arabic), *laiton* (French) or *laton* (Spanish) meaning brass, used in the early medieval period for a mixed copper alloy with significant zinc content. Blair and Blair (1991, 84) conclude that the term suggests the deliberate inclusion of zinc in its composition.
leaded brass	brass containing an addition of lead to improve casting properties
liquation	the separation of silver from copper in lead-rich ores by heating the ore to above the melting point of lead, whilst the copper (having a higher melting point) remained solid. Silver, which has an affinity for lead, poured off with it, leaving only copper.
lost-wax	a term used to describe a method of casting a detailed metal image based on a wax model. There are two basic approaches, the direct, where model is formed wholly of wax or is modelled around a core. The indirect method is to form the wax in sections inside a mould and then assemble the wax and fill the hollow with core material. In both cases the model is then sprued and gated, then invested with clay or plaster, dried/fired and then the wax melted out ('lost') and molten metal poured in. Where a core is used, core pins have to be used to maintain the separation between mould and core. The method is also often referred to by its French name, *cire perdu*
ormolu	'or moulu', a bright alloy, around 10% zinc, used for gilding, to give a gold-like appearance at lower cost than gold.
paktong	a ternary (three-part) nickel brass alloy, containing about 40-65% copper, 20-50% zinc and 5-20% nickel, which is equivalent to modern nickel brasses and nickel silvers (also known as German silver). 'Paktong' derives from Chinese words meaning 'white copper'.
plastic deformation	permanent deformation of the brass, in a particular direction or plane, as a result of its forging.
Princes' metal	cementation brass, about 75% copper to 25% zinc. It included small but varying amounts of added distilled zinc, and was bright yellow in colour. It was named after Prince Rupert of the Rhine.
quaternary	a word describing, an alloy containing considerable percentages of major elements. In the context of this work the elements will be copper, zinc, lead and tin.
reduction	the use of carbon – in the form of carbon monoxide derived from heated charcoal or coal – to remove oxygen
refined	removing unwanted elements (such as sulphur) from brass that has already been alloyed. Refining involved re-melting, usually to a lower temperature than before. The process might be repeated many times in order to achieve a purer brass, often adding calamine (or metallic zinc) to replace zinc that had vaporised off.

reverberatory	a word describing a furnace with an outer domed cavity, to carry hot smoke and gases around the exterior of an inner chamber, which contained charged brass objects to be annealed, or crucibles containing material to be heated.
sublimation	transition direct from the solid to the gas phase, without passing through the liquid phase. Most elements have three distinct temperatures at which they become solid, liquid or gas, but a few, including zinc, may readily pass direct from solid to gas
ternary alloy	an alloy containing considerable percentages of three elements, at least one of which is a metal
volatile	easily evaporated at normal temperatures, so liable to change readily from solid to gas

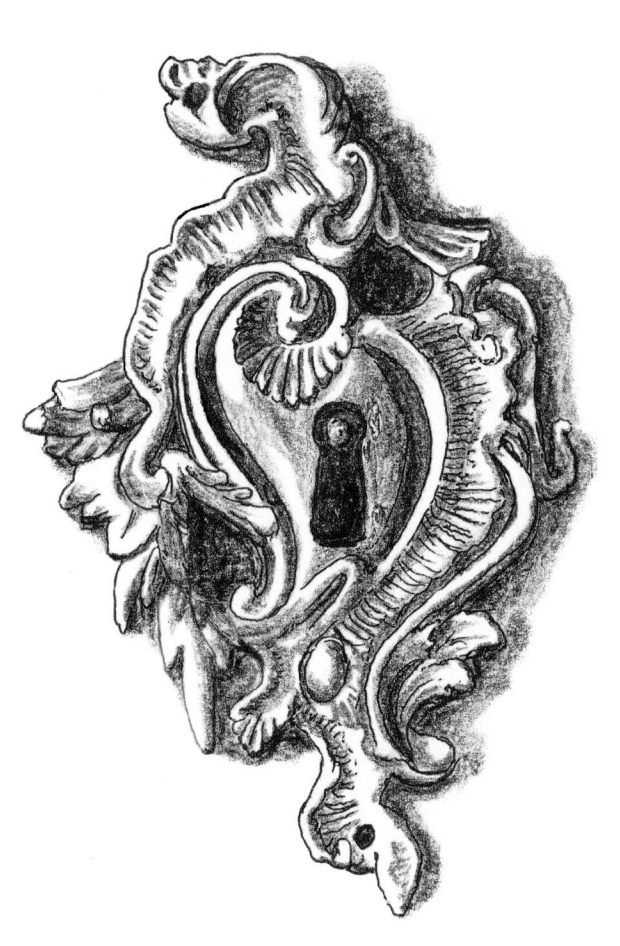

Bibliography

Abesadze, Tsisana 1969 (Georgian edition). *Metal production in the Transcaucasus during the third millennium BC*, (trans Gilmour), Tbilisi, Georgian National Museum

Ade Ajayi, J. F. and M. Crowder 1985. *Historical Atlas of Africa*, Harlow, Longmans.

Aiken, John 1795. *Description of the Country from 30-40 Miles round Manchester*, London, Stockdale.

Aitchison, L. 1960. *A History of Metals*, vols. 1 and 2, London, Macdonald and Evans.

Aitken, W.C. 1866. Brass and Brass Manufactures, in Samuel Timmins, *The Resources, Products and Industrial History of Birmingham*, London, Hardwicke.

Alcock, Anthony 1998. *A Short History of Europe*, London, Macmillan.

Alcuin of York AD 804. Propositiones Alcuini Doctoris Caroli Magni Imperatoris ad Acuandes Juvenes Opositis de disco pensante Book XXX, Problem 7. Translation in Singmaster, David and John Hadley, (1995). *Problems to sharpen the young*, London, South Bank University.

Allan, James 1979. *Persian Metal Technology, 700-1300 AD*, London, Ithaca Press.

Allan, James 1999. *Islamic metalwork: The Nuhad es-Said Collection,* London, Philip Wilson.

Allan, James 2002. *Metalwork Treasures from the Islamic Courts*, Doha, Museum of Islamic Art

Allison, Warren, Samuel Murphy and Richard Smith 2010. An early railway in the German mines of Caldbeck, in Grahame Boyes, *Early Railways 4, Papers from the fourth International Early Railways Conference*, Sudbury, Six Martlets

Alpers E.A. 1976. Gujarat and the trade of East Africa, c.1500-1800, *International Journal of African Historical Studies*, 9/1, 1-45

Amin, Muhamad and Abu Usaman Al-Arabi bin Razduq 2010. website, islamhouse.com; translation of Al-Nawawi, Abu Zakania, (c. 1260-70). *Riad-us-Saliheen,* chapter 116, paragraphs 775 and 777.

Amman, Gert 1990. *Silber, Erz und weisses Gold; Bergbau in Tirol.* Exhibition catalogue, Tiroler Landesmuseum Ferdinandeum, Innsbruck, 20 May - 28 October, 1990.

Anthony, David 2009, The Sintasha Genesis: the roles of climate change, weather and long-distance trade, in Bryan Hanks and Kathryn Linduff, *Social Complexity in Prehistoric Eurasia: monuments, metals and mobility*, Cambridge University Press, 47-73.

Aschauer, Josef 1953. Das Messingwerk Reichraming: ein Beitrag zur oberösterreichischen Wirtschaftsgeschichte. *Oberösterreichische Heimatblätter* 7, 313-326

Asmus, Bastian 2018. The Harz Mountains and some thoughts on the copper trade, in Nicolas Thomas and Pete Dandridge, *Cuivres, bronzes et laitons mediévaux,* Agence Wallone du Patrimoine, Études et Documents, Archéologie 39, 25-35

Arnold, Denis 1983. *The new Oxford Companion to Music*, Volume 1, Oxford University Press

Atkins, J. 1737. *A Voyage to Guinea, Brazil and the West Indies in his Majesty's ships the Swallow and the Weymouth*, London, Ward and Chandler

Aubrey, John, 1671, (1975 edition). *Natural History and Antiquities of Surrey* 1, containing John Ogilby, map of Surrey, Dorking: Surrey Archaeological Society, 14

Aubrey, John 1718. *The Natural History and Antiquities of Surrey: a Perambulation of Surrey begun 1673, ended 1692*, vol. 1, London, Curll; containing 'Mr Evelyn's letter to Mr Aubrey' (1676) between pages xlviii and 1

Aurich, Hermann 1906. Messingwerk, in *Die Industrie dem Finow Kanal*, Eberswalde, 117-130

Backowski, Krzysztof 2002. Handelsbezieungen zwischen Krakau und Oberungarn (der Slowakei) im 15 und 16 Jahrhundert (c.1471-1526). In Rainer Goimmel and Markus Denzel, 2002. *Weltwirtschaft und Wirtschaftsordnung: Festschrift fur Jürgen Schneider zum 65 Geburtstag*, Stuttgart, Frans Steiner, 15-24

Baber, Zaheer 1996. *The Science of Empire, Scientific Knowledge, Civilization and Colonial Rule in India*, State University of New York Press

Balon, Joseph 1863. Études franques, part 1, Namur, Godenne

Banks, Sir Joseph 1898. Journal of an Excursion to Eastbury and Bristol in May and June 1767, *Proceedings of the Bristol Naturalists Society* (1898) 9/1, 22

Barba, Alvaro Alonso 1640 (1817 edition). *El Arte de los Metales*, Book 4, Lima, los Huérfamos,

Barbot, Jean 1732. A Description of the Coasts of North and South Guinea, vol. 5, translation in A. Churchill 1746. *Collection of Voyages and Travels*, third edition, London, Lintot and Osborn

Barker, T.C. 1966. *Pilkington Brothers and the Glass Industry*, London, Allen and Unwin, 33-54

Barnes, J. W. 1973. Ancient clay bars from Iran, *Historical Metallurgy* 7/2,(1973), 8-17

Barnet, Peter 2006a. Beasts of every land and clime, in Peter Barnet and Pete Dandridge (eds), *Lions, Dragons and other beasts: Aquamanilia of the Middle Ages, Vessels for Church and Table*. New Haven, Yale University Press, 2-17

Barnet, Peter and Dandridge, Pete 2006b. Aquamanilia, in Peter Barnet and Pete Dandridge, *Lions, Dragons and other beasts: Aquamanilia of the Middle Ages, Vessels for Church and Table*. New Haven, Yale University Press, 66-176

Barratt, W. 1789. *The History and Antiquities of the City of Bristol*, Bristol, Pine

Baxter, T. and A. da Silva Rego 1964. *Documents on the Portuguese in Mozambique and Central Africa,* National Archives of Zimbabwe and Malawi, Lisbon, Centro de Estudos Históricos Ultramarinos

Bayley, Justine 1984. Roman Brassmaking in Britain, *Historical Metallurgy* 18/1, 42-43

Bayley, Justine 1990. The production of brass in antiquity, with particular reference to Roman Britain, in P. Craddock, 2000. *Years of Zinc and Brass*, London, British Museum occasional paper 50, 7-26

Bayley, Justine 1992. Copper alloys, in Justine Bayley, *Anglo-Saxon Non-Ferrous Metalworking from 16-22 Coppergate*, Council for British Archaeology, 807-810

Bayley, Justine 2014. Copper alloy working, in Justine Bayley, *Non-Ferrous Metalworking from Coppergate. The Archaeology of York, The Small Finds* 17/7 Council for British Archaeology, for York Archaeological Trust, 803-810

Bayley, Justine and Sarnia Butcher 1980. Variations in alloy compositions of Roman brooches, *Revue d'Archéometrie*, supplement III, Actes du XX symposium International d'Archeometrie, Paris, 1980, 29-36

Bayley, Justine and Sarnia Butcher 2004. *Roman brooches in Britain, a technological and typological study based on the Richborough collection*, Society of Antiquaries of London

Bayley, Justine and Kerstin Eckstein 2004. Roman and medieval litharge: structure and composition, in *Proceedings of the 34th Symposium on Archaeometry*, 3-7 May 2004, Zaragoza, Spain, 145-153

Bayley, Justine and Thilo Rehren 2007. Towards a functional and typological classification of crucibles, in Susan la Niece, Duncan Hook and Paul Craddock, *Metals and Mines: studies in archaeometallurgy*, London, Archetype, 46-55

Bayley, Justine, Jonathon Cotton, Thilo Rehren and Ernst Pernicka 2016. A Saxon brass bar ingot cache from Kingsway, London. In J.Cotton, J. Hall, J. Kelly, R. Sherris and R. and R. Stephenson, *Hidden Histories and records of antiquity: essays on Saxon and medieval London,* for John Clark, Curator Emeritus, Museum of London. London, LAMAS (special paper 17), 2014, 121-128

Bedell, John 2001. Delaware archaeology and the revolutionary eighteenth century, *Historical Archaeology* 35/4, 83-104

Beddies, Thomas 1996. *Becken und Geschütze: der Harz und sein nördliches Vorland als Metallgewerbelandschaft in Mittelalter und frühen Neuzeit*, Frankfurt am Main. Peter Lang

Beeley, Peter 2001. *Foundry Technology*, 2nd edn. Oxford, Butterworth, 353-4, 491, 525

Begemann, Friedrich, Sigrid Schmitt-Strecker, Ernst Pernicka (1992). The metal finds from Thermi III-V: a chemical and lead-isotope study, *Studia Troica* 2: 219-240.

Beiler, Rosalind, 2008. *Immigrant and Entrepreneur: the Atlantic World of Caspar Wistar 1696-1750*, University Park Press, Pennsylvania State University

Belényesy, Károly 2018. Cannon foundry workshop in late medieval Buda (Hungary) at the turn of the 15th-16th centuries, in Nicolas Thomas and Pete Dandridge, *Cuivres, bronzes et laitons médiévaux,* Agence Wallone du Patrimoine, Études et Documents, Archéologie 39, 155-167

Bentley-Smith, Dorothy 2004. *A Georgian Gent and Co: the Life and Times of Charles Roe*, Ashbourne, Landmark

Bequet, Alfred 1893. Les Bagues Franques et Mérovingiennes. *Annales de la Société Archéologique de Namur* XX, 209-240

Berg, Torsten and Peter Berg 2001. *R.R .Angerstein's illustrated travel diary 1753-1755: industry in England and Wales from a Swedish perspective*, London, Science Museum

Bettge, Gotz 1987. *Iserlohn-Lexicon*, Iserlohn, Hans-Herbert Mönnig

Biddle, Martin and David Hinton 1990. Book-clasps and page-holder, in Martin Biddle, *Object and Economy in Medieval Winchester: Artefacts from Medieval Winchester,* Oxford, Clarendon Press, vol.2, 755-758

Biringuccio, Vanoccio 1540. 1959 translation by Stanley Smith and C. Gnudi. *The Pirotechnia of Vanoccio Biringuccio, 1540*, New York, Basic Books

Bischoff, Bernard 1981. *Mittelalterliche Studien: Ausgewählte Aufsätze zur Schriftkünde und Literaturgeschichte* 3, Stuttgart, Nietsmann

Biswas, Arun Kumar 1993. The primacy of India in ancient brass and zinc metallurgy, *Indian Journal of History and Science* 28/4, 309-330

Biswas, Arun Kumar 1996. *Minerals and Metals of Ancient India*, Vol.2, Indigenous literary evidence, New Delhi, Printworld

Blair, Claude, John Blair, Roger Brownsword 1986. An Oxford brasiers' dispute of the 1390s: evidence for brass-making in medieval England, *The Antiquaries Journal* 66, 82-9

Blair, Claude and John Blair 1991. Copper Alloys, in John Blair and Nigel Ramsey, *English Medieval Industries*, London, Hambledon, 81-106

Blair, Claude and Angus Pattison 2000. 'Surrey' Enamels reattributed: part 2, an illustrated list of known types. *Journal of the Antique Metalware Society* 14, 10-19

Bocarro, Antonio 1644. 1937 reprint: Livro des plantas de todas as fortalezas, cidades e povoacoes do Estado da India Oriental. In A. B. Pereira da Braganza, *Arquivo Portugues Oriental.* Tome 4, vol.2, part 1, 352-4

Bollingberg, Haldis 2005. Origin and distribution of Scandinavian and other European Perlrandbecken as revealed by spectrochemical elemental analysis, in H. J. Hassler (ed). Studien zur Sachsenführung 15, *Neue Forschungsergebnisse zur nordwesteuropäischen Fruhgeschichte unter besonderer Berücksichtigung der altsachsischen Kultur im heutigen Niedersachsen,* Oldenburg, 487-505

Bondeson, Gustaf 1965. *Gusums Bruk*, Göteborg, Wezäta

Bonson, Anthony 1995. The History of Bosley Works, Cheshire, in *Wind and Water Mills* 14, Midland and Water Mills Group, 17-22

Borst, Arno 2006. *M G H Schriften zur Komputistik in Frankenreich von 721 bis 818*, vol. 3, Hannover, Hahnsche Buchhandlung

Bourgarit, David and Fanny Bauchau 2010. The ancient brass cementation processes revisited by extensive experimental simulation, *Archaeotechnology* 62/3, 51-57

Bourgarit, David and Benoit Mille, Thierry Borel, Pierre Baptiste and Thierry Zéphir 2003. A millennium of Khmer bronze metallurgy: analytical studies of bronze artifacts from the Musee Guimet and the Phnom Penh National Museum, in Paul Jett (ed) *Scientific Research in the field of Asian Art: proceedings of the first symposium at the Freer Gallery of Art*, (2003). London, Archetype 103-126

Bourgarit, David and Nicolas Thomas 2011. From laboratory to field experiments: shared experience in brass cementation, *Historical Metallurgy* 45/1, 8-16

Bourgarit, David and Nicholas Thomas 2012. Late medieval copper alloying practices: a view from a Parisian workshop of the 14th century AD, *Journal of Archaeological Science*, 3052-3070

Bowen, Emanuel and Thomas Kitchen 1777. *Large English Atlas*, London, Bowles

Bowman, S. G. E., Michael Cowell and Joe Cribb 1989. Two thousand years of coinage in China, an analytical survey, *Historical Metallurgy* 23/1, 25-9

Boyce, Helen 1920. *Mines of the Upper Harz from 1514 to 1589*, Wisconsin, George Bauta

Brandon, Peter 1984. *The Tillingbourne River Story*, Shere, Gomshall and Peaslake Local History Society, 84

Brookshaw, Dominic 2019. *Hafiz and his contemporaries. Poetry, Performance and Patronage in Fourteenth Century Iran.* EPUB ebook

Brautigam, Uwe and Yonghee Lee-Kalisch 2007. *Tibet, Klöster öffnen ihre Schatzkämmer,* Exhibition catalogue, Museum für Asiatische Kunst, Berlin, Munich, Hirmer

Breen, T. H. 1994. The meanings of things: interpreting the consumer economy in the eighteenth century, in J. Brewer and M. Porter, *Consumption and the World of Goods*, London, Routledge, 249-260

Brown, Josiah 1761. *Reports of Cases upon Appeals and Writs of Error in the High Court of Parliament, 1701-1779*. London, H.M. law printers, 296-7, 522, 525

Browne, Frank 1916. *Journal of the Royal Society of Arts* No. 3318, Vol. 64, 576

Brownsword, Roger 1986. A penannular brooch from Braithwaite, Cumbria, *Transactions of the Cumberland and Westmoreland Antiquarian and Archaeological Society* 86, Note 12, 264-6

Brownsword, Roger 1997. Evidence for a medieval English 'brass' industry, *Journal of the Antique Metalware Society* 5, 14-16

Brownsword, Roger. and E. Pitt 1983. Alloy composition of some cast 'latten' objects of the fifteenth to sixteenth centuries, *Journal of Historical Metallurgy* 17/1, 44-9

Brownsword, Roger and John Hines 1994. The analysis of a sample of Anglo-Saxon great square-headed brooches, *The Antiquaries Journal* 73, 1-10

Brunnthaler, Adolf 2000. *Reichraming*, Gemeinde Reichraming

Buchanan, B. 2000. The Africa Trade and the Bristol Gunpowder Industry, *Transactions of the Bristol and Gloucestershire Archaeological Society* 118, 133-56

Buckeridge, N. 1654. (1973, edition) J. R. Jensen. *Journal and Letter-book of Nicholas Buckeridge, 1651-1654*, Minneapolis, 32

Burke, P. 1994. Res et Verba: conspicuous consumption in the early modern world, in J. Brewer and M. Porter, *Consumption and the World of Goods*, London, Routledge, 148-161

Burks, Jean 1986. *Birmingham Brass Candlesticks.* Charlottesville, University of Virginia

Butzer, Paul, Max Kerner and Walter Oberschalp (1997). *Karl der Grosse und sein Nachwirken, 1200 Jahre Kultur und Wissenschaft in Europa*, Colloquium Carlus Magnus, 1995, Aachen, 251-275

Calcutt, Vince 2000: A brief early history of brass, in Introduction to Brasses, part 1 of *Innovations*, January 2000, online newsletter of the Copper Development Corporation, 1

Caley, Earle 1955. On the existence of chronological variations in Roman brass, *Ohio Journal of Science* 55, 137-140

Caley, Earle 1963. Investigations on the origin and manufacture of orichalcum. In M. Levey, *Archaeological Chemistry*, Pennsylvania, American Chemical Society, 67-8

Calver, Clare, 1990. Magno cum artificio: medieval metallurgy and the monumental brass industry, unpublished MSc dissertation, Newnham College, Cambridge, appendix 1, metallurgical tables, folio 68 ff

Camden, William 1637. *Britannia*, trans. Philemon Holland, London, Crooke, 767 and map.

Camden, William 1695. *Britannia*, 1971 reprint, Newton Abbot, David and Charles, 822

Cameron, Peter 2013. Identifying ID-with a groove above, *Journal of the Antique Metalware Society* 21, 2-13

Campbell, Gordon 2006. *The Grove Encyclopedia of Decorative Arts*, volume 1, New York, Oxford University Press

Capes, Michael and Diane Chamberlain 1998. Stirling Castle 1703, in Mensun Bound, 1998, *Excavating Ships of War*, Oswestry, Nelson, 125-141

Catulla, Tommaso Antonio 1833. *Elementi di minerologia applicati alla medicina e alla farmacia,* 2

Chakrabarti, Dilip K. and L. Nayanjot 1996. *Copper and its alloys in Ancient India*, Delhi, Munshiram Manoharlal

Chalenor, W. H. 1953. Charles Roe of Macclesfield, 1715-81, an eighteenth-century industrialist, 1. *Transactions of the Lancashire and Cheshire Antiquarian Society* 62, 133-156.

Chaptal, Jean and Antoine-Claude 1807. *Chemistry applied to arts and manufactures*, Vol 2, London, Philips
Chaucer, Geoffrey 1380s, in Fred N. Robinson (1957). *The Complete Works of Geoffrey Chaucer*, Cambridge University Press
Chaudhuri, Kirti N. 1990. *Asia before Europe: Economy and Civilisation of the Indian Ocean from the rise of Islam to 1750*. Cambridge University Press
Chernykh, Evgenii Nikolaevich 1992, translator Sarah White, *Ancient Metallurgy of the USSR*, Cambridge University Press
Clark, P. and L. Murfin 1995. *A History of Maidstone*, Stroud, Allan Sutton
Clark, John, Geoff Egan and Nick Griffiths 2004. Harness fittings, in John Clark (ed), *The Medieval Horse and its Equipment, c. 1150-1450*, London, Boydell, 43-74
Clifford, Helen 1990. Colonel Shorey, citizen and pewterer of London. *Journal of the Pewter Society* 7, 130-2
Cohn, Norman 1970. *The Pursuit of the Millennium: revolutionary millennians and mystical anarchists of the Middle Ages*, Oxford University Press, 258
Coleman-Smith, R. 2002. Excavations in the Donyatt Potteries, *Post Medieval Archaeology* 36, 118-172
Collingwood, W. G. 1912. *Elizabethan Keswick: Extracts from the Original Ancient books 1564-1577 of the German Miners, in the Archives of Augsburg*, Kendal, Cumberland and Westmorland Antiquarian and Archaeological Society, tract series 8,
Connor, R. 1987. *The Weights and Measures of England*, London, HMSO
Corin, Carl Frederik 1978. Mässingsbruket, in Sven Ljung, *Arboga Stads Historia från 1500-talets mitt till 1718*, Arboga, Stad Arboga, 166-169
Corse, Christopher de 2001. *An Archaeology of Elmina: Africans and Europeans of the Gold Coast, 1400-1900*, Washington, Smithsonian Institute Press
Cowell, Michael, Paul Craddock, Alistair Pike and A. M. Burnett 2000. An analytical survey of Roman provincial copper-alloy coins and the continuity of manufacture in Asia Minor. In B. Kluge and B. Weisser (eds), *XII International Numismatischer Kongress, Berlin, 1997,* Berlin, Mann-Verlag, 670-77.
Cowell, Michael, Susan La Niece, and Jessica Ranson 2003. A study of later Chinese metalwork, In Paul Jett, 2003. *Scientific Research in the Field of Asian Art, Proceedings of the first Forbes symposium at the Freer Gallery of Art*, London, Archetype, 80-89
Craddock, Paul 1977. The Composition of copper alloys used by the Greek, Etruscan and Roman civilisations. *Journal of Archaeological Science* 4, 103-124.
Craddock, Paul 1978. The composition of the copper alloys used by the Greek, Etruscan and Roman civilisations, 3: The origins and early use of brass. *Journal of Archaeological Science* 5/3, 1-16
Craddock, Paul 1979. The copper alloys of the Medieval Islamic World: inheritors of the classical tradition. *World Archaeology* 11/1. 68-79
Craddock, Paul 1984. How zinc was smelted in Ancient India. *New Scientist*, 1403, March 1984, 23
Craddock, Paul 1986. The metallurgy and composition of Etruscan bronzes, *Studi Etruschi* 52, 211-271
Craddock, Paul 1995. *Early Metal Mining and Production*, Washington, D. C., Smithsonian
Craddock, Paul, A. Burnett and K. Preston 1980. Hellenistic copper-base coinage and the origins of brass; in W. A. Oddy, *Scientific Studies in numismatics*, British Museum Occasional Paper 18, 53-64

Craddock, Paul, I. C. Freestone, L. K. Gurjar, A.P. Middleton and L. Willies 1990. Zinc in India, in Paul Craddock, *2000 Years of Zinc and Brass*. London, British Museum Occasional Paper 50, 29-73

Craddock, Paul, Susan La Niece and Duncan Hook 1998. Brass in the Medieval Islamic World. In Paul Craddock, *2000 Years of Zinc and Brass*, London, British Museum Occ. Paper 50, 73-113

Craddock, Paul and Michael Cowell 2000. Appendix 3, The Isleworth Sword, a note on the brass foils. In Ian Stead, *British Iron Age Swords and Scabbards*. Appendix 3, 123-5

Craddock, Paul and K. Eckstein 2003. Production of brass in Antiquity by direct reduction. In P.T. Craddock and J. Lang (eds), *Mining and Metal Production through the Ages,* London, British Museum. 216-230

Craddock, Paul, Michael Cowell and Ian Stead 2004. Britain's First Brass, *Antiquaries Journal* 84, 339-4

Crawford, Andrew 2006. Casters, dredgers and Muffineers, 1680-1840, *Journal of the Antique Metalware Society* 14, 1-9

Crone, Gerald 1937. *The Voyages of Cà da Mosto and other documents of western Africa in the fifteenth century*, London, Hakluyt Society

Curtin, Philip 1975. *Economic change in Pre-colonial Africa: Senegambia in the era of the Slave Trade*. Wisconsin, Madison

Curtis, John, Miroslaw Kruszynsky and Alistair Pike 2002. *Ancient Caucasian and related Metalwork in the British Museum*, British Museum Occasional Paper 121

Custenson, Paul 2011. Brasses in Meissen Cathedral, Germany, *Bulletin of the Monumental Brass Society* 116, 316-7

Daaku, Kwame. 1970. *Trade and Politics on the Gold Coast 1600-1720: a study of the African reaction to European trade,* Oxford University Press

Dai Zhiqiang and Zhou Weirong 1992. Studies of the alloy composition of more than two thousand years of Chinese coins, *Journal of the Historical Metallurgy Society*, 26, 45-55

Daim, Falko 2010. Byzantine belt ornaments of the seventh and eighth centuries in Avar contexts. In Chris Entwhistle and Noel Adams, *Intelligible Beauty: recent research on Byzantine jewellery*. British Museum Research Publications 178, 61-71

Dan Berg, Sven, and George Hassell 1992. *The Geddy Foundry*, Williamsburg, Virginia, Colonial Williamsburg Foundation

Dandridge, Pete 2018. The Hildesheim font; a window into medieval workshop practices, in Nicolas Thomas and Pete Dandridge, *Cuivres, bronzes et laitons médiévaux,* Agence Wallone du Patrimoine, Études et Documents, Archéologie 39, 25-35

van Danzig, A and A Jones 1987, *Description and Historical Account of the Gold Kingdom of Guinea*, translation of Pieter de Marees (1602). New York: Oxford University Press

Dapper, Olfert. 1686. *Naukeurige Beschryvinge der Afrikaensche Gewesten*, Wolfgang, Amsterdam

Davey, Christopher, J. 2009. The early history of lost-wax casting. In Jianjun Mei and Thilo Rehren, *Metallurgy and Civilisation: Eurasia and beyond*. International Conference on the Beginnings of the Use of Metals and Alloys, 2006. Beijing, China. London, Archetype Press, 147-154

Davies, K. and Williams, C. J. 1986. *The Greenfield Valley*, Holywell, Town Council

Davies, Norman 1997. *Europe, a History*, London, Pimlico

Davis, John 2015. The Chetwode quadrant: a medieval unequal-hour instrument, *British Sundial Society Bulletin* 27 (ii) 2-6

Davis, John 2017. A royal English medieval astrolabe made for use in Northern Italy, *Journal for the History of Astronomy* 48 (1): 3-32

Davis, Mary and Adam Gwilt 2008. The Seven Sisters Hoard: relationships between technology, style and function, in Duncan Garrow and Chris Gosden (eds), *Rethinking Celtic Art*, Oxford, Oxbow (146-178)

Dawkins J.M. 1950. *Zinc and Spelter*: Notes on the History of Zinc, from Babylon to the 18th century, compiled for the curious. Oxford, Zinc Development Association

Day, Joan 1973. *Bristol Brass: the History of the Industry*, Newton Abbot, David and Charles

Day, Joan 1984. The Continental Origins of Bristol Brass, *Industrial Archaeology Review* 7, 32-36

Day, Joan 1998. Brass and Zinc in Europe from the Middle Ages until the mid-nineteenth century, in Craddock, Paul. *2000 Years of Zinc and Brass*, London, British Museum Press, 133-158

Defoe, Daniel 1725 (2006 edition), *A Tour through the Whole Island of Great Britain*, vol.2, London, Folio Society

Dennert, Herbert, 1979. *Quellen zur Geschichte des Bergbaus und des Huttenwesens im Westharz von 1524-1631*, Claustal-Zellerfeld

Diderot, Denis, and Jean d'Alembert 1758. *Encyclopédie ou Dictionnaire Raisonnée des Sciences, des Arts et des Métiers*, Paris, le Breton

Dietz, Alexander 1925. *Frankfurter Handelsgeschichte,* part 2, Frankfurt, Knauer

Diodorus Siculus 60-30 BC. *Biblioteca historia*, book 5

Dollinger, Philippe 1968. *The German Hansa*, Stanford, Stanford University Press

Donnan, E. 1930-33. *Documents illustrative of the Slave Trade to Africa,* Washington, Carnegie Institute

Douxchamps-Lefèvre, Cécile 2005. L'art du laiton dans le pays mosan, in Jacques Toussaint (ed), *Art du Laiton, Dinanderie*, Namur, Société Archéologique, 9-25

Drescher, Hans 1985. Metallwerk des 8-11 Jahrhunderts in Haitabu auf Grund der Werkstattabfälle... In Herbert Jankuhn (ed) *Das Handwerk in vor und frühgeschichtlicher Zeit*: [Bericht über die Kolloquien der Kommission für die Altertumskunde Mittel- und Nordeuropas in den Jahren 1987 bis 1980], Göttingen, Vandenhoeck and Ruprecht

Dresser, Madge 2001. *Slavery Obscured: the Social History of the Slave Trade in an English Provincial Port*, London, Continuum

Drewal, Henry and Enid Schildkrout 2010. *Kingdom of Ife: sculptures from West Africa*, London, British Museum

Duczko, Wladyslaw 2004. *Viking Rus: Studies on the Presence of Scandinavians in Eastern Europe*, Leiden, Brill

Duhamel du Monceau, Henri-Louis 1764. De la Fonte et de l'Affinage du Cuivre et du Potin à Villedieu-les-Poêles en Normandie. In Jean-Gafin Galon and Henri-Louis Duhamel du Monceau, 1764. *L'Art de convertir le Cuivre Rouge ou le Cuivre Rosette en Cuivre Jaune,* Paris, Académie des Sciences, 57-78

Dungworth, David 1996. The production of copper alloys in Iron Age Britain, *Proceedings of the Prehistoric Society*, 62, 399-413

Dungworth, David 1997. Roman Copper Alloys: Analysis of artefacts from Northern Britain, *Journal of Archaeological Science* 24, 901-9

Dungworth, David 2010. Newent Glasshouse, Newent, Gloucestershire: Examination of glass and glassworking waste, Resesearch Department Report 6/2010, Portsmouth, English Heritage

Dungworth, David and H. White 2007. Scientific examination of zinc-distillation remains from Warmley, Bristol. *Historical Metallurgy* 4/1 77-83

Dungworth, David and Roger Wilkes 2008. *Taynton Brassmill, Newent, Gloucestershire: early 17th-century brass manufacture. Technology report.* English Heritage Research Department Report Series 28-2010

Dungworth, David, Paul Belford and Rob Ixer 2010. The examination of crucibles, copper ore and slag: technology report, *English Heritage Research Department Report Series*, November 2010

Dungworth, David and Roger Wilkes 2010. Taynton, Gloucestershire, early 17th-century brass manufacture, *English Heritage Research Department Report*, Series 28, 6

Dunn, Richard 2006. Touching and cleaning, the routine work of an east London instrument supplier, *Bulletin of the Scientific Instrument Society* 89, 21-6

Eiwanger, Josef 1996. Barrenmessing, ein mittelalterliches Handelsgut zwischen Europa und Afrika? *Beiträge zur Allgemeinen und Vergleichenden Archäologie* 16, 215-226

Ekström, Frans 1985. *Vattholma och Trollbo*, Tierp, Lindquist

Ellacombe, H.T. 1881. *The History of the Parish of Bitton*, Exeter, William Pollard

d'Elvert, Christian 1866. *Zur Kultur-Geschichte Mährens und Oest Schlesiens*, vol.1, Brünn, Ritsch, 164-7

Eniosova, Natasha, Lyibov Pokrovskaya, Victor Singh and Olga Tarabardina 2014. Evidence concerning jewellery production and workshops from the Lyudin End of Medieval Novgorod. *Medieval Copper, Bronze and Brass, Dinant-Namur, 2014*, conference report, Namur, La maison du patrimoine médiéval mosan

Entwhistle, Chris 2010. Notes on selected recent Acquisitions of Byzantine Jewellery at the British Museum. In Chris Entwhistle and Noel Adams, *Intelligible Beauty: recent research on Byzantine jewellery*, London, British Museum Research Publication 178, 20-26

Ercker, Lazarus 1598. *Beschreibung aller furnehmesten mineralischen Erszt und Bergwercks Arten*, Frankfurt, Zunners/Feyrabend

Erixon, Sigurd 1957. *Skultuna Bruks Historia 1607-1860*, Västerås, Svenska Metallverke

Erixon, Sigurd 1969. Investigation of an Industry and its production: an ethnological programme, In John Geraint Jenkins, *Studies in Folk Life*, London, Routledge, 293-301

Erixon, Sigurd 1978. *Mässing: svenska manufakturer och konsthantverksprodukter under 400 År*, Lund, Walter Ekstrand

Evans, E.W. and David Richardson 1995. Hunting for Rents: the Economics of Slaving in Pre-colonial Africa, *Economic History Review* 48, 665-86

Eveleigh, David 1995. *Brass and Brassware*, Princes Risborough, Shire

Eyll, K. van 1998. Die Kupfermeister im Stolberger Tal zu wirtschaftlichen Aktivität einer religiösen Minderheit, in *Mühlen, Hammerwerke und Kupferhöfe im Tal der Vicht und ihre Besitzer*, Stolberger Geschichte 3, 83-104

Fabijanec, Sabine 2018. L'exploitation du cuivre en Europe Centrale et dans les Balkans et sa disribution commerciale à travers les territoires Croates aux 15e et 16e siècles, in Nicolas Thomas and Pete Dandridge, *Cuivres, bronzes et laitons médiévaux*, Agence Wallone du Patrimoine, Études et Documents, Archéologie 39, 51-64

Fang, Jui-Lien and Gerry McDonnell 2011. The colour of copper alloys, *Historical Metallurgy* 45 (1), 52-61

Fechner, Hermann 1903 (2013 reprint). Berg und Huttenpolitik in *Geschichte zur Schlesischen Berg- und Huttenwesens in der Zeit Friederich der Grossen, Friedrich Wilhelm II und Friedrich Wilhelm III, 1741bis 1806*, part 1, Potsdam, Becker

Fenney, David and Joseph Collier 1765. *A System of Geography*, vol 2,

Fiala, Claudia 1998. Erzbergbau – Segen und Niedergang. In Claudia Fiala, Reinhard Dörnig, Fritz Gratius and Silke Leisner, *Ilmenau: Beiträge zur Geschichte einer Stadt*, Hildburghausen, Frankenschwelle, 149-172

Fischbach, Friedrich 1786. Das grosse und weltberümtes Königliche Messingwerk bey Heegermühle, in *Statisch-typographische Städte Beschreibung der Mark Brandenburg*, Berlin, 80-83

Forbes, R.J. 1972. *Metallurgy in Antiquity*. E.J.Brill, Leiden.

Forgeng, Jeffrey 1999. *Daily Life in Medieval Europe*, Greenwood, Westport, Connecticut

Forsberg, Karin 1953. *Gusums bruks historia 1653-1953*, Stockholm

Forsgren, Nils 2010. *Mässingsbruk och Bruk av Mässing*, Nifo, Borgholm

Fox, R. and J. Barton 1986. Excavations at Oyster Street, Portsmouth, 1968-71. *Post Medieval Archaeology* 20, 31-256

Frachetti, Michael 2006. Ancient nomads of the Androvnovo Culture: the globalisation of the Eurasian Steppe during Prehistory, in Claudia Chang, *Of Gold and Grass: Nomads of Kazakhstan*, exhibition catalogue, 21-28

Franklin, M. 1977. translation of Abu al Fazl ibn Mubarak (1596). 'Ain I Akhbari', in *Representing India*, 5, part 1, London, Routledge, 43-4

Fraser-Liu, Sylvia 2012. *Burmese Crafts, Past and Present*, Oxford University Press, Kuala Lumpur

Frifelt, Karin 1991. *The island of Umm an Nar, 1, Third Millennium Graves*, Jutland Archaeological Press publications XXVI: 1, Aarhus University Press, Aarhus

Frobenius, Leo 1913. *The Voice of Africa*, translated Rudolph Blind, London, Hutchinson, vol. 1. 292-308, vol 2, plate opposite 492

Frumento, Armando 1963. *Il ferro milanese tra 1450 e 1796: Imprese lombardo nella storia della siderurgia italiana*, vol 2, Milan

Furia, Luigi 2012. *Le Miniere di Piombo e Zinco della Bergamasca*, Bergamo, Bolis

Gadd, Jan 2007. Skultuna brass manufactury, Sweden: *Journal of the Antique Metalware Society* 15. June 2007, 2-19

Gadd, Jan 2010. The curious case of the Tolånga candelabrum. *Journal of the Antique Metalware Society* 18, 2-7

Gale, Noel, Zofia Stos-Gale, I. Panyotov, A. Raduncheva, I. Ivanov, P. Lilov, T. Todorov 2003. Early Metallurgy in Bulgaria, in Paul. Craddock and J. Lang, *Mining and Metal Production through the Ages*, London, British Museum,122-173

Galon, Jean-Gafin 1764. L'art de convertir le cuivre rouge ou cuivre de rosette en laiton ou cuivre jaune, in H-L Duhamel du Monceau and J-G Galon, *Déscription des Arts et Métiers*, Académie de Paris, 1-56

Garenne-Marot, Laurence 2009. 'Fils á double tête and copper-based ingots: copper money-objects at the time of the Sahelian empires of Ghana and Mali, in Catherine Eagleton, H. Fuller and J. Perkins (eds), *Money in Africa*, British Museum publication 171: 1-8

Garenne-Marot, Laurence and Michael Wayman 1994. Early copper and brass in Senegal, in Terry Childs, *Society, Culture and Technology in Africa*, University of Pennsylvania Museum of Archaeology and Anthropology, 45-62

Garonne-Marot, Laurence, Caroline Robion, Benoit Mille 2003. Cuivre, alliages de cuivre et histoire de l'empire du Mali: à propos de trois figurines animales d'un tumulus du delta intérieur du Niger (Mali), *Techné* 18, 74-85

Garenne-Marot, Laurence and Benoît Mille 2007. Copper-based metal in the Inland Niger delta: metal and technology at the time of the Empire of Mali. In Susan La Niece, Duncan Hook and Paul Craddock, *Metals and Mines: Studies in Archaeometallurgy*, London, Archetype, 159-169

Gaur, Ram Chandra 1983. *Excavations at Atranjikhera: Early Civilisations of the Upper Ganga Basin. New Delhi*, Motil Barsidass, Algarth Muslim University

Gechter, Michael 1993. Römischer Bergbau in der Germania Inferior: eine Bestandsaufnahme. In Heiko Steuer and Ulrich Zimmermann (eds), *Archäologie in Europa, Montanarchäologie*, Sigmaringen, Jan Thorbecke, 161-165

Gentle, Rupert, and Rachel Field 1975. *English Domestic Brass 1680-1810 and the History of its Origins*, London, Paul Elek

Gérard, Edouard 1958. *Dinant, Ville d'art: la Dinanderie*, Dinant, Heraldic

Gericke, Hans Otto 2008. *Das priviligierte sachsische Messingwerk Niederauerbach i. Vogtland: die Geschichte eines bedeutendes Huttenwerkes von 1593 bis 1926*. Plauen-Jössnitz: Vogtland Verlag.

Ghambashidze, Davit 1919. *Mineral Resources of Georgia and Caucasia. Manganese Industry of Georgia*, London, Allen and Unwin

Giachi, Gianna, M. Colombini, M. Lippi, J. Lucejko, P. Pallechi, E. Ribechini and A. Romaldi 2013. Ingredients of a 2,000-year-old medicine revisited by chemical, mineralogical and botanical investigations. *Proceedings of the National Academy of Science of the United States of America*, 110/4, 1193-6

Giertz, Wolfram and Sebastian Ristow 2013. Goldtessellae und Fensterglas:Neue Untersuchungen und Nutzung von Glas im Bereich der karolingischerzeitlichen Pfalz, Aachen. *Antike Welt* 5/13, 59-66

Gilbert, Keith 1969. *Science Museum: a Descriptive Catalogue of the Collection illustrating Fire Fighting Appliances*, London, HMSO

Gilmour, Brian 2000. Dutch composite ordnance of the early 17th century, *Royal Armouries Yearbook* 5, 85-105

Gilmour, Brian and Eldon Worrall 1995. Paktong, the trade in Chinese brass to Europe, in Duncan Hook and David Gaimster, *Trade and Discovery: the scientific study of artefacts from post-medieval Europe and beyond* London, British Museum Occasional Paper 109, 259-287

Giumlia-Mair, Alessandra 2001. Technical studies on the copper-based finds from Emona, *Berliner Beiträge zur Archäometrie* 18, 5-42

Glamann, Kristof 1977. Japanese copper on European market in the 17th century, in Hermann Kellenbenz (ed), *Schwerpunkte des Kupferproduktions und des Kupferhandels in Europa 1500-1650*, Vienna, Bohlau, 280-289

Glauber, Johann Rudolf 1656a. *Pharmacopaea Spagyricae* 2. Jansson Amsterdam, 16-17, 62

Glauber, Johann Rudolf 1656b. *Prosperitas Germaniae* 2. *de Zinco quod his Speauter Belgis.* Amsterdam, 6-7

Golas, P.J. 1999. Chemistry and Chemical Technology, section 36a, vi, Zinc. In Joseph Needham, *Science and Civilisation in China*, 5/5, Cambridge University Press, 136-9

Gomes, A, 1648 1959 reprint. Viagem que faz o Padre Antonio Gomes da Compa de Jesus ao Imperio de Manomotapa, *Studia* 3, Lisbon, 155-242

Goodison, Nicholas 1999. *Ormolu: the work of Matthew Boulton*, London, Phaidon

Goodsall, Robert 1957. Watermills on the River Len, *Archaeologia Cantiana* 71, 106-129

Goss, Elbridge Henry 1891. *The Life of Colonel Paul Revere*, volumes 1 and 2

Granger, James 1775. *A Biographical History of England: from Egbert the Great to the Revolution*, 2nd edition, Vol 4, London, Davies

Grassi, Elisa 2015. Roman metalworking in northern Italy between archaeology and archaeometry: two case studies, in Andreas Hauptmann and D. Modaressi-Ferani, *Archaeometallurgy in Europe*, 3, Bochum, 155-164

Green Christopher and Roderick Butler 2011. Domestic Metalware in a seventeenth-century Parsonage: from the journal of Giles Moore, *Journal of the Antique Metalware Society* 19, 18-29

Gunther, Robert T. 1932. *Astrolabes of the World*, Vol. 1, Oxford University Press, translation of Severus Sabokt, by Jessie Margouliouth, 82-103

Gunther, Robert T. 1933. *The Greek Herbal of Dioscorides*, trans. Goodyer (1655), book V, chapter 84, 623-5, Oxford University Press

Hachenberg, Karl 1998. The complaint of the Markneukirchen brass-instrument makers about the poor quality of brass from the Rodewisch foundry, 1787-1795, *Historic Brass Society Journal* 10, 116-145

Hachenberg, Karl and Helmut Ullwer 2013. *Messing nach dem Gallmeifahren: drei Handschriten des 18. Jahrhunderts experimetell erläutert*, Hamburg, Disserta

Haedecke, Hanns-Ulrich. 1970. Trans. V. Menkes, *The Secret History oif the Decorative Arts*, London, Weidenfeld and Nicholson

du Halde, Jean-Baptiste 1735. *Description géographique, historique, chronologique, politique et physique de la Chine et de la Tartarie chinoise*, vols.1, 2 and 3, Paris, le Mercier

Hallen, William 1885. *An Account of the family of Hallen or Holland from AD 1280 to AD 1885*, Edinburgh, Neill

Hamilton, Alexander 1727. *A New Account of the East Indies*, vols 1 and 2, Edinburgh

Hamilton, Elizabeth 1996. *Technology and Social Change in Belgic Gaul: Copper-working at the Titelberg, Luxembourg, 125 BC-AD 300*, Philadelphia, Museum for Applied Science centre for archaeology

Hamilton, Henry, 1926 (1967 edition). *The English Brass and Copper Industry to 1800*, London, Frank Cass

Han Rubin and Ko Tsun 2000. Tests on brass items found at Jianzhai, Phase 1, In Linduff, Katheryn, Han Rubin and Sun Shuyun, *The Beginnings of metallurgy in China*. Lewiston, Mellen Press, 195-9

Harris, J. R. 2003. *The Copper King: A Biography of Thomas Williams of Llanidan*. Ashbourne, Landmark Publishing

Hart, W. A. 1987. Masks with metal-strip ornament from Sierra Leone, *African Arts* 20/3, 68-74

al-Hassan, Ahmad and D.R. Hill 1986. *Islamic Technology, an Illustrated History*, Paris, Unesco, 248-260

Hauptmann, Andreas. and Ernst Pernicka 2004. *Die Metallindustrie Mesopotamiens von den Anfangen bis zum 2 Jahrtausend v.Chr.* Orient-Abteilung des Deutschen Archäologischen Instituts, Verlag Marie Leidorf

Hawthorne, J. and C. S. Smith 1963. *On Divers Arts: The Treatise of Theophilus* (Book III, chapters 64-6) University of Chicago Press, 140-144

He Tangkun 1987. Metallurgy in Chinese Academy of Science, Institute of Natural Sciences (ed). *Ancient China's Technology and Science*, Beijing, Foreign Languages Press, 392-407

Hedeagar, Lotte 2007. Scandinavia and the Huns: an interdisciplinary approach to the migration Era. *Norwegian Archaeological Review* 40/1, 42-58

Heinricher, Alois 2006. Die einstigen Betriebe an der Leisacher-Lienzer Drauwiere. *Östtiroler Heimātblätter* 74/10, 1-16

Hekscher, Eli 1954. *An Economic History of Sweden*, vol.19, Harvard

Helmfrid, Björn 1965. *Norrköpings Historia*, vol.II (1568-1655). Stockholm

Henckel, J.F. 1737. De Zinco, in *Acta Physico-Medica 1, Academiae Caesariae Leopoldino-Carolinae Naturae Curiosum*, Nuremberg, Endteri

Heppe, Dorothea 1996. *Industriedenkmal Messinghof*, Kassel: Stadt Kassel

Herbert, Eugenia 1984. *The Red Gold of Africa*, Madison, University of Wisconsin Press

l'Heritier, Maxime and Florian Téreygoel (2010). From copper to silver: Understanding the Saigerprozess through experimental liquation and drying, *Historical Metallurgy* 44 (2), 136-152

Heyd, Volker 2008. In the Copper Age, *British Archaeology* 101, 19-27

Higham, Robert 2007. 300 years of the Warming Pan, *Journal of the Antique Metalware Society* 15, 40-43

Hildenbrand, Hanswerner 1983, *Die Iserlohner Industrie: vom Panzerhemd zur Ankerkette*, Iserlohn, Zeitungsverlag

Hildebrandt, Reinhard 2002. '...geet es hie vil anderst zue als Ir euch daussen Zuversteen gebt ...'. Informationsprobleme im interkontinentalen Handel des 16. Jahrhunderts. In Rainer Gömmel and Marcus Denzel, *Weltwirtschaft und Wirtschaftsordnung: Festschrift fur Jurgen Schneider zum 65 Geburtstag*, Stuttgart, Franz Steiner, 57-67

Hill, Donald 1974. *The book of knowledge of mechanical devices,* Dordrecht, Reidl; translation of Al-Jazarī, Ibn al-Razzāz (1204-6) 15-279

Hollunder, Christian Furchtegott 1824. *Tagebuch einer metallurgisch-technologischen Reise durch Mähren, Böhmen, einen Theil von Deutschland und der Niederlande*

Holmes, Helen 2005. Stone Brassworks and some memories of Brass Works Farm and Cottages, *Stone Historical and Civic Society Yearbook,* 2004-5, 8-18

Holtmeyer, Alois 1923. *Die Bau- und Kunstdenkmaler im Regierungsbezirk Cassel*, Marburg

Hoover, L. H. and L. R. Hoover 1950. *De Re Metallica, Georgius Agricola*, Dover, New York

Horst, L. 1972. *The Radical Brethren*, Nieuwkop, B. de Graaf, 31, 36-47

Houghton, John, 1693 (1728 edn). *Husbandry and Trade Improv'd, a collection of letters for the improvement of husbandry and trade*, vol. 1.13,109, 192; vol. 2. 175, 183-198, 203-4, London, Woodman and Lyon

Huey, Lois 2009. *American Archaeology uncovers the Dutch Colonies*, Cavendish Square digital publishing

Hulme, Edward W. 1896. History of the Patent System under the Prerogative and at Common Law, *Law Quarterly Review* 12, 141-154

Hussennether, Baptiste 2010. Forward, in Mertel, Rudolf, *Nürnberg-Laufamholz: ehemaliges Fabrikgut-Hammer*, Arbeitskreis Kultur und Geschichte des Vorstadtvereins Nürnberg-Laufamholz

Hutton, William 1835. *The History of Birmingham,* London, Berger, 187-91

Hyde, Charles K. 1998. *Copper for America,* Tucson, University of Arizona Press

Ingalls, Walter R. 1903. *The Metallurgy of Zinc and Cadmium,* New York Engineering and Mining Journal, 306, 393

Istenič, Janka. and Šmit, Ž. 2007. The beginning of the use of brass in Europe with particular reference to the southeastern Alpine region. In Susan LaNiece, Duncan Hook, Paul Craddock. *Metal and Mines: Studies in Archaeometry*, London, Archetype, 140-147

Isikili, Mehmet and Altunkaynak, Gulsah 2014. Some observations on relationships between South Caucasus and North-Eastern Anatolia based on recent archaeometallurgical evidence.*In Problems of Early Metal-Age Archaeology of Caucasus and Anatolia. Proceedings of International Conference 19-23 November 2014*, Tbilisi, 73-93

Istenič, Janka, 2015. *Roman Stories from the Crossroads,* Lubljana, National Museum of Slovenia (Narodni Muzei Slovenje)

Ivimey, J. 1811. *A History of the English Baptists, including an investigation of the History of Baptism in England*, London, privately printed, vol.2

Jackson, Ralph and Paul Craddock 1995. The Ribchester hoard: a description and technical study. In B. Rafferty, *Sites and Sights of the Iron Age*, Oxford, Oxbow, 75-102.

Jacobs, Edward 1889. Johann Christian Ruberg, *Allgemeine Deutsche Biographie*, 268-270

Janin, Valentin and Petr Gajdukov 1998. Ein Schatzfund aus Novgorod mit westeuropäischen und byzantinischen Münzen, in Anke Wesse, *Studien zur Archäologie des Ostseeraumes: von der Eisenzeit bis zum Mittelalter. Festschrift für Michael Müller*, Neumunster, Wachholz, 345-358

Jansson, Alfred 1960. *Nacka: kring Nacka ström 1557-1887*, Nacka, Nacka Kulturamnd

Johnson, Samuel 1774. *Diary of a Journey into North Wales*, London, Jennings

Jones, G. 1968. *History of the Vikings*, Oxford,

Jones, Henry Leonard 1929. Strabo, *the Geography VI*, book 13; 1, 56, page 115.

Jovandovič, Borislav 1990. Die Vinca-Kultur und der Beginn der Metallnutzung auf dem Balkan, in Dragoslav Srejovič and Nicola Tasič. *Vinča and its world: the Danubian Region from 6000 to 3000 BC*, Beograd, Serbian Academy of Sciences and Arts, 55-60

Jürgen, Adolf 2012. *Zur Schleswig-Holsteinischen Handelsgeschicht des 16 und 17 Jahrhunderts*, Bremen, Unikum. Reprint of 1914 edn, Berlin, Kurt Curtiss

Jurriča, A. R. J. 2010. Taynton, in Jurriča, A. R. J. ed., *Victoria County History of Gloucestershire*, vol. 12, Newent and May Hill section, London, Boydell and Brewer

Karsten, Carl J. B. 1827. Entwurf einer Geschichte der Schlesischer Bergwerks-Verfassung vor dem Jahr 1740. In C. J. B.. Karsten. *Bergbau und Huttenwerken* vol 16. Berlin, Reimer, 227-401

Kashkai, Mir Ali and Isa Rizaevich Selimkanov 1973. *Iz istorii drevnei metallurgii kavkaza [uz ucтoрии древний металлиргий Кавказа]*, Baku, Elm

Kavtaradze, Georgi Leon 1999. The importance of metallurgical data for the formation of Transcaucasian technology, in Andreas Hauptmann, Ernst Pernicka, Thilo Rehren and Ünsal Yalçin (eds), *The Beginnings of Metallurgy*, Bochum, Deutsches Bergbau Museum, 67-101

Keagle, Matthew 2009. The 'Brass Caps': Germanic military headgear from the American War of Independence, *Journal of the Antique Metalware Society* 17, 36-43

Kealhofer, Lisa 2005. *The Archaeology of Midas and the Phrygians: recent work at Gordion*, Philadelphia, University of Pennsylvania Museum of Archaeology and Anthropology

Kellenbenz, Hermann 1977. Europäisches Kupfer. In Hermann Kellenbenz (ed), *Schwerpunkte der Kupferproduktion und des Kupferhandels in Europa 1500-1650*, Bohlau, Vienna. 291 -351

Kent, Eliza 2004. *Converting Women: Gender and Protestant Christianity in Colonial South India*, Oxford, Oxford University Press,

Kershaw, Jane 2013. *Viking Identities, Scandinavian Jewellery in England*, Oxford, OUP

Kharakwal, Jayshee 2012. *Zinc production in Ancient India*, website article, http://www.infinityfoundation.com/mandala/t_pr/t_pr_khara_zinc.htm, 1-7

Kharakwal, Jayshee. and L.K.Gurjar 2006. Zinc and Brass in Archaeological Perspective. *Ancient Asia* 1, www.ancient-asia-journal.com/article/view/aa.06112/23, 1-11

Kilson, Marion 1969. Libation in Ga ritual, *Journal of Religion in Africa* 2, 161-178

Klein, H.S. 1999. *The Atlantic Slave Trade*, Cambridge University Press, Cambridge, 47-60

Kisluk-Grosheide, Daniëlle 1988. Dutch tobacco boxes in the Metropolitan Museum of Art: a catalogue, *Metropolitan Museum Journal* 73, 201-231

Klingström, Lars 2009. *Holmen: en resa in fyra sekel*, Stockholm, Holmen

Klostermann, Rolf 1996. *Der Bergbau in Iserlohn*, Schacht-Audorf, Köller

Knoke, Horst 1990. *Die Mühlen in Bettenhausen*, www.agathof.de/6-Mühlen_in_Bettenhausen-1.cfm

Kock, Friedrich 1981. *Beiträge zu einer Chronik der Gemeinde Bäk*, Bäk local council

Köfler, Werner 1962. Das Messingwerk Achenrain. In Hanns Bachmann (ed), 1972, *Das Buch von Kramsach*, Innsbruck, Universitätsverlag Wagner, 367-95

Kopytoff, Igor 1986. The cultural biography of things: commodities as process. In Appadurai, A., *The Social Life of Things*, Cambridge University Press

Kroll, Stephan, Gunter, C., Hallweg, U., Roaf, M. and Zimansky, P. 2010. Introduction, in S. Kroll et al. *Biainili Urartu:* Proceedings of the Symposium held at Munich, 2007. Leuven, Peeters,1-38

Kumlien, Kjell. 1977. Staat, Kupfererzeugung und Kupferausfuhr in Schweden, 1500-1650, In Herman Kellenbenz *Schwerpunkte der Kupferproduktion und des Kupferhandels in Europa*, Köln, Böhlau, 226-241

Labouchère, Rachel 1988. *Abiah Darby, 1716-1793*, New York, William Sessions

Laharnar, Boštjan 2009. The Zerovišček Iron Age Hillort site near Bločice in the Notranjska region: *Archeološki vestnik* 60, 97-157

Lamberg-Karlovsky, Clifford, and Peter Magee 2001. *Excavations at Tepe Yahya, Iran, 1967-1975: the Third Millennium*, American School of Prehistoric Research bulletin 45. Cambridge, Mass. Peabody Museum of Archaeology and Ethnology

Lammers, Dieter 2009. *Das Karolingisch-ottonische Buntmetallhandwerker-Quartier auf dem Plattenberg in Soest*, Soest, Westfalischen Verlag

La Niece, Susan 2003. Medieval Islamic metal technology, in Paul Jett, 2003. *Scientific Research in the Field of Asian Art, Proceedings of the first Forbes symposium at the Freer Gallery of Art*, London, Archetype, 90-96

La Niece, Susan 2018. Brass in the medieval Islamic world and contact with Europe, in Nicolas Thomas and Pete Dandridge, *Cuivres, bronzes et laitons médiévaux,* Agence Wallone du Patrimoine, Études et Documents, Archéologie 39, 387-393

La Niece, Susan and G. Martin 1987. The technical examination of Bidri Ware, *Studies in Conservation* 32, 97-101

Latham,R. 1958 *The Travels of Marco Polo.* Translation of Marco Polo, 1298. Penguin Books, London, 31-2

Leone, M. 1988. The Georgian order as the order of merchant capitalism in Annapolis, Maryland, In M. Leone and P. Potter, *The Recovery of Meaning: Historical Archaeology in the Eastern United States*, Smithsonian Institute Press, 235-61

Lester, Ayala 2004. Glass and metal objects, in Yizhar Hirschfeld, 2004. *Excavations at Tiberias,1989-94,* Jerusalem, Israel Antiquities Authority, 59-68

Levtzion, Nehemia and J. Hopkins 2000. *Corpus of early Arabic Sources for West African History*, Princeton, Markus Wiener, 62-87

Lewis, William 1759. *The Chemical Works of Caspar Neumann*, Johnston, London

Li Ch'iao-p'ing 1948. *The Chemical Art of Old China*, Easton, Pennsyvania, Journal of Chemical Education, 44

Linne, Carl von, 1734, (1953 edition). Massingsbruk wid Bjurfors. In Arvid Hj Uggla (ed), 1953, *Linnes Dalaresa jamte utlandsresan och bergslageresan och Bergslageresan.* Stockholm, Hugo Gebers, 377-383

Lisch, Georg 1842. Geschichte der Eisengewinnung in Meklenburg aus inlandiscen Rasenerz, in *Jahrbuch für Geschichte* 7, Verein für meklenburgische Geschichte und Altertumkunde, 52-155

Liu Haiwang, Bao Wenbo, Chen Jianli, Han Rubin, Li Yanxiang, Sun Shuyun, Wu Xiaohong and Yuan Dongshan 2007. Preliminary multidisciplinary study of the Miaobeihou zinc-smelting ruins at Yangliusi village, Fengdu county, Chongqing, in Susan La Niece, Duncan Hook and Paul Craddock, *Metals and Mines: Studies in Archaeometallurgy*, London, Archetype, 170-178

Lockyer, C. 1711. *An Account of the Trade in India*, London, Crouch

Löhneyss, Georg Engelhard von (1690). *Bericht vom Bergwerck*

Lyson, S. and D. 1806. *Magna Britannia,* Cadell and Davies, London, 1, 198-9; and 2, 427, 725

Magnusson, Lars 2000. *Swedish Economic History*, London, Routledge

Makarov, N. 2009. Rural settlement and trade networks in northern Russia AD 900-1250. In M. Mango, *Byzantine trade, 4th-12th centuries: the archaeology of local, regional and international exchange.* Farnham, Ashgate publishing, 443-461

Malmsten, Karl 1939. Den svenska mässingsindustriens uppkommst. *Med hammare och fackla* X, 132-178

Malmsten, Karl 1942. En Industrieplanerung under Vasatid. *Mere hammare och fackla* XII, 1-99

Malynes, G. 1629. *Lex Mercatoria*, London, Islip, 265-6

Mansi, Giovanni Domenico 1759. Ex concilio Rhemensi (notes on the canons of the Council of Rheims c.625-30 AD), in *Sacrorum conciliorum nova et amplissima collectio*, Vol. X, chapter VI, 591-603

Marggraf, Andreas Sigismund 1746. Expériences sur la manière de tirer le zinc de sa véritable minière, c'est à dire de la pierre calaminaire, *Histoire de l'Académie Royale des Sciences et Belles Lettres de Berlin*, 49-57

Markoe, Glenn 2000. *Phoenicians*, London, British Museum Press.
Marmol, Eugene del 1859. Fouilles dans un cimetière de l'époque Franque, à Samson, *Société Archéologique de Namur, Annales* 6, 345-391
Marsette, Marcel 2003. An encounter in the Bronze Age, *Historical Archaeology* 37/4, 29-39
Marshall, John 1951. *Taxila: an illustrated account of archaeological excavations*, vol 2, Minor Antiquities, Cambridge, Cambridge University Press
Martinon-Torres, Marcos. And Thilo Rehren 2002. Agricola and Zwickau: Theory and Practice of Renaissance Brass Production in SE Germany, *Historical Metallurgy* 36/2. 95-111
May, W.E. 1973. *A History of Marine Navigation*, Henley-on-Thames, Foulis
Mei, Jianjun 2009. Early metallurgy in China: some challenging issues in current studies. In Jianjun Mei and Thilo Rehren. *Metallurgy and Civilisation: Eurasia and beyond*. International Conference on the Beginnings of the Use of Metals and Alloys, 2006, Beijing. London, Archetype, 9-16
Melikian-Chirvani, Assadullah 1982. *Metalwork from the Iranian World, 8th to 11th century*, London, HMSO
Mellaart, James 1967. *Çatal Hüyük: a Neolithic town in Anatolia*, London, Thames and Hudson
Meller, Simon 1925. *Peter Vischer der Älter und seine Werkstatt*, Leipzig, Insel Verlag
Mende, Ursula 2006. Late Gothic Aquamanilia from Nuremberg, in Peter Barnet and Pete Dandridge, *Lions, Dragons and other beasts: Aquamanilia of the Middle Ages, Vessels for Church and Table*. New Haven, Yale University Press, 18-33
Merkel, Stephen 2018. Archaeometallurgical investigation of a Viking brass ingot hoard from the Hedeby Harbor in northern Germany. *Journal of Archaeological Science Reports*, 293-302
Metcalf, Michael and Peter Northover 1987. Coinage in ninth-century Northumbria, in Melinda Mays, *Celtic coinage in Britain and beyond*. Proceedings of the tenth symposium on monetary history, 1-21
Meyer, Gunther 2006. Mulich Familie, in *Biographischen Lexikon für Schleswig-Holstein und Lübeck*, vol. 12, Neumünster, Wachholz, 321-7
Mikkelsen, Egil 1998. Islam and Scandinavia during the Viking Age, in Elisabeth Piltz, *Byzantium and Islam in Scandinavia: Acts of a Symposium at Uppsala University, June 15-16, 1996*, Jonsered, Aström, 39-51
Mitchener, M. B., C. Mortimer and Mark Pollard 1988. The alloys of continental copper-base jettons, *Numismatic Chronicle* 148, 117-128
Molbech, Christian 1820. *Brief über Schweden im Jahre* 1812, vol.2. 66-69 (translated from Breve fra Svenske i Aaret 1812, Altona, Hamerich
Monod, T. 1969. Le Ma'den Ijâfen: une épave caravanière ancienne dans la Majâbat al-Koubrâ. *Actes du premier colloque international d'archéologie africaine*. Fort-Lamy, Chad, 286-320
Montero-Ruis, Ignacio and Alicia Perera 2007. Brasses in the early metallurgy of the Iberian Peninsula, in Susan La Niece, Duncan Hook, and Paul Craddock, *Metals and Mines: Studies in Archaeometallurgy*, London, Archetype, 136-139
Moore, Francis 1739. *Travels into the Inland Parts of Africa*, London
Moreland, John 2001. *Archaeology and Text*, London, Duckworth

Morton, John 1985. The Rise of the Modern Copper and Brass Industry in Britain, 1690-1750; unpublished PhD thesis, Birmingham University Library

Morton, Vanda 2009. The Brass Revolution, unpublished PhD thesis, University of Bristol

Mosselman, E. 1825. Fusion des Minerais de Zinc en Angleterre. *Annales des Mines: receuil de mémoires sur l'exploitation des mines*, 10. Paris, Conseil général des mines, 485-90

Murgatroyd, L. 2003. *Mill walks and industrial yarns: a history of the mills of Congleton and district*, booklet,

Musschenbroek, Peter van 1762. *Introductio ad Philosophicam Naturalem*, vol.2, Utrecht, Lugd. Bat

Musty, J. 1975. A brass sheet of first century A.D.-date from Colchester (Camulodunum), *Antiquaries Journal* 55, part 2, 409-11

Mutschlechner, Georg 1990. Bergbau auf Silber, Kupfer und Blei, in Gert Amman, *Silber, Erz und weisses Gold; Bergbau in Tirol*, exhibition catalogue, Tiroler Landesmuseum Ferdinandeum, Innsbruck, 20 May-28 October, 1990, 231-267

Needham, Joseph 1974. *Science and Civilisation in China* vol. 5/2. Cambridge, Cambridge University Press, 208, 212, 220

Neri, A., 1611 1662 translation by C. Merrett). *L'arte vetraria distinta, in libri sette*, London, Pulleyn, 239 ***

Newbury, Brian, Jon Almer, Dean Haeffner, Michael Notis, Brian Stephenson, Bruce Stephenson 2003. Synchroton Analyses of high-zinc brass Astrolabes, in *Archaeometallurgy in Europe*, 2. Proceedings of International Conference, Milan, September 2003, Associazione Italiana di Metallurgia, 359-368

Newbury, Brian, E. Dale and Michael Notis 2005. Revisiting the zinc composition limit of cementation brass. *Historical Metallurgy* 39/2, 75-81

Norden, John, 1593 1723 edition). *Speculum Britanniae:Middlesex*. London, Browne, 41

Northover, P. 1987 Metallurgical analyses, batch 2, Metcalf, D. M. and Peter Northover 1987. The Northumbrian royal coinage in the time of Æthelred II and Osbehrt, pp 187-218, in D.M. Metcalf (ed.), *Coinage in ninth-century Northumbria*, Proceedings of the 10th Oxford Symposium on coinage and monetary history, Oxford, BAR series 180, 187-218, Plates 14-21

Noel-Hume, Ivor 1969. *Guide to Artifacts of Colonial America*, New York, Knopf

Norris, Malcolm, 1978. *Monumental brasses: the craft*, London, Faber

North 2005, Brass mounted hangers from the Hounslow sword mill, *Journal of the Antique Metalware Society* 13, 24-27

Northover, Peter 1992. Materials issues in the Celtic coinage, in Melinda Mays, *Celtic Coinage: Britain and Beyond*. Eleventh Symposium on Coinage and Monetary History, Oxford, BAR British series 222. 235-299

Northover, Peter 1997. Appendix, in Houshang Mahboubian, 1997. *The Art of Ancient Iran: Copper and Bronze*, London, Philip Wilson, 325-41

Norton, J. and J. Robey 1985. Jacob Momma and the Ecton Copper Mines. *Bulletin of the Peak District Mines Historical Society*, 9/3, 195-196

O'Brien, William 2015. *Prehistoric Copper Mining in Europe: 5500-500 BC*, Oxford Universty Press

Ogier, Charles 1656. *Caroli Ogerii ephemerides, sive iter Danicum Sveciacum*, Paris, Luet, 134-5

Oldeberg, Andreas 1966. *Metallteknik under vikingatid och medeltid*, Stockholm, Seelig

Özdogan, Asti 1999. Çayönü, In Mehmet Özdogan and Nezih Ba gela, *Neolithic in Turkey: the cradle of civilization,* Istanbul, Arkeoloji ne Sanat Yayinlari, 35-63

Pal, Pratapaditya 1975. *The Bronzes of Kashmir*, Graz, Academic Press

Pal, Pratapaditya 1995. *Jain Art from India*, London, Thames and Hudson

Palme, Rudolf 2000. *Das Messingwerk Mühlau bei Innsbruck :ein Innovationsversuchs Kaiser Maximilians I: aus den Quellen dargestellt.* Hall in Tirol, Berenkamp

Parkinson, C.N. 1952. *The Rise of the Port of Liverpool,* Liverpool University Press

Paterson, Caroline 1998. Insular belt fittings from the pagan Norse graves of Scotland: a reappraisal in the light of scientific and stylistic analysis, in Justine Bailey, *Pattern and Purpose in Insular Art: Proceedings of the fourth International Conference on Insular Art*, held at the National Museum and Art Gallery, Cardiff, 3-6 September 1998, Oxford, Oxbow, 125-132

Pennant, Thomas 1796. *History of the parishes of Whiteford and Holywell*

Petruszka, Andrew 2011: Artifacts of Exchange: the muscular approach to Maritime Archaeology at Elmina, anthropology dissertation, paper 87, Syracuse University, New York

Pettus, John 1670. *Fodinae Regales: or the History, Laws and places of the Chief Mines and Mining Works in England etc.*, London, Basset, 25, 32-5, 72-3, 188

Pettus, John 1686. *Fleta Minor,* London, Bateman, Part 1, 86, 104, 285-6; Part 2, BR, BU, CU, GL,

Piersig, Wolfgang 2011. Kupferreport aus dem Jahre1820, Kupfer, Metall der Antike, Gegenwart, Zukunft: Fonds fur Technik Kultur, Kunst. *Beitrag zür Technik Geschichte* 17. GRIN Verlag, 29-31

Pike, Alistair 2002. Appendix, in Curtis, J., *Ancient Caucasian and related Material in the British Museum*, 87-99, London, British Museum Press

Pinkerton, John 1808-10). *A General Collection of the best and most interesting Voyages*, vol.6, London, Kimber and Conrad

Plot, Robert 1686. *The Natural History of Staffordshire*, Oxford, Halton, 165

Plumier, Jean 2014. Des ateliers de dinandiers à Bouvignes, in Nicolas Thomas, I Leroy and J.Plumier, *L'or des dinandiers, fondeurs et batteurs mosans au Moyen Age*, Bouvignes- Dinant,cahiers de la maison du patrimoine médiéval mosan 7, section I

Pohl, H. 1977. Staat, Kupfergehandel und Kupferarbeitung im Aachen-Stolberger Raum, in H. Kellenbenz, *Schwerpunkte der Kupferarbeitung und das Kupferhandels in Europa*, Böhlau, Cologne, 226-240

Pollard, Mark 2008. The Chemical Study of Metals, in Mark Pollard and Carl Heron, *Archaeological Chemistry*, Cambridge, Royal Society of Chemistry 198-234

Pöllnitz, K. L. von, 1737/1998. *Amusemens des Eaux d'Aix la Chapelle*, Amst

Ponting, Matthew, 1999. East meets west in post-classical Bet She'an: the archaeometallurgy of culture, change, *Journal of Archaeological Science* 26, 1311-1321

Ponting, Matthew, 2003. From Damascus to Denia: scientific analysis of three groups of Fatimid period metalwork, *Historical Metallurgy* 37/2, 85-105

Ponting, Matthew and I. Segal, 1998. ICP-AES analyses of Roman military copper-alloy artefacts from the excavation of Masada, Israel. *Archaeometry* 40/1, 109-123

Postlethwayt, Malachy, 1746. *The national and private advantages of the African Trade*, London, Knapton

Priesner, Claus, 1997. *Bayerisches Messing*, Stuttgart, Franz Steiner Verlag

Price, Neil, 2000. Novgorpod, Kiev and their satellites: the City-State model and the Viking Age: Polities of European Russia, in Hansen, Mogens Horman,(ed), *A Comparative study of thirty City-State Cultures,* Konlige Danske Videnskabernas, Selskad, 263-275

Quanya Wangan and Saschia Priewe, 2013. Scientific analysis of a Buddha attributed to the Yongle period of the Ming dynasty, *British Museum Technical Research Bulletin* 7: 61-68

Rackham, H. 1961. *Pliny Natural History*, vol.9, London, Heinemann, 126-7, translation of Pliny the Elder, Natural History, Book 35, chapter 2, dated c.77-9AD

Rao, S. R. 1991. Further excavations of the submerged city of Dwarka, In S. R. Rao, *Recent Advances in Marine Archaeology: Proceedings of the second Indian Ocean conference on Marine Archaeology of Indian Ocean Countries*, January 1990, Goa, Nio, 51-59, http://drs.nio.org/drs/handle/2264/3290

Radivojević, Miljana' Thilo Rehren, Shahina Farid, Ernst Pernicka, Duygu Camurcuoğlu, 2017. Repealing of the Catahöyük extractive metallurgy: The green, the fire and the 'slag' *Journal of Archaeological Science*, 101-122

Ralph, Elizabeth 1985. Bishop Secker's Diocese Book, in *A Bristol Miscellany*, Bristol Record Society 37, Bristol, 33-63

Ratelband, K. 1953. *Fijf Dagregisters van het Kesteel Sao Jorge da Mina* (Elmina aan der Goudkust), 1645-47, Hague, Martinus Nijhoff

Rattray, R.S. 1923. *Ashanti*, Oxford University Press

Rau, Reinhold 1975. *Fontes ad Historiam Regni Francorum Aevi Karolini illustrandum (Quellen zur Karolingischen Reichsgeschichte),* part 3. Darmstadt, Wissenschaftliche Buchgesellschaft

Rawley, J.A. 1981. *The Transatlantic Slave Trade: a history,* New York, Norton, 171-91

Ray, P. 1956. *History of Chemistry in Ancient and Medieval India,* Calcutta, Indian Chemical Society

Rayfield, Donald 2013. *Edge of Empires: a History of Georgia,* London, Reaktion Books

Rea, Alexander 1969. *South Indian Buddhist Antiquities,* Varanasi, Ideological Bookhouse

Reade, Julian 1991. *Mesopotamia,* London, British Museum Press

Reedy, Chandra 1997. *Himalayan bronzes: technology, style and choices,* Newark, University of Delaware

Reedy, Chandra 2003. New evidence for the historical context of Buddhist bronzes from Swat Valley, Northern Pakistan, In Paul Jett, *Scientific Research in the Field of Asian Art, Proceedings of the first Forbes symposium at the Freer Gallery of Art,* London, Archetype, 133-145

Rees, William 1968. *Industry before the Industrial Revolution,* Cardiff, University of Wales Press

Rehren, Thilo, Karl Deutmann, Andreas Hauptmann and Egon Lietz 1993. Excavation by the watch-tower and the eastern medieval city walls of Dortmund, in Heiko Steuer and Ulrich Zimmermann, *Montanarchäologie in Europa*, Simaringen, Jan Thorbecke, 303-314

Rehren, Thilo 1999. Small size, large scale. Roman brass production in Germania Inferiore. *Journal of Archaeological Science* 26, 1083-7

Rehren, Thilo 2002. Metallanalysen an romischen fibeln aus der CUT, appendix to Ulrich Boelicke, Die Fibeln aus dem Areal der Colonia Ulpia Traiana. *Xanten Bericht* 10, 146-151

Rehren, Thilo and Marcos Martinón-Torres 2008. Naturam ars imitate: European brassmaking: between craft and science, in Marcos Martinón-Torres and Thilo Rehren, *Archaeology, History and Science: Integrating approaches to ancient materials*, Walnut Creek, California, Left Coast Press, 167-188

Reynolds, Susan 1962. Isleworth Hundred, Mills, *Victoria County History, Middlesex* 3, Oxford University Press, 112-14

Richards, John F. 1993. *The Mughal Empire*, Cambridge, Cambridge University Press

Richardson, Colin and Roger Brownsword 1986. A roman Brooch from Braithwaite, Cumbria, Transactions of the Cumberland and Westmoreland Antiquarian and Archaeologicall Society, note 12, 264-6

Richardson, David 1986. *Bristol, Africa and the Eighteenth-Century Slave Trade to America*, 1, 1698-1729, The Years of Expansion, Bristol, Bristol Record Society

Riederer, Josef 1982. Die Berliner Datenbank von Metallanalysen, Metallanalysen Nūrnberger Statuetten aus der Zeit der Labenwolf Werkstatt . *Berliner Beiträge zur Archäometrie* 7, 157-202

Riederer, Josef 1983. Die Berliner Datenbank von Metallanalysen, Metallanalysen an Erzeugnissen der Vischer-Werkstatt. *Berliner Beiträge zur Archäometrie* 8, 88-90

Riederer, Josef 2000. Die Berliner Datenbank von Metallanalysen Kulturgeschichtlicher Objekte, II, Objekte aus Kupferlegierungen des 17/18 Jahrhunderts, der Renaissance und des Mittelalters *Berliner Beiträge zur Archäometrie* 17, 143-216

Riederer, Josef 2001. Die Berliner Datenbank von Metallanalysen Kulturgeschichtlicher Objekte, III. Römische Objekte. *Berliner Beiträge zur Archäometrie* 18, 139-259

Riederer, Josef 2002. Die Berliner Datenbank von Metallanalysen Kulturgeschichtlicher Objekte, IV, Objekte der Mitteleuropaischen Bronzezeit sowie etruskische, sardische, greichische, ägyptische, vorderasiatische Objekte, *Berliner Beiträge zur Archäometrie* 19, 72-226

Riederer, Josef 2003. Die Berliner Datenbank von Metallanalysen, Kulturgeschichtliche Objekte V, Objekte aussereuropäischer Kulturen, *Berliner Beiträge zur Archäometrie* 20, 133-199

Roderburg, Andreas von 1927. Aus der Geschichte des Stolberger Messinggewerbes. In A.Roderburg (ed), *Stolberg, Rheinland: Beiträge zur Geschichte und Kultur der alten Kupferstadt*. Düsseldorf, Otto Fritz, 9-16

Rösler, Balthasar 1700. *Speculum Metallurgiae Politissimum, oder hell-polierter Berg-Bau-Spiegel*. Winckler, Dresden.

Roslund, Mats 1997. Crumbs from the rich man's table: Byzantine artefacts in Lund and Sigtuna c. 980-1250. In Hans Andersson, Peter Carelli and Lars Ersgård, *Visions of the Past: Trends and Traditions in Swedish Medieval Archaeology*. Stockhom, Lund Studies in medieval Archaeology 239-297

Rouse, Paul 2015. *Sport and Ireland: a history*. Oxford University Press

Roxburgh, Marcus, S. Heeren, H. Huisman, and B. van Os. Early Roman copper-alloy brooch production: a compositional analysis of 400 brooches from Germania Inferior. *Journal of Roman Archaeology* 29, 2016, 411-421

Rudder, Samuel 1779. *A New History of Gloucestershire*, Cirencester, privately printed

Sachau, Edward 1989, 2003 reprint). *Alberuni's India*, Dehli, Lowprice Publications

Sagona, Antonio 2018, *The Archaeology of the Caucasus: from earliest settlements to the Iron Age*, Cambridge University Press

Sahlin, Carl 1954. *Valsverk inom den svenska metallurgiska industrien intill början av 1870-talet*, Stockholm, Jernkontoret

Saint-Amand, Pascal 2014. Jean Josès de Dinant, maître fondeur de laiton au XIVe siècle, In Nicolas Thomas, I Leroy and J.Plumier, *L'or des dinandiers, fondeurs et batteurs mosans au Moyen Age*, Bouvignes- Dinant, cahiers de la maison du patrimoine mediéval mosan 7, section VII

Sas, Kathy 2004. Military bracelets in Oudenberg: troop movements, origins and relations in the Litus Saxonicum in the 4th century AD, in Frank Vermeulen, Kathy Sas and Dhaeze Wouter (eds) *Archaeology in Confrontation: aspects of Roman Military Presence in the Northwest. Studies in honour of Prof. Emeritus Hugo Thoen*, Ghent, Academia, 343-378

Saussus, Lise and Benoit Wery 2014. Un cadran solaire portative dans le puit du chateau de Logne, in Nicolas Thomas, I. Leroy and J. Plumier, *L'or des dinandiers: Fondeurs et batteurs mosans au Moyen Age*, cahiers de la maison du patrimoine mediéval mosan 7, Bouvignes- Dinant, section VIII

Schiffer, Peter and Nancy 1978. *The Brass Book*, Exton, Pennsylvania, Schiffer

Schleicher, Karl 1974. *Geschichte der Stolberger Messingindustrie*, Stolberg/Rheinland, Stolberger Metallwerken KG

Schmauder, Michael and Frank Willer 2004. Römische Kastenbeschläge aus Buntmetall im Römisch-Germanishen Museum, Köln, *Kölner Jahrbuch* 37, 137-221

Schmauder, Michael and Frank Willer 2010. Ein spätantiker Schlossbeschlag im Römisch-Germanischen Museum, Köln, *Kölner Jahrbuch* 43, 675-694

Schulte, Wilhelm 1937. *Iserlohn: die Geschichte einer Stadt*. Iserlohn town council

Segers-Glocke, Christiane, and Harald Witthöft 2000. *Aspects of Mining and Smelting in the Upper Harz mountains up to the 13th/14th century*, St Katharinen, Scripta Mercaturae

Seib, Gerhard 1977. Studien zur Geschichte der Industriearchitektur in Hessen (iii), der Messinghof in Kassel-Bettenhausen, *Hessische Heimat* 27/4, 170-175

Shalev, Sariel and Mira Freund 2002. Using a Traditional metal workshop in modern Cairo as a ready-made' lab for studing aspects of early Islamic metallurgy, *Bulletin of the Israeli Centre in Cairo* 25, 21-30

Sharma, Deo Prakash 2000. *Early Buddhist Metal Images of South Asia*. Delhi, Bharstija Kala Prakashan

Sharp-Joukowsky, Martha 1996. *Early Turkey: an introduction to the Archaeology of Anatolia from Prehistory through the Lydian period*, Dubuque, Iowa, Kendall Hunt

Shaw, T. 1970. *Igbo Ukwu: an account of archaeological discoveries in Eastern Nigeria*, vols. 1 and 2, London, Faber and Faber

Shepard, Jonathon 2008. The Viking Rus and Byzantium, in Stafan Brink and Neil Price, *The Viking World*, London, Routledge, 496-516

Sheppard, Mubin 1971. *Taman Indera: Malay Decorative Arts, Past and Present*. London, Oxford University Press

Sindbaek, Søren 2001. An object of exchange: Brass bars and the routinization of Viking Age long-distance exchange in the Baltic area, *Offa*, 58, 49-60

Slafski, Heinz 1962. *Der Jungere Peter Vischer*, Nürnberg, Hans Carl

Smith, E.A. 1918. *The Zinc Industry*, London, Longmans Green

Smith, Richard 1994. An overview of the principles of copper metallurgy and the practice at Keswick, 1567-1602. *Bulletin of the Peak District Mines Historical Society* 12/3, 116-123

Song Yingxing 1968. *T'ien Kung Ka'ai Wu: Chinese Technology in the Seventeenth Century*, University Park, Pennsylvania

Souza, George 1991. Ballast goods: Chinese maritime trade in zinc and sugar in the seventeenth and eighteenth centuries, In Roderich.Potak and Dietmer Rothermund, *Emporia, Commodities and Entrepreneurs in Asian Maritime trade, c.1400-1750*, Stuttgart, Steiner Verlag, 291-315

Sperber, Erik 1989. The weights found at the Viking Age site of Paviken, *Forvannem 84*, 129-34

Srinivasan, Sharada 1999. "Preliminary insights into provenance of south Indian copper alloys and images using a holistic approach of comparisons of their lead isotopes and chemical composition with slags and ores", in Suzanne Young, Mark Pollard, Paul Budd and Robert Ixer, *Metals in Antiquity*, BAR International Series 792. Oxford, Archaeopress, 200-211

Srinivasan, Sharada 2008. Megalithic and Early Historic metalwork in southern India: some issues of technology, emergence and transmission. In: Gautam Sengupta and Sharmi Chakraborty, *Archaeology of Early Historic South Asia.* New Delhi: Pragati Publications with Centre for Archaeological Studies and Training, Eastern India, 375-392

Stahlschmidt, Rainer 1970. Das Messinggewerbe im Spätmittelalterlichen Nürnberg, *Mitteillungen des Vereins für Geschichte der Stadt Nürnberg 57*, 124-148

Stead, Ian 1986. The Brooches, in Ian Stead and Valery Rigby, *Baldock: the excavation of a Roman and pre-Roman settlement,* London, Society for the Promotion of Roman Studies, 108-125

Steel, Louise 2004. *Cyprus before History: from the earliest settlers to the end of the Bronze Age,* London, Duckworth

Steuer, Heiko 1999. Handel und Wirtschaft in der Karolingenzeit, In Christoph Stiegemann and Matthius Wemhoff, *Kunst und Kultur der Karolingenzeit*, Mainz, Phillip von Zabern, 406-416

Stos-Gale, Zofia 1992. The origin of metal objects from the early Bronze Age site of Thermi on the island of Lesbos. *Oxford Journal of Archaeology* 11 (2) 155-77

Stos-Gale Zofia 1993. The origin of metals from the Roman-period levels of a site in south Poland. *Journal of European Archaeology* 1/2, 101-130

Straume, E.M., Bollingberg, H. J. and Christensen A. E. 2005. Pearl-edged basins from the late Roman and Migration Periods in Norway: Spectrochemical elemental analysis of the Norwegian basins and of some related basins from Belgium and Germany, in H. J. Hassler, *Studien zur Sachsenführung 15*, Oldenburg, 457-484

Strieder, J. 1933. Negerkunst von Benin und Metallexportgewerb im 15 und 16 Jahrhundert, *Zeitschrift für Ethnologie 1/3*, 249-259

Stringer, Moses 1713. *Opera Mineralia Explicata*, London, Brown

Styles, J. 1994. Manufacturing, consumption and design in eighteenth-century England, in J. Brewer and M. Porter, *Consumption and the World of Goods*, London, Routledge, 527-554

Suttor, Marc 2014. La Dinanderie, "fille" de la Meuse? In *L'or des Dinanderies, fondeurs et batteurs mosans au moyen Age*, exhibition catalogue, Maison du Patrimoine Médiéval Mosan, March-November 2014, Bouvignes- Dinant

Swedenborg, Emanuel 1734. *Principia*, 3. *Tomo de Regnum Subterraneum sive Minerale de Cupro et Orichalco*, Dresden, Hekel, 352-4

Talbot-Rice, David 1965. *The Dark Ages: the making of European Civilizations*, London, Thames And Hudson

Thévenot, Jean de, 1683 1687 translation by A. Lowell). *Travels of Mr de Thévenot into the Levant*, Vol.1/3, The Relation of Indostan, the New Moguls and of the People and Countries of the Indies

Thiessen, Tamara, 2016. *Borneo, Sarawak, Brunei*, Bradt Travel Guide, third edition, Connecticut, Globe Pequot,

Thomas, Nicolas 2007. Quand Melle enterrait ses metallurgists: étude de creusets lutés découverts à Melle et à Niort en context funéraire mediéval (Deux-Sèvres, France), *ArchéoSciences, revue d'archéometrie*, 30, 45-59

Thomas, Nicolas 2010. L'industrie du cuivre au bas moyen âge, *Histoire et Image Medievale* 34, 32-37

Thomas, Nicolas and David Bourgarit 2006. Une industrie medieval du bronze. *La Recherche* 403, 56-58

Thomas, Nicolas and David Bourgarit 2014. Les techniques de production des batteurs et fondeurs mosans au Moyen Age (XIIe-XVIe siècles), in Nicolas Thomas, I Leroy and J.Plumier, *L'or des dinandiers, fondeurs et batteurs mosans au moyen Age,* Cahiers de la maison du patrimoine mediéval mosan 7, section V, Bouvignes- Dinant

Thomas, Nicolas, Bastien Asmus, David Bourgarit, Jean Plumier and Marie Verbeek 2013. Commerce et techniques métallurgiques: les laitons mosans dans le marché européen du Moyen Âge (XIIIe-XVIe siècles), in, Stephanie Thiébault et Pascal Depaepe, *L'archéologie au laboratoire*, Paris, La Découverte, 169-182

Thomas, Nicolas and Françoise Urban 2014. Du côté du marché : uniformité et diversité des productions en alliage à base de cuivre au Moyen Age, in Nicolas Thomas, I Leroy and J.Plumier, *L'Or des Dinandiers, fondeurs et batteurs mosans au moyen Age*, Cahiers de la maison du patrimoine mediéval mosan 7, section VI

Thompson, A., F. Drew and J. Schofield, 1984. Excavations at Aldgate, 1974, *Post Medieval Archaeology* 18, 1-148

Thornton, Christopher and Christine Ehlers, 2003. Early Brass in the Ancient Near East. *Bulletin of the Institute for Archaeo-Metallurgical Studies* 33, 3-8

Thornton, Christopher, 2007. Of brass and bronze in prehistoric South-West Asia. In Susan La Niece,Duncan Hook and Paul Craddock, *Metals and Mines: Studies in Archaeometallurgy*, London, Archetype. 123-135

Tizzoni, Marco, 1996. Condensatori per la produzione dell'osso=ido di zinco da Conca del Naviglio a Milano. *Notizie Archaeologiche Bergomensi* 4, 111-120

Tizzoni, Marco, 1999. Le scorie di rame, in Costanza Cucini and Marco Tizzoni, *La miniera perduta: cinque anni di ricerche archaeometallurgiche nel territorio di Bienno*, Bienno, Comune di Bienno, 195-198

Toločko, Petr, 1998. Die Rolle der Handels- und Handwerkszentren des 9 Jahrhunderts bis zum Anfang des 11 Jahrhunderts bei der Entstehung der altrussischen Städte, in Anke Wesse, *Studien zur Archäologfie des Ostseerauns. Festschrift für Michael Müller-Wille*, Neumunster, Wachholtz, 213-217

Toussaint, Jacques, 2005. *Art du Laiton, Dinanderie*, booklet 7, Société Archéologique de Namur

Treixler, G., 1933. *Musik und Heimatfest, Graslitz*, Kraslice, booklet

Trinder, Barry and Jeff Cox, 1980. *Yeomen and Colliers in Telford*, London, Phillimore

Tucci, Ugo, 1977. Rame dell'economia veneziana del secolo XVI, In Hermann Kellenbenz (ed), *Schwerpunkte des Kupferproduktions und des Kupferhandels in Europa 1500-1650*, Bohlau, Vienna. 95-116

Turner, Anthony, 1987. *Early Scientific Instruments, Europe, 1400-1800*, London, Philip Wilson

Twilley, John, 2003. Scientific investigations of an enthroned Buddha figure, In Paul Jett, 2003. *Scientific Research in the Field of Asian Art, Proceedings of the first Forbes symposium at the Freer Gallery of Art*, London, Archetype, 140-148

Tylecote, Ronald, 1972. A contribution to the metallurgy of 18th and 19th-century brass pins, *Post Medieval Archaeology* 6, 183-190

Tylecote, Ronald and Joan Day, 1991. *The Industrial Revolution in Metals*, London, Institute of Metals]

Ucik, Friedrich Hans, 2002. Messing in Österreich: die Herstellung und die wirtschaftliche Bedeutung unter besonderer Berücksichtigung der ehemaligen Messinghütte Möllbrücke, *Carinthia* 11, 192/112, Klagenfurt, 161-188

Unger, Eike, 1966. Nürnbergs Handel mit Hamburg im 16. und beginnenden 17. Jahrhundert, *Mitteilungen des Vereins für Geschicht der Stadt Nürnberg* 54, 1-193

Ure, Andrew, 1839. *A Dictionary of Arts, Manufactures and Mines*, London, Longman

Vaccari, Ezio, 2007. From Tyrol to Venice: the papers of Giovanni Arduino 1714-1795 as valuable sources for the history of Mining and Geology. *GeoAlp, special issue* 1/5, 155-164

Vainker, Shiela, 1971. *Chinese Pottery and Porcelain from Prehistory to the Present*, London, British Museum

Verbeek, Marie, 2014. Les sources archéologiques du travail du laiton: les ateliers de Dinant, in N.Thomas, I.Leroy, and J. Plumier, *L'or des Dinandiers: fondeurs et batteurs mosans*, Bouvignes/Dinant, cahiers de la Maison du Patrimoine Médiéval Mosan No. 7, section IV

Vincent, Brice, David Bourgarit and Paul Jett, 2012. Khmer bronze metallurgy during the Angkorian period: technical investigation of a new selected corpus of artefacts from the National Museum of Cambodia, Phnom Penh, in *Studies of Ancient Asian Metallurgy using Scientific Methods, Proceedings of the fifth Forbes Symposium at the Freer Gallery of Art*, Washington D.C. (28-29 October, 2010, Archetype, 124-153

Vogt, J., 1969. *Portuguese Rule on the Gold Coast, 1469-1682*, Athens, USA, University of Georgia Press, 69

Vries, J. de, 1994. Between purchasing power and the world of goods: understanding the household economy in early modern Europe, in J. Brewer and M. Porter (eds.), *Consumption and the World of Goods*, London, Routledge, 85-132

Wakelin, Peter and Joan Day, 1982. Joseph Harris in Bristol, 1748, *Bristol Industrial Archaeology Journal* 15, 8-12

Waldbaum, Jane, 1983. *Metalwork from Sardis: the finds through 1974*, Harvard University Press, Cambridge, Massachusetts

Ward, R., Susan La Niece, Duncan Hook, and R. White, 1995. Veneto-Saracenic metalwork: an analysis of the bowls and incense-burners in the British Museum, in D. Hook and D. Gaimster, *Trade and Discovery: the scientific study of artefacts from post-medieval Europe and beyond*. Occasional paper No. 109, 235-58.

Warner, Richard, 1801. *The History of Bath*, London, Robinson

Warncke, Johannes, 1938. Die Tremse Mühle aus der Geschichte eines Lübecker Gewerkebetriebes, *Zeitung des Vereins für Lübecke Geschichte und Altertums* 14, 207-224

Watson, Richard, 1786. *Chemical Essays*, London, Evans, vol. 4, 1-84

Webster, John, 1671. *Metallographia or a History of Metals*, London, Kettilby, 250, 337-99

Weeks, Lloyd, 2003. *Early Metallurgy of the Persian Gulf: Technology and Trade in the Bronze Age World*, Leiden, Brill

Weisgerber, Gerd, 2007. Roman brass and lead ingots from the Western Mediterranean, in Susan La Niece, Duncan Hook, Paul Craddock, *Metals and Mines: studies in archaeometallurgy*, London, Archetype, 148-156

Welter, Jean-Marie, 2003. The zinc content of brass: a chronological indicator? *Techné* 18, 27-36

Werner, Otto, and Frank Willett, 1975. The composition of brasses from Ife and Benin, *Archaeometry* 17/2, 141-156

Westermann, Ekkehard, 1971. *Eislebener Garkupfer und seine Bedeutung fur den Europäischen Kupfermarkt von 1460-1560*, Köln, Bohlau

Westermann, Ekkehard, 2002. Kupferhalbfabrikathandel vor dem Tor zur Welt. In Rainer Gömmel and Marcus Denzel (eds), *Weltwirtschaft und Wirtschaftsordnung: Festschrift fur Jurgen Schneider zum 65 Geburtstag*, Stuttgart, Franz Steiner Verlag, 85-100

Weston, Tony, 2009. English Roasting-Jacks, Weight-driven Jacks of the 17th century, *Journal of the Antique Metalware Society* 17, 14-26

Weston, Tony, 2011. English Roasting-Jacks, Non-standard weight-driven Spit Jacks of the eighteenth century, *Journal of the Antique Metalware Society* 19, 60-76

Wiersma, Lisa, 2014. abstract 30, in *Medieval Copper, Bronze and Brass*, handbook, International Symposium, Dinant-Namur, May 2014

Wiersma, Lisa 2018. Monumental Dinanderie: achievement and tradition of metal sculpture in the Low Countries in the late Gothic and Renaissance period, in Nicolas Thomas and Pete Dandridge, *Cuivres, bronzes et laitons médiévaux*, Agence Wallone du Patrimoine, Études et Documents, Archéologie 39, 377-386

Whisker, James Biser, 1924. *Pennsyvania workers in brass copper and tin, 1681-1900*, Lewiston, NY, Edwin Mellen Press

White, D. P., 1977. The Birmingham button industry, *Post Medieval Archaeology* 11, 67-79

Wiegleb, Johann Christian, 1789. *A General System of Chemistry*, vol.2, trans.G. Hopson, from *Handbuch der Allgemeine Chemie*, London, Robinson

Willems, J, 1973. Le quartier artisanal gallo-romain et merovingien de ‚batta' à Huy, *Archaeologia Belgia* 148, 1-64

Willett, Frank and E. V. Sayre, 2000. The elemental composition of Benin memorial heads, *Archaeometry* 42, 159-188

Williams, Gareth, Peter Penitz and Matthius Wemhof, 2014. *Vikings, Life and Legend*, London, British Museum

Wilson, Andrew, 2002. Machine, Power and Ancient Economy, *The Journal of Roman Studies* 92, 1-32

Wittek, Ansgar 1984. Mühle, Hammerwerk, Kraftwerk, Freilichtmuseum: der Nürnberger Vorort Laufamholz. In Ansgar Wittek, *Mitteilung des Vereins für Geschichte der Stadt Nürnberg* 54, 158-170

Woiwode, C., 1996. Das Messingwerk bis zum Jahre 1863. In *Messingwerk – eine historische Industriesiedlung am Finowkanal*. Berlin, Technische Universität, 21-32

Woolrich, A. P. and Alex den Ouden (translators) 1987. *Ferrner's Journal: an Industrial Spy in Bath and Bristol [1760]*, Eindhoven, De Archaeologische Pers

Xu Li, 1998 (trans. J. O. Petersen). Traditional zinc-smelting technology in the Guma district of Hezhang county, In Paul Craddock, *Two Thousand Years of Zinc and Brass*. London, British Museum Occasional Paper 50, 115-131

Yante, Jean-Marie, 2014. Dinant et Bouvignes, pôles majeures de la batterie mosane (XIIIe-XVIe siècles), in Nicolas Thomas, I Leroy and J.Plumier, *L'or des dinandiers: Fondeurs et batteurs mosans au moyen Age*, Dinant-Bouvignes, cahiers de la Maison du Patrimoine Médiéval Mosan No.7, section 1

Young, Rodney, 1981. *Three Great Early Tumuli*, Philadelphia, University of Pennsylvania Library

Zebrowski, Mark, 1997. *Gold, Silver and Bronze from Mughal India*, London, Alexandria Press,

Zechi, Helke, 2010. Theatre of War: two enamelled copper snuff-boxes with images, *Journal of the Antique Metalware Society* 18, 48-55

Zhou Weirong, 2007. The origin and invention of zinc smelting technology in China, In Susan La Niece, Duncan Hook and Paul Craddock, *Metals and Mines: Studies in Archaeometallurgy*, London, Archetype. 179-186

Zhou Wenli 2012. Distilling Zinc in China: the technology of large-scale zinc production in Chonqing during the Ming and Qing dynasties (AD 1368-1911, unpublished PhD thesis, Institute of Archaeology, University College, London

Zhou Wenli 2016. *The Technology of large-scale Zinc Production in Ming and Qing China*, British Archaeological Reports, international series 2835, Oxford, BAR publishing

Zhou Wenli, Marcus Martinón Torres, Jianli Chen, Haiwang Liu, Yangxiang Li, 2012. Distilling zinc for the Ming Dynasty: the technology of large scale zinc production in Fengdu, south-west China, *Journal of Archaeological Science* 39 (4), 908-921

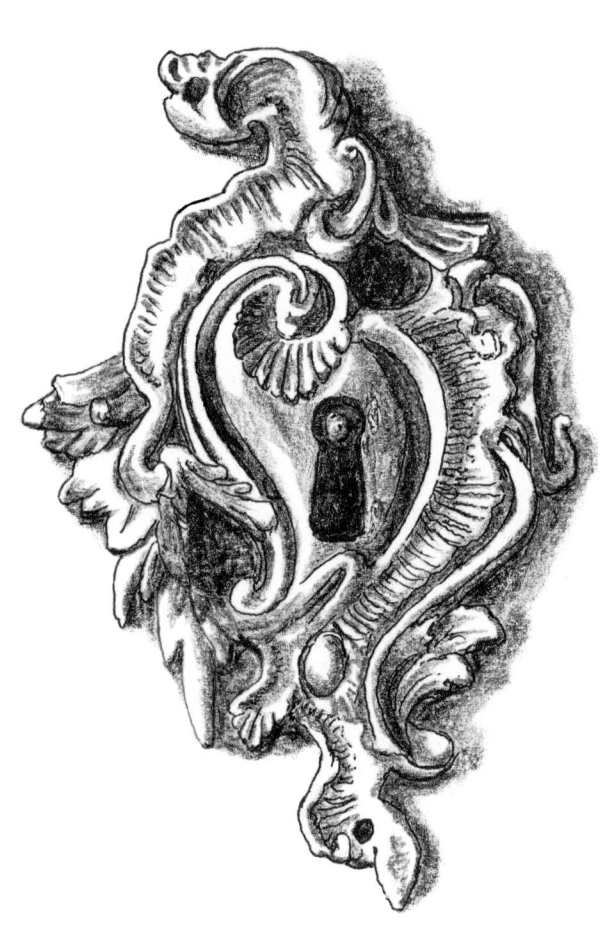

Appendix
Metallurgical tables relevant to individual chapters

Metallurgical tables are avaliable online at https://bit.ly/2HgrCwi

Index

Aachen/Aix-la-Chapelle, 22, 45–53, 78–93, 106–118, 130–175, 188–207, 231
Abbasid, 35, 59–65
Aberdulais, 123
Abinger, 163–164
Abu Dhabi, 9
Accra, 155
Achenrain, 215–217, 253–257, 302
Aden, 143
Adlerwald, Carl, 293
Æthelred, king, 49
Afghanistan, 8, 30–43, 102–103
Africa, 41, 63–69, 91–108, 139–153, 188–207, 224–239, 268–287
Africa (East), 98
Africa (North), 65, 91
Africa (West), 63–69, 97–107, 139, 153, 189–190, 224–239, 281–282
Agadès, 2
Agricola, (Georg Bauer), 119–120
Agra, 72, 103, 183
Akan, 108, 154–155
Akhbar, emperor, 103–104
Akkadia, 10
Albacete, 19
Albania, 44
Albany (U.S), 198
alchemy, 40
Alchester, 25
Alcuin, 51
Alderley Edge, 283
Alesia, 21
Alexander the Great, 15–19

Algeria, 191
al Jazarī, 43
alloy, 1–65, 79–85, 101, 114–126, 140–144, 167–186, 211, 227–235, 266–269, 301–305
Almoravid, 63–65
Altenberg or Vieille Montagne, 22, 45–55, 77–95, 114–131, 145–152, 288, 300
America, north, 80, 281
America, south, 144, 221
American Revolutionary War, 296, 304
Amoy, 184, 221
Amsterdam, 132, 144–167, 185–198, 226, 274
Amu Darya, 10
anabaptist sect, 166–167
Anatolia, 2–21, 43, 86
Andalucia, 64
Andeira/Andreida, 17–20
Andhra Pradesh, 29–39, 104, 269
Angkor, Angkor Wat, 75
Anglesey, 283–289
Anglo-French war, 239
Anglo-Saxon, 49, 264
Anglo-Scandinavian, 49
Annamaboe, 189
Anthée, 22
Antioch, 41
Antwerp, 83–106, 131–132, 144–157
Aphrodisias, 3
Arab, Arabia, 9, 42, 60–67, 76, 96–107, 142
Aravalli hills, 16, 38, 73

Arboga, 94, 132–144, 293–294
Archangel, 247
Arda/Alada, 189–190
Arduino, Giovanni, 280
Armenia, 3–12, 68, 102–103
Arnold, John, 265
Arnstadt, 169, 213, 253
Aschaffenburg, 110
Ashanti, 155–156, 189, 225, 268
Asia Minor, 8–26, 44
Assyria, 13
astrolabe, 42, 65–68, 85–86, 114, 140
Athens, 17–19
Atranjikhera, 14
Auertal, 167, 213, 253
Augsburg, 26, 87–132
Aurangzeb, emperor, 181–183, 223
Aureillon family, 246
Aurichalcum, 17, 51
Auronzo, 216, 256
Australia, 267
Austria, 2, 90, 109–123, 170–174, 247–257, 298–304
Avar, 43–46
Avesta, 200–202
Avoca, 284
Avon Mill, 237
Awdaghust, 63–65
Ayawaro, 155
Ayyabid dynasty, 43
Ayen, 297
Azelick, 2
Azerbaijan, 12–13
Baber's Tower, 241–242
Babylon, 18
Baghdad, 8, 59–65

Baldock, 24
Balikesir, 17
Bäk, 128–129, 150, 205–206
Balkans, 2, 18–21, 40, 52, 87
Ball, 192
Balikesir, 17
Baltic, 21, 44–95, 147, 203–205
Bangladesh, 15, 34
Bantam/Bantem, 144
Banten, 144
Baptist Mills, 167, 188, 233–242, 274–285
Barbary, 97, 126–127
Bark, Lambrecht, 134
Batavia/Jakarta, 183, 221
Bateshwar, 72
Bath, 237–239, 303
battery or trip hammer, 88–93, 112–133, 147–175, 188–216, 229–262, 278–303
batteur, 84
Bavaria, 2, 54, 86, 109, 215, 257, 302
Becks, Gerlach, 188
Beijing, 144, 180, 267
Belgium, 22, 78, 117–126, 207, 281, 297
Benares/Varanasi, 30, 222, 269
Bengal, 34–40, 71–72, 103, 221–223, 269
Benin, 107, 154–155, 189–190, 224–226, 239, 268
Berber, 65, 97
Bergamo, 18–22, 52, 91
Berlin, 206–214, 242–247, 282, 297–298
Berry, William, 166–167
Bettenhausen, 211–212, 251, 304
Beuthen/Bytom, 115–116, 169–170, 252
Bhir Mound, 15
Biainili, 12
bidri, 140–141, 182

Bienno, 91

Bihar, 15, 34

Biringuccio, Vanocchio, 109–119, 196, 244

Birmingham, 194, 211, 235–243, 272–290

Bisham, 188, 229–231, 287

Bitton, 279–280

Bivort, Henri de, 297

Bjurfors, 199–201, 251–260, 291–292

Black Sea, 3–21, 41–67

Bohemia, 57, 85–94, 112–120, 132–136, 167–170, 213–214, 258, 300–301

Bolzano/Bozen 44, 90, 109, 215–216, 254

Bombay/Mumbai, 210, 302

Bordeaux, 297

Bosley, 284

Boston, Massachusetts, 78, 175, 198, 229, 304

Böttger, Johann F., 281

Boulle, Charles, 185, 297

Boulton, Matthew, 272–290

Bouvignes, 57–58, 78–82

Bradley, James, 266

Brandenburg, 115–116, 152–157, 169, 206, 245–247, 298

Braunschweig (Brunswick), 54–55, 78–94, 121

Brazil, 153

Bremen, 118, 130

Breslau, 86, 116, 172, 297–299

Bridgewater, 186

Brilon, 130, 169, 212–213

Bristol, 80–81, 138, 161–167, 188–199, 225–244, 266–290

Bristol Channel, 80, 239

Britain, 20–26, 41–57, 79, 107, 125–126, 157, 221–242, 256–257, 281–305

Brittany, 118, 151, 233

Brixlegg, 215–216, 302

Brocklesby, Peter, 163–164

Brode, John, 125–127

Bronze Age, 3–12

Brookhouses, 281

Browall, Carl, 259, 292–293

Bruges, 77–86, 105–106

Brunei, 105

Brunswick, Braunschweig, 54–55, 78–94, 121

Braunschweig, 54–55, 78–94

Brussels, 83–86, 111, 172

Bulgaria, 2

Bundheim, 120–121, 207

Burgundy, 83–93

Burma/Myanmar, 40, 74, 141, 182, 224, 268

Bustleholme, 289–290

Byfleet, 232–236

Byron, John, 164–166, 194

Bytom/Beuthen, 115–116, 169–170, 252

Byzantine, 41–63, 264

Cadiz, 19, 274

Caesarea (Cairo), 43, 63-65, 95–97

Calabar, 227, 288

calcining, 16, 134, 237, 250, 284–299

calamine, 5–29, 45–58, 77–175, 191–261, 274–303

Cambay, 143, 181–182

Cambodia, 75–76, 142

Cameroon, 107, 221–227

Campania, 22

Canterbury, 24

Canton, 105, 179, 221–222, 266

Cape Corso, 190, 225–226

Caribbean, 175, 189–190, 207, 227–239, 274–281, 304

Carinthia/Kärnten, 90, 112–120, 174, 216–217, 254–256, 303

Carlisle, 79

Carolingian dynasty, 61

Carpathians, 61

Caspian Sea, 3–13

Castello di Parre, 18

casting/foundry work, 55, 165, 243

Catal Höyük, 2

Caucasus, 3–20, 58, 102–103

Cavustepe, 13

Çayönü, 2

Celtic, 20–26

cementation, 5–54, 74–77, 95–107, 121–130, 142–184, 215, 236, 250, 271–302

Central Asia, 29–44, 61–76, 103

Ceuta, 67, 97

Chamen, Aegidius, von der, 83, 117

Champion, John, 285–286

Champion, Nehemiah, 236–242, 266–285

Champion, William, 241–242, 277–280

Champs du Boule, les, 151

Chandragupta, 15

charcoal, 5–16, 36, 54, 70–173, 192–216, 250–259, 277–303

Charlemagne, 50–53

Charles I, 157–166

Charles II, 185–196

Charles V, emperor, 111–115

Chaucer, Geoffrey, 86

Cheadle, 281–290

Chernykh, 3

Chew, 237

Chile, 152, 300

Chilworth, 162–163

China, 2–4, 30–33, 65–77, 89, 101–105, 118, 139–144, 161, 179–184, 221–224, 241–242, 257–267, 305

Chongqing province, 102, 141–143

Chongzhen, emperor, 143–144

Clarke, Samuel, 188

Claudius, emperor, 25–26

Coalbrookdale, 197, 225, 242–243

coins, 15–37, 50–60, 101–102, 143, 159–170, 222

Colchester, 24–26

Cologne (Köln), 47–57, 77–85, 118–119, 152, 171

Colombo, 223

colour of brass, 140, 151, 176, 186, 214

Connecticut, 280

Constantine, emperor, 40–44

Constantinople, 17, 40, 53, 117, 210

Coptic, 41, 264

Cordoba, 50–63

Cornelimunster, 84

Cornwall, 46, 80, 122–128, 195, 230–233, 279, 300

Corsica, 22

Coromandel, 181

Cotswold, 161–167

Croatia, 169

Cromwell, Oliver, 157, 193–196

Cronebane mine, 284

Cronström, family, 198–201, 260

crucible, 6–7, 24, 36–57, 79, 109, 132–151, 192, 209–212, 232, 247–259, 277–280, 303

Cumbria, 57, 77, 122

Cyprus, 2–17

Czech, 44–46, 90, 112–131, 167–170, 213, 252, 297
Dalversin, 10
Damascus, 43, 65–67, 95–96
Dannemora, 132
Danube/Donau river, 21, 61, 254
Danzig, 95, 116
Darby, Abraham, 233-244
Darius, 14–15, 38
Daulagala, 223
Deccan, 14, 39–40, 71, 182–183
Defoe, Daniel, 240–248
deformation-working, 184
Delaware, 229
Delphi, 19
Demetrius, Daniel, 191–192
Denia, 63
Denmark/Danish, 58–61, 157, 189, 201–206 297
Deptford, 184, 266
Derbyshire, 165–166, 195, 235, 282–290
Devon, 126–128, 195, 232
Dillinger, Bergrath, 281
Dinant, 22, 47–98
Diodorus Siculus, 17
Dioscorides, 19, 38
distillation (metal), 16, 36–39, 73–75, 102, 140–143, 179–184, 214–224, 241–242, 274–285
Dobunni tribe, 21
Dockwra, William, 191–192
Dognascau, 254
Döllach, 281
Dony, Daniel, abbot, 281
Dortmund, 51–54, 118, 248
Dresden, 197, 214, 301

Düppengiesser family, 145–149
Düna, 52–55
Dutch, 107–108, 122–209, 221–259, 271–274, 299
Duwall family, 292
Earnshaw, Thomas, 265
Ebener, Erasmus, 88, 120–121
Ebenau, 113, 214, 257, 302
Eberswalde, 214, 245–246, 297
Ecton mine, 191, 283
Edo, 224, 268
Edward II, 57
Edward III, 79–86
Edward IV, 85
Egypt, 2–22, 40–53, 65, 102, 264
Ehrenstrahl, David, 202
El Amajaro, 19
El Oualadji, 66
Elamite, 12
Ellefeld, 169, 213–214, 253
Elizabeth I, 122–125
Elmina, 106, 153–156, 239
Ember, 164–166, 192–195, 232
Emerson, James, 280–281
England/English, 21, 49–57, 77–86, 101, 117–128, 147–288, 178–209, 229–248, 264–280, 292–304
Enlightenment, 30–34, 69–70, 98, 208–209
English Civil War, 123, 157–192
Enköping, 145
Ercker, Lazarus, 121
Ergeni/Ergeninski, 8–9
Erik XIV, 134
Erith, 162
Erzgebirge (Krušné Hory), 46, 89–95, 120, 167–170, 214, 253, 300–301

Esfahan/Isfahan, 42, 65–70, 139–140

Esher, 188–192, 231–242

Etruscan, 18

Euphrates, 8

Ewuakpe, oba, 224

Falun, 77, 93–94, 117, 130–157, 200–203, 258, 293

Fatimid, 43, 63–65

Ferdinand I, archduke, 111–116

Ferdinand III, 215

Fez, 64

Ficker, Peter, 167–169, 213

Finow-Eberswalde, 214, 245–246, 297

Flanders/Flemish, 56–57, 78, 83–85, 97, 107, 119, 145–162, 180–207, 225, 240–243, 259

Flemming, Jakub von, 252

Flintshire, 237–240, 274–290

Florence, 86, 101, 215

Fock, Hermann, 206

forging or hammering, 1–13, 43, 56–63, 84–94, 175–184, 198–214, 223, 232, 251–254, 289, 301–302

Forsberg, Jacob, 260–261

Forster family, 251, 301

France/French, 22, 53–64, 78–85, 107–130, 146–165, 181–212, 222–259, 269–304

Franco-Roman, 28, 48–49

Frankfurt, 77–97, 110–114, 168

Franks, 45–58

Frauental, 254

Freiburg, 241–252

French revolutionary wars, 296

Friedrich I, 247

Friedrich Wilhelm I, 248

Friedrich Wilhelm II, 299

Frosse, Ulrich, 123

Fugger merchants, 87–95, 107–131

Fulda, Marcus 253

Funck, Isak, 260

furnace, 17, 36, 55–56, 74–95, 116–132, 151–179, 195–215, 229–303

furnace accretions, 55, 92–95, 121, 169, 207, 252

Galicia, 20

Ga people, 155

Gambia, 97, 189, 225–227, 239

Gandhara, 29–32

Ganges, 14, 72, 222, 269

Gaul, 19–23, 44–45

Gdansk/Danzig, 95, 116

Geddy foundry, 272

Geer, Louis de, 146–150, 201

Gela, 17

Genoa, 53, 119

Georgia, 3–20

German/Germany, 26, 52–62, 78–169, 191–216, 225–235, 240-259, 271–281, 301–304

Germania Inferiore, 22, 44

Ghana, 106–108, 153–156, 189, 225, 239

Ghanaian empire, 63–67, 156, 198

Ghantasâlâ, 40

Ghent, 85–86

Glauber, Johann, 186, 281

Gloucestershire, 158, 194

Goa, 101–105, 143, 182

Golconda/Telengana, 86, 181–183

Telangana, 181

Gomsez, Petter de, 259

Gordion, 13

Gorno, 18–22, 52, 91, 119

Gorodishche, 58–62

Goslar, 53–56, 85–95, 120–121, 151, 169, 212

Gothenburg, 199, 222

Gotland, 60–63

granulation, 236

graycober, grey copper, 79–80

Great Northern War, 259–261

Greece, Greek, 18, 44, 89

Greenfield/Maes Glas, 285–287

Gressenich, 45–46

Grünau, 169, 253

Grüne (Iserlohn), 51, 78, 92–96, 118, 169–178, 208, 248–250, 298–300

Guadalajara, 42, 65

Guangdong, 3, 141, 179, 221

Guinea Coast, 67, 97–107, 153–157, 176–207, 225–239, 266–284, 304

Gujarat, 14, 29, 71–73, 104–105, 139, 182–183

gunmetal, 13–14, 54–56

Gussmann, Dominik, abbot, 256

Gustav I, 130-133

Gustav Adolf I, 148-149

Gusum, 199–203, 259–260, 294

Habsburg, 109–136, 150, 169–172, 214, 247–258, 300–303

Hackney, 186–188, 229

Haitabu, 58–63

Hallan Çenü, 2

Hallen family, 162, 197–198, 243

Hallstatt, 20

Hamburg, 46, 79–86, 118–130, 150, 168–172, 206

Hammer, Laufamholz, 88–93, 113, 171–172, 209–210, 251, 301–302

Hanham, 280

Hanseatic, 78–95, 118, 131–135, 147

Harappan, 14

Harrison, John, 265

Harz, 52–61, 77, 94–95, 120–134, 151, 168, 207, 245–246

Haugk family, 213-214

Havannah, 285

Hegermühle, 206, 245–248, 297–298

Hemer, 250, 299

Hemmoor, 46

Henckel, Jean Frédéric, 241–252

Henry III, 57, 122

Heidelberg, 46, 227

Heinlein, Paul, 113

Herat, 37, 102

Heroldsberg, 88

Hesse, 120–121, 211–212, 251, 304

Herwick, Abraham van, 127–128

Hildesheim, 54–55, 94

Himalaya, 15, 30–34, 67-69, 98, 102, 222, 305

Hitchcock, John, 195, 232–235

Hittite, 17

Höchstetter merchants, 110–124

Hockerode, 169, 213, 253

Hod Hill, 23–25

Hohenberg, 43–46

Hohenzollern, 116–121, 152, 206, 246–253, 297–298

Höjen, 134–144

Holstein, 129–130, 205–213, 246–258, 297

Höltenklacken, 129

Holywell, 242, 281–287

Honafwer, Hans, 198

Hongwu, emperor, 76

Hoote, Peter, 191
Horni Slavkov/ Schlackenwald, 131
Hounslow, 161–162
Huangji Hall, 144
Huguenot, 203–212, 233, 294
Humfrey, William, 124–125
Hun, 44
Hungary, 41–44, 90–95, 110–117, 129–130, 145, 174, 191–203, 216, 292–300
Huy, 50–56, 85
Huygens, Christiaan, 209
Iberia, 18, 63
Ife, 67
Ilmenau, 95, 116–120
Ilsenburg, 120–121, 168, 245
Indebetou, Carl, 294
India, 8–17, 29–42, 65–86, 98–105, 139–144, 157, 176–198, 210–242, 269, 291, 305
Indonesia, 105, 183
Indus Valley, 14–15
Inn Valley, 90, 122, 215–216, 302
Innsbrück, 92, 109–110, 122, 216
Iran, 2–15, 30–43, 70–71, 86, 102, 140–141, 182
Iraq, 8–10, 37, 70
Ireland/Irish, 47, 81, 234, 284
Iron Age, 20–23
Iser, Matthius, 261
Iserlohn, 51, 78, 92–96, 118, 169–178, 208, 248–250, 298–300
Isfahan, 42, 65–70, 139–140
Isleworth, 20, 125–127
Italy, 18–25, 44–53, 89–91, 109, 175, 215–216, 248–256, 280, 303
Ivory Coast, 107, 239
Jablonkov pass, 95

Jáchymov, 120
Jägerndorf/Krnov, 112–116, 169, 252
Jahangir, emperor, 104
Jahore/Singapore, 143
Jakarta/Batavia, 183, 221
Jakobswalde/Kotlarnia, 214, 252, 298
Jalor (Zawar), 16–17, 37–39, 72–74, 102, 140–142, 181–183
Japan, 142, 179–183, 221, 267–268
Järle lake, 204
Jerusalem, 18, 53
Jesuit, 180
Jiajing, emperor, 101
Joar/Jawahar, 227
Johansson, Jacob, 132–145
Jordan, 41–43
Julich, duchy, 152, 233, 247–248
Julius Caesar, 21–25
Jung, Ulrich, 112
Kalmeter, 195, 235–236
Kalmykia, 8–9
Kamänget, 62
Kanchipuram, 72
Kandy/Maha Nuwara, 223
Kanishka, 30
Karakota dynasty, 32
Kargaly, 2
Karl IX, 133–146
Karl Gustav X, 202
Karl XI, 201–204
Karl XII, 201, 259–262
Karnataka, 39, 269
Kärnten /Carinthia, 90, 112–120, 174, 216–217, 256, 303
Kashmir, 31–37

Kassel, 120–121, 211–213, 251
Kaufungen, 120–121, 212
Kautilya, 15–16
Kazakhstan, 8
Keele, 197
Kelston, 279
Kent, county, 157–162, 229–232
Kent, Walter, 232
Kerman/Kirman, 10, 35–36
Keswick, 57, 77–79, 122–125
Keynsham, 237
Keys/Keyes/Keysar family, 92, 124, 166, 273–291
Khachbulag, 13
Khorasan, later Khurasan, 30-42, 71, 102
Khmer, 75
Kierman, Gustav, 291–292
Kiev, 58–62
Kings Lynn, 78–86
Kish/Kiš, 8
Köberer, Wolfgang 112
Köberer, Hanns 173
Kock, Isaac, 200
Kogelmann, Peter, 120
Koi Gourrey, 66
Kolkata/Calcutta, 210, 302
Körber family, 169, 213
Kosala, 29
Kotlarnia/Jakobswalde, 214, 252, 298
Krakow, 94–95, 117, 252, 281
Krasliče, 168–170, 214, 258, 300–301
Kristina, queen, 147–160, 201
Krnov/Jägerndorf, 112–116, 169, 252
Krušné Hory (Erzgebirge), 46, 89–95, 120, 167–170, 214, 253, 300–301

Kshemagupta, 32
Kushan dynasty, 30
Kushlik, 36
Kumasi, 108, 155–156
Kussinger, Ulrich, 112
Küstrin, 116
Kostrzyn, 116
Ladakh, 32
La Tène, 20-21, 35
Laufamholz, 88–93, 113, 171–172, 209–210, 251, 301–302
lead, 2–121, 140–143, 158–181, 207–224, 252, 274–288, 301
Leen valley, 164–165, 244
Leers, Matthius, 150, 205–206
Leipzig, 83–86, 117, 169, 212–214, 301
Lesbos, 7, 19
Lhasa, 34
Liège, 55–57, 83–84, 132, 146–162, 196–207
Lienz, 112, 173–174, 216–217, 256, 303
Lincoln, 49
liquation process, 90, 120, 207
Lisbon, 97–106, 247, 274
Liverpool, 231–239, 271–287
Loket, 301
Loire river, 50, 297
London, 20, 57–61, 78–85, 102, 118–127, 157–248, 266–279, 301
lost-wax casting, 20, 43, 107
Lothal, 14
Lübeck, 62, 78–97, 121–135, 147–151, 168, 205–206, 258, 292–297
Lucca, 91
Lyon, 24, 189, 279
Maastricht, 56–57

Macao, 142, 179
Macclesfield, 281–289
Macedonia, 15–25
Madeira, 234
Madeley, 243
Ma'den Ijâfen, 64
Madhya Pradesh, 73, 269
Maes Glas/Greenfield, 285–287
Magdeburg, 55, 92–94, 247
Maghreb, 63–65, 108
Maharatha, 181–183, 223, 269
Maharashtra, 223
Malabar, 142
Malacca, 105, 142, 221–223
Malay, 105
Mali, 66–67, 97
Malines, 162
Mamluk, 95–103
Maidstone, 157–158, 277
Manchu, 139–144, 179, 267
Manila, 142–143
Mannhagen, 128, 150, 206
Mansfeld, 90, 118–120, 152–169, 300–301
Manstein, Leonard, 112–113, 173
Manyika, 143
Mansfeld, 90, 118–120, 152–169, 300–301
Marees, Pieter de, 107–108
Marggraf, Andreas, 242
Maria Theresa, empress, 252, 303
market, 17, 50–63, 78–107, 133, 146–259, 277–304
Markneukirchen, 300–301
Marlow, 188
Marseilles, 101, 247

Martin, Richard, 124
Masada, 40
Massachusetts, 175
Massa Marittima, 91
Matheson family, 259, 293
Mauretania, 63–64
Mauryan dynasty, 15, 38
Maximilian I, 92, 109–111, 152
Maximilian II, 170
Mayer, Hans, 113
Mecklenburg, 128–129, 150, 205–206
Mediterranean, 12–20, 40–67, 95–107, 305
Mell, 53
Mendip hills, 123–128, 158–159, 191–193, 233–242, 274
Menzlin, 58
Meissen, 92, 281
Merovingian, 47–50
Mesopotamia, 10–18, 65
Meuse (Maas) river, 22–24, 44–69, 82–83, 118, 210, 250
Meyer, Christian, 206
Michell, Humfrey, 125
Midas, king, 13
Middle East, 4–11, 27–42, 63–73, 96, 142, 305
Milan, 19, 91, 109–119, 215
Mineral and Battery Works, society of, 124–128, 157–166, 189–193
Ming dynasty, 76, 101–102, 139–143
Mithradates VI, 20
Möllbrücke, 174, 217, 303
Momma family, 153, 194–204
Mongelas, hotel de, 81
Mongol, 74, 103

Monomotapa, 105–107, 143
Moresnet, 22, 45, 77–83, 114–131, 152
Morocco, 44, 63–67, 97, 126, 191
Moscow, 247
Mosul, 37
Mosto, Alvise da Ca' da, 97
Motala river, 147, 202, 294
Mozambique, 105, 143
Mughal, 102–104, 139, 181–183, 223, 269
Mühlau, 109–112
Müller, 246
Mumbai, 210, 302
Muslim, 35–50, 107, 223, 269
Myanmar/Burma, 40, 74, 141, 182, 224, 268
Myrvälde, 60–64
Nacka, 148–150, 199–205, 260–261
Nacken, Winandt, 145–150
Nāgārjuna, 29
Namazga, 9
Namur, 28, 46–57, 83, 250, 297
Nantes, edict, 211, 296
Nasafjall, 292
Nassereith, 110, 216
Native American, 175, 198, 269
Native Canadian, 222
Naugatuck, 280
Near East, 15, 36–41, 53, 85–106, 210
Neath, 26, 123, 284
Nepal, 31–34, 70
Nero, emperor, 26
Netherlands, 111–158, 171–175, 195–240, 282
Neubrunn/Sachsenbrunn, 88, 168
Neumann, Caspar, 252
Neustadt in Mecklenburg, 128, 206

Nevis, 190
Newcastle-under-Lyme, 197
Newfoundland, 184
New York, 270–273
Niger, 2, 66, 288
Nigeria, 67, 154, 190
Niger river, 66, 288
Nightingale, James, 190
Nikolas I of Schleswig, 57, 94
Niort, 53
Norfolk, 86, 98
Normandy, 151, 188–192, 296
Norrköping, 147–148, 199–203, 259–262, 294–296
Norse, 58–60
North Sea, 61, 81, 122–128, 201
Norway, 46–47, 62
Nostitz, duke of, 214
Nottingham, 164–166
Novgorod, 58–63
Nubia, 2
Nuli, 9
Nuremberg, 58, 85–132, 168–172, 186–197, 209–214, 230–233, 251, 300–302
Nuzi, 10
Nyköping, 146–148, 199–202, 261–262, 294
Oberalm, 113, 214, 257, 302
Ockendon, William, 287
Offenhauser, Leonhard, 109–110
Oder river, 26, 116, 206
Ofin river mine, 108
Oker, 207
Oldesloe, 130, 213
Onitsha, 225
Orange, William of, 132

oreichalkos, 17
Osterode, 52–55
Ostragoth, 41, 92
Ottoman, 246
Oxus river, 10
Oxford, 25, 68–80, 264
Pasch, Elias, 296
Pachmann family, 174
Pagan, 40, 75, 264
Pamunuwa, 223
Pa Niu, 3, 141
Pakistan, 15, 30–31, 102–104, 140
Palestine, 41–42
Pannonia, 41–46
Pacquet, Vergil, 132–134
Paracelsus, 119
Parina valley, 119
Paris, 57, 79–85, 111–119, 208, 304
Parthia, 30
Parys mine, 283–286
Patten, Thomas, 281–287
Pegnitz river, 87–93, 172
Pepin, king, 50
Persia, 15, 35–42, 69–71, 98–103, 139–141, 181
Persian Gulf, 8–10
Peru, 152, 300
Pflach bei Reutte, 110–111
Philadelphia, 227–229, 269–270
Philimalawa, 223
Philip I of Spain, 93
Phoenicians, 17–18
Phophnar, 38–39
Phrygia, 13
Pless/Pszczyna, 281

Pliny, 22
Pločnik, 2
Plutarch, 19
Plymouth, 265–266
Poland, 57, 116–117, 130–145, 195–202, 246–260, 300
pollution, 1, 242–246, 287, 305
Pompeii, 27
Polo, Marco, 36, 70
Populonia/Piombino, 18
Porter, Abel and Levi, 280
Portsmouth, 235
Portuguese, 96–108, 142–157, 179–189, 223
Potsdam, 246, 298
Prague, 136, 214
Prakash, 14
Prettau/Predoi, 90, 109–112, 216
Prussia, 110–116, 246–252, 297–299
Punjab, 71
Qāytbāy, sultan, 95
Quebec, 222, 271
Qing dynasty, 179–181, 222, 267
Radmer, 113, 254–256, 303
Rajasthan, 16–17, 38, 72–73, 102, 181, 223, 269
Rammelsberg, 53–55, 94, 120
Rantzau, Barbara, 129–130
Ratzeburg, 128–129, 150–151, 205–206
Redbrook, 184–191, 236
Reenstierna, Abel, 203–205, 259–261
Reeves, John, 165
refining, 195–200, 232, 269, 284
refractory clay, 57, 119
Reichraming, 112–113, 172–173, 217–218, 254–257, 303–304

Reinmann, Paul, 170
retort, 16, 74, 101–102, 143, 179, 242, 281
reverberatory furnace, 160–171, 195–196, 288
Revere, Paul, 304
Rhaetia, 120
Rheims, Council of, 49
Rhine river, 22–26, 44–63, 118, 186
Rhine-Meuse area, 24, 44–63, 118
Ribchester, 24
Ribe, 58–61
Richter, Kaspar, 131–132
Rieser family, 217–218, 254
Rinman, Sven, 293–294
Rodewisch, 167–169, 213, 253, 300–301
Roe, Charles, 283–285
rolling mill, 192–204, 230–244, 260, 279–284, 304
Roman, 2, 18–55, 69, 89, 109–119, 136, 215, 256–257, 305
Romania, 252–254
Romano-British, 23–26
Rome, 22–26, 40–47
Røraas, 300
Rosenheim, 253–258, 302
Rotherhithe, 127–128, 195, 232
Rouen, 85, 296
Rudolf II, 136, 152, 170
Ruhberg, 281
Rupert, prince, 186
Russia, 58–61, 117, 199, 246, 259–261, 297–298
Sachsenbrunn/Neubrunn, 88, 168
Safavid dynasty, 102
Sahara, 65–67, 97
Saint Helens, 287

Saint Joachimstal/Jachymov, 120, 170
Saint Malo, 188
Saltford, 239
Salzburg, 113, 214–218, 257, 302
Samson, 28, 47–49
sand-casting, 42–43, 67, 230
Sarawak, 105
Sardis, 41
Sary Tepe, 12–13
Sauerland, 46–51, 169
Saxony/Saxon, 49-61, 89–95, 124, 152, 167–170, 213–214, 241–253, 264, 298–301
Scandinavia, 50–63, 117, 248, 300
Schissler, Christophe, 111
Schleswig, 57, 93–94, 128, 297
Schutz, Cristofer, 123–131, 206
Schwabia, 151
Schwaz, 90, 123, 253–256, 302
Schweina, 116
Schwerte, 51–52
Scotland, 60
Scythia, 44
Seitenstetten, abbey, 254–255, 303
Senegal, 63–65, 97, 189, 226
Senuwar, 15–16
Seven Years War, 251, 269, 297
Sforza, second duke, 109, 215
Shan states, 141
Sherbro river, 190, 225
Shaanxi province, 3
Shorey family, 187–188, 229–241
Sicily, 17, 42–44
Siena, 109, 215
Sierra Leone, 189, 225, 239, 268

Sigismund I, 133

Sigismund III, 135

Silesia/Śląsk, 57, 94–95, 115–120, 169–170, 207–214, 246–252, 281, 297–298

Sintiou Bara, 65

Skultuna, 144–150, 198–205, 244, 259, 292–293, 306

slave trade, 69, 154, 189–221, 237, 281–305

Slavic, 50

Slovakia, 94, 115, 129, 145, 174, 292

Slovenia, 21, 304

Smethwick, 290

Smith, William 227

Soest, 51–52, 78, 118

Sofala, 107

Soghun valley, 10

Solingen, 162

Songhai empire, 107, 189

South-East Asia, 34–42, 73–76, 141–144, 179, 221–222

Spain/Spanish, 18–19, 37–65, 85–93, 107–132, 144–162, 185, 230–250, 272, 299–304

Spalding family, 259–260, 294

Sri Lanka, 223

Staffordshire, 159, 191, 281–289

Stapleton (Bristol), 166, 194–195

Stapleton, John, 194–195

Steel, 81, 110, 147, 165, 257, 277, 305

Steel, Joshua 277

Steere, Thomas, 162

Steyr, 90, 112–113, 173, 217–218, 255–256

Stockholm, 131–149, 198–204, 260–262, 291–296

Stolberg, 22, 46, 79, 118, 130, 145–175, 188–207, 225–235, 247–257, 300–305

Stone, 9–10, 53, 70, 82, 117, 133, 147, 200–207, 234–239, 252–255, 278, 290–291, 304

Stourbridge, 197, 236–243, 276–282

Strabo, 17

Štříbrná, 170, 214, 301

Styria/Steiermark, 123, 174, 254

sublimed zinc vapour, 35, 207

Sumatra, 105, 224

Sumeria, 8, 12

Surat, 143, 181–182, 221–222

Surrey, 128, 162–164, 185–192, 232–239

Sus, 63–66

Swab, Anton, 242

Swansea, 284–290

Swat Valley, 30–37

Sweden/Swedish, 46–64, 77, 94, 117, 129–166, 176, 189–215, 221, 236–262, 282–306

Switzerland, 120, 215–216, 303

Syria, 12, 36–43, 68, 97–103

T'ai Hang mountains, 3

Tamerlane, 103

Tamil, 223

Tamil Nadu, 72, 269

Tarnowitz, 115–116, 169–170

Taxila, 15–16

Tegdaoust, 65

Telebi, 9

Telangana/Golconda, 86, 181–183

Temeswar, 252

Temne, 268

Temple Mills, Hackney, 186–188, 229–241

Temple Mills, Bisham, 188, 230–241, 287

Tepe Yahya, 10

Terengganu, 105
Tern forge (Attingham), 244
Teutoburger forest, 78
Tewkesbury, 158
Thailand, 3, 75–76
Thames river, 20–22, 79, 125–128, 162–164, 188–192, 229–232, 287
Thames Ditton, 164, 192
Theodosius, emperor, 40
Theophilus, 54
Theopompus, 17
Theophrastus, 19
Thermi, 7
Thiller, 188, 231–233
Thirty Years War, 117, 136, 148–185, 205–216
Thun, Henrik, 204
Thuringen, 120, 169, 213, 252–253
Thurzó, James, 95, 115
Tiberias, emperor, 42
Tiberias, site, 42
Tibet, 31–37, 70–77, 141
Ticino river, 19
Tigris river, 10, 37
Timbuctu, 67, 97
Tintern, 123–127, 158–167, 184
Titelberg, 21
Tongeren, 78
Toruń, 95
Tournai, 83–85
trade fair, 50, 78–89, 97–98, 114, 139, 213
Trave river, 93, 128–130, 205
Trems, 130, 205, 258, 297
Tresoigne, Jean-François, 297
Trinovantes, 21

trip- or battery-hammer, 88–93, 112–133, 147–175, 188–216, 229–262, 278–303
Trondheim, 300
Troy, 7–8
Try, Henrik de, 203
Tuher, Thomas, 170
Turkey, 12–17, 40–44, 70, 217, 303
Turkmenistan, 9
Turners' Brass House, 288–289
Tuscany, 18, 119
Twelve-Brothers house, 87
Tyrol, 44, 78–89, 109–120, 151, 173, 215–218, 253–257, 302–303
Ugarit, 12
Umm an Nar, 9-10
Ummayad, 43, 65
Upper Hungary/Slovakia, 93–95, 115–117, 129–130, 145, 174, 191, 203, 216, 292–300
Uppsala, 132–133
Ur, 8
Urartu, 12–13
Urbino, 215
Utrecht, treaty, 230, 299
Uttar Pradesh, 14, 73
Uzbekistan, 10
Vakataka dynasty, 38–39
Val del Riso, 18
Valencia, 119
Vällinge, 148–149, 199–204, 259
Van Herwick, Abraham, 127–128
Varanasi/Benares, 222, 269
Varna culture, 2
Vasa dynasty, 109, 130
Västerås, 134–145

Vattholma, 132–134

Veneto-Saracenic, 53–54, 96

Venice, 53, 67, 86–110, 172, 210–218, 248, 303

Verbiest, Ferdinand, 180

Verdun, 56

Verona, 215, 280

Versailles, 185

Vieille Montagne/Altenberg, 22, 45–55, 77–95, 114–131, 145–152, 288, 300

Vicenza, 280

Vienna, 86, 119, 136, 170–174, 214, 218, 256-257

Villedieu les-Poêles, 151, 296

Viking, 60–67

Vinča, 2

Virginia, 220–228, 270–276, 304

Vischer family, 88–98

Visigoth, 45

Vitsch, Paul, abbot, 255

Volga river, 59–61

Volkamer family, 209–210, 251, 301

Wales/Welsh, 26, 123, 157–158, 184–187, 233–244, 267, 274–290

Walloon, 146–158

Wandsworth, 162, 196–197, 243

Wanli, emperor, 101, 142–143

Warmley, 242, 267–280

Wenzel, king, 86

Wernigerode, 120, 207

Westerberg, Maria, 259–260, 294

West Indies, 195, 230, 274–283

Weston, 237

Wethered, Thomas, 232

Westphalia, 52, 169, 205, 248

Whydah/Ouida, 189, 225, 239

Wiesloch, 46

Williamsburg, 220–228, 270–276, 304

Winchester, 81

Winkelhofen, Andree von, 216

wire, 1, 24, 48–60, 78–94, 106–215, 227–303

Wistar, Caspar, 227, 270

Wolin, 58–62

Woodborough, 237

Wotton, 163–164

Wudang mountains, 76–77

Wutai mountains, 141

Xanten, 24, 46

Xuande dynasty, 76

Xuanzang, 29

Yangliusi, 102, 143

Yangtze Valley, 102, 179

Yongle, emperor, 76, 98

Yongzheng, emperor, 222

York, 49–51, 85, 270–273

Yuan, dynasty, 76

Yunnan, 74, 179

Zawar/Jalor, 16–17, 37–39, 72–74, 102, 140–142, 181–183

Zerovnišček, 21

Zimbabwe, 105, 143

zinc accretions, 95

zinc carbonate/calamine, 4–30, 43–54, 76–101, 118–125, 152, 170–178, 254, 280

zinc metal/metallic zinc, 3–4, 16–19, 36–39, 69–74, 86, 92–102, 118–121, 139–144, 161, 179–195, 207–253, 274–298, 305

zinc oxide, 5–55, 73–74, 95–102, 119–121, 141–142, 207–224, 278–285

zinc sulphide, 4–16, 55, 74, 96, 278–285

Zwickau, 94–95, 128, 167, 253